W9-CVA-099

CECIL AND IDA GREEN

Cecil Howard Green (1900–) and Ida Flansburgh Green (1903–1986)
Photo courtesy of the *Dallas Morning News*, 1983

CECIL AND IDA GREEN
Philanthropists Extraordinary

Robert R. Shrock

The MIT Press
Cambridge, Massachusetts
London, England

© 1989 Massachusetts Institute of Technology

All rights reserved. No part of this book may be reproduced in any form by any electronic or mechanical means (including photocopying, recording, or information storage and retrieval) without permission in writing from the publisher.

This book was set in Palatino by Graphic Composition, Inc., Athens, Georgia and printed and bound by Halliday Lithograph in the United States of America.

Library of Congress Cataloging-in-Publication Data

Shrock, Robert Rakes, 1904–
  Cecil and Ida Green.

  Includes index.
  1. Green, Ida M. 2. Green, Cecil Howard, 1900– . 3. Philanthropists—United States—Biography. 4. Science—Study and teaching—United States—Finance—History. 5. Medical education—United States—Finance—History. I. Title.
Q183.3.A1S47   1989   361.7'4'0922 [B]   88-26853
ISBN 0-262-19276-4

To Ida
For sixty years
Cecil's devoted wife and partner
in all his achievements

CONTENTS

ILLUSTRATIONS

ACKNOWLEDGMENTS

I started this biography of Cecil and Ida Green, of Dallas and La Jolla, as a labor of love with the primary purpose of making known to the general public the nature and extent of their remarkable philanthropy. I have used the adjective extraordinary in the title because it best describes the program of giving that Cecil and Ida have followed together for the past thirty-five years. Not only have they devoted an impressive amount of time, effort, and thought in deciding how much and to whom each gift should be made, but they have also exercised rare imagination and innovation in selecting the recipients of their individual and joint donations. The results have been dramatically generous and timely benefactions to educational, medical, and civic institutions. These benefactions, largely in the form of funding for buildings or endowments for teachers and students, will continue to benefit humankind far into the future.

When I expressed the wish to write the story of their married life together, with particular emphasis on their philanthropy, they graciously granted me the permission I requested and assured me of their fullest cooperation. Suffice it to say that I could not have written the story that appears in the following pages without that cooperation. They allowed me to record on tape many hours of questioning and informal conversation; made available their diaries and much of their correspondence; shared with me the logbook of their philanthropy; and by correspondence and telephoning kept me informed of their most recent donations and future commitments, as well as of the awards and honors accorded them—cooperation that has continued even while my manuscript is being prepared for publication. In addition to the aforementioned tapes, I have had access to many of the lectures, speeches, and informal commentaries that Cecil has delivered during the past twenty-five years, and I have been granted permission to quote from them as I wished. Comments attributed to Cecil and Ida that are not enclosed in quotation marks are based on my memory and notes of the particular conversation.

During the early stages of writing, Cecil's longtime Dallas secretary, Dolores Hall, and mine at MIT, Pauline Richmond, patiently kept open the lines of communication between Cecil and me by letter and telephone. They deserve special credit for the readable copy they each extracted from the taped recordings assigned them for transcription.

I am deeply grateful to the many friends of the Greens who without exception promptly sent me the materials or information requested and gave me permission to use them. Individuals and organizations are specifically credited in the text, but if I failed in any case to give appropriate credit I ask forebearance, for the lapse was surely unintentional. And I would be remiss if I did not mention a few of those friends who gave me especially important assistance.

Conversations with Cecil's longtime business partners Eugene McDermott and J. Erik Jonsson were most helpful earlier on, and I am especially grateful to Jonsson for his permission to quote at length from his informative and authoritative in-house account of the early history of Geophysical Service Inc. (GSI) and Texas Instruments Inc. (TI). Dot Adler, supervisor of special projects and editor of GSI's monthly magazine *Grapevine*, has kept me abreast of the activities and accomplishments of the staff of GSI and TI by sending me photographs, news items, and other material, as well as seeing that I received every issue of the magazine.

From academia abroad has come indispensable assistance from two scholars associated with Oxford's Cecil H. Green College. Sir Richard Doll, who worked closely with Cecil in creating Green College and who served as its first warden, was most helpful in supplying information about the new college, and graciously showed my wife and me about the premises when we visited the college in 1980. Sir John Walton, the present warden, has been equally helpful by sending both photographs and informational booklets and granting permission for me to use them in the biography. His personal tribute to Ida, at the recent Memorial Service, is a moving and heart-warming statement of the gratitude and admiration that the scholars of Oxford accord Ida and Cecil, one of England's native sons. Members of the faculty and staff of the University of British Columbia warmly welcomed me when I visited Vancouver in quest of firsthand information about Cecil and the results of his and Ida's benefactions to his first alma mater. Rosanne Rumley and Bel K. Nemetz kindly provided the list of distinguished people who have contributed so importantly to the Vancouver–Victoria community as Green Visiting Professors. Again from outside the United States, the Greens' and my irrepressible friend Harry Messel, founder and longtime head of Australia's University of Sydney's School of Physics, has sent me successive copies of *The Nucleus* and otherwise kept me informed of the timely donations Cecil and Ida have made to his school and the University of Sydney. Messel also made important alterations in the final proof.

In the United States, help has come from so many institutions and people that I can mention only a very few of them, but all deserve my sincerest gratitude. Orlo E. Childs, while president of the Colorado School of Mines, kept me informed of the Greens' developing interest in that institution, and J. Edward White, the first Charles Henry Green Professor of Geophysics at CSM repeatedly provided information that I desired. Charles C. Sprague, president of the University of Texas Health Science Center in Dallas (UTHSCD), and his staff deserve mention for providing information about the Greens' impressive benefactions, especially excellent photographs of the buildings Cecil and Ida have helped to fund; also Andrew D. Suttle, Jr., of the University of Texas Medical Branch at Galveston (UTMBG), who not only gave Cecil exceedingly important assistance with the arrangement of the two international conferences on biological imaging but also sent me the information I sought on several matters at UTMBG. James S. Triolo, assistant to the president of the Scripps Clinic and Research Foundation (SCRF), graciously gave me permission to quote from his dramatic in-house report on the Greens' impressive donations to SCRF. Walter Munk, J. Freeman Gilbert, and their associates in the Institute of Geophysics and Planetary Physics (IGPP) of the University of California at San Diego (UCSD) have been most helpful in keeping me updated on the Greens' different academic benefactions in the San Diego community. And of course I should not omit our appreciation of Roger Revelle's thoughtful remarks at the dedication of MIT's Cecil and Ida Green Building for the Earth Sciences. Former heads of department and then deans—the late Richard H. Jahns and Allan V. Cox—George A. Thompson, a more recent department head in geophysics, and most recently Associate Dean Dudley C. Kenworthy have repeatedly helped me to stay up to date on the Greens' impressive benefactions to Stanford, especially in geophysics and to the library system. Finally, I have been able to keep informed of the rapid development of TAGER (The Association for Graduate Education and Research of North Texas), one of the Greens' most impressive educational efforts, through the timely and enthusiastic reports of Fred Baus and his associates of AHE (Association for Higher Education of North Texas). Also in Dallas, Robert Decherd of the *Dallas Morning News* never failed me when I asked for a special photograph or other item involving Cecil and Ida, and David V. Hicks, headmaster of St. Mark's School, has kept me abreast of the latest of Cecil's activities at St. Mark's.

During preparation of the successive drafts of my manuscript, quite another group of people and institutions provided the assistance that made its completion possible. From the very beginning, some six years ago, the Massachusetts Institute of Technology has permitted me to continue using a small part of the space in the Green Building that I occupied as a geology professor before retiring in 1970. In addition the Institute has provided working materials, given me an IBM Selectric II typewriter, and made available part of the time of Dorothy Frank, an expert at word processing, who patiently and cheerfully rendered my handwritten and typed composition into amazingly accurate printed form. My longtime friend John I. Mattill gave up some of his free time while editor of MIT's *Technology Review* to edit a portion of the early printed copy, greatly improving it in structure and style and adamantly refusing any recompense. I happily record my gratitude to him and regret that he could not have done more. I am equally grateful to two other longtime MIT colleagues—Vincent A. Fulmer, at the time vice president and secretary of the Institute, who always provided the answers desired to the questions I asked about MIT, and Warren A. Seamans, director of the MIT Museum, who, together with his efficient staff, produced the photographs and other historical material I sought. Finally, I thank former MIT President Julius A. Stratton, who more than anyone else brought the Greens' philanthropy to MIT when he assured Cecil, his 1923 classmate, and Ida that he would approve, for support of the earth sciences, the first major donation they offered the Institute.

When the first draft was completed and in word-processed form and its revision began, my whole family came to my assistance. Our son, Robert E., spent most of his 1985 year-end vacation carefully reading the entire manuscript. Our daughter, Wendolyn T., with an impressive memory for details, spent many hours critically revising the manuscript from beginning to end, greatly improving it in almost every way. In the final stage of revision, my wife, Theodora, spent many hours using her skill in composition, an effort that greatly improved the main part of the manuscript. The assistance that my three family members gave me was the driving force that kept me going, for their admiration and esteem for Cecil and Ida are equal to mine. I cannot thank them enough for their encouragement and assistance and consider them my true co-editors even if their names do not appear on the title page. And I am deeply grateful to the manuscript editor, Caroline Birdsall, and to the editorial staff of The MIT Press—Helen Osborne, managing editor, Dana Andrus, editor, and their efficient associates—for the skillful editing that further improved the final manuscript.

I come now to my last expression of gratitude—my sincerest thanks to Cecil and Ida for trusting me to write about them. At no stage did they suggest that I change my story in any substantial way. Ida read much of the earliest version of the manuscript and was told before she succumbed to her last illness that the book would be dedicated to her. Cecil read all of the manuscript except the appendixes and the captions for the illustrations and called on his remarkable memory again and again to supply the name of a person or a place, the date of an important event, the details of an amusing happening, even the name of the street and the number of the house where they lived at different times. I alone, however, assume full responsibility for the contents of this work, having made every effort to be accurate and fair in my story of the two persons for whom I have had the greatest admiration and affection since the day I first met them. I salute them now with the same words I wrote fourteen years ago:

To think seriously of giving to help others is commendable:
To give is the essence of humaneness and nobility;
To give generously and with deep purpose is the greatest
act of all because it requires thought, effort, and
discrimination of the highest order.

CECIL AND IDA GREEN

PROLOGUE

AN INTERNATIONAL TRIBUTE TO CECIL AND IDA GREEN

On the evening of November 9, 1978, a group of 225 distinguished academicians and their spouses gathered at the National Academy of Sciences in Washington, D.C., to participate in *An International Tribute to Cecil and Ida Green*. The reception and the dinner that followed were designed to honor the Greens for their extraordinarily creative philanthropy. The black-tie celebration, possibly unique in the annals of academic assemblies, brought together representatives of thirty-two leading educational, medical, and scientific institutions of the English-speaking world. From the introduction of the first guests in the receiving line to the reluctant departure of friends who lingered for a final word with the Greens, the celebration was marked by a pervasive air of friendliness and an impressive outpouring of admiration and affection for the honorees. Many times since, Cecil and Ida have remarked that this celebration was their greatest party ever.

Presidents, vice presidents, chancellors, deans, directors, chairmen of regents—from abroad they came from England (Oxford), Australia (University of Sydney), and Canada (University of British Columbia). From the United States they came from the President's Office of Science and Technology, the National Academy of Sciences, the American Association for the Advancement of Science, the American Association of University Women, and the Smithsonian Institution. And from Massachusetts Institute of Technology, Woods Hole Oceanographic Institution, Suffolk University, the University of Texas System (including the Health Care Center in Dallas), Austin College, Bishop College, Southern Methodist University, St. Mark's School of Texas, TAGER Educational Television, Texas Christian University, and Texas Instruments Foundation. And from Colorado School of Mines, Scripps Clinic and Research Foundation, Stanford University, California Institute of Technology, and the University of California at San Diego (including Scripps Institute of Oceanography and the Institute of Geophysics and Planetary Physics). All but four of the institutions had received munificent benefactions from the Greens.

1   Participants in An International Tribute to Cecil and Ida Green, a festive occasion hosted by the National Academy of Sciences in Washington, D.C., November 9, 1978. *Above:* Ida and Cecil Green on the receiving line. Photo courtesy of the *Washington Post*. *Opposite page:* Dignitaries who delivered remarks during the dinner (*from left to right*): Phillip Handler, president of the National Academy of Sciences; Frank Press, director of the Office of Science and Technology Policy in the White House on leave as Robert R. Shrock Professor of Geophysics at MIT; Peter S. Bing, president of the board of trustees at Stanford University; Cecil and Ida Green; Marjorie Bell Chambers, president of the American Association of University Women; Harry Messel, head of the School of Physics at the University of Sydney, Australia; Charles A. LeMaistre, chancellor of the University of Texas System, and Jerome B. Wiesner, president of MIT, at the lectern. MIT Photo by Calvin Campbell.

An International Tribute to Cecil and Ida Green

After the dinner, convener Jerome B. Wiesner, president of MIT, extolled the Greens for their innovative and generous philanthropy and emphasized its far-reaching influence. Then Frank Press, President Carter's science adviser, spoke of Ida's philosophy of life—to try to make the world a better place to live in—a philosophy fully shared by Cecil, who also gave great importance to cooperation and collaboration in human endeavor. Other dignitaries on the stage (figure 1) lauded the Greens for their immensely generous actions, which had benefited so many thousands of people. Cecil and Ida spoke last, expressing their deep gratification at the appreciation they had received. The celebration ended on a sustained note of admiration and affection for the extraordinary couple it was held to honor.

Who are Cecil and Ida Green? Why should they receive such an unprecedented tribute from so many leading academic and scientific institutions, and under the sponsorship of the National Academy of Sciences?

The answer to the foregoing questions can be found in the eloquent essay, "An Appreciation," prepared by Vincent A. Fulmer, a member of the planning committee for the program booklet of the International Tribute. The essay is reproduced here wih the author's kind permission.[1]

IN APPRECIATION
Few Americans have given as much time and energy to the exercise of creative philanthropy as have Cecil and Ida Green.

Quietly and unintentionally, but with total dedication and consummate skill, they have become internationally prominent as

[1] An informative and well-illustrated program booklet, titled *An International Tribute to Cecil and Ida Green* (Washington, D.C.), written by Vincent A. Fulmer, secretary of MIT and member of the planning committee for the Tribute, and printed by the MIT Design Services, was distributed to participants as a memento of the festive gathering. Copies are included in the Greens' memorabilia to be preserved in due course in a suitable archive. An excellent commentary on the Tribute appeared in the Style (F1–2) section of the *Washington Post* on Saturday, November 11, 1978.

philanthropists. They are probably the leading private individuals providing support to the geophysical sciences in the United States.

Their encouragement of education and basic research in advanced fields is reminiscent of the strategic role private individuals played in fostering the Age of Exploration centuries ago. The difference is that the Greens themselves have been actively involved in the work-a-day world of science, whereas the patrons of world exploration in earlier times rarely became involved on a personal basis.

Practical necessity led the Greens to champion the marriage of science and industry—first as practitioners and later as advocates and sponsors. As members of the founding group of Texas Instruments Incorporated and its predecessor company, Geophysical Service Inc., their pioneering use of the reflection seismograph in the search for mineral resources led to broader interests in the whole domain of the natural sciences, and in improving the instrumentation available for measurements. They and their associates in Texas Instruments helped to foster an electronic revolution. Their later interest in the biological and medical sciences and in improved health care was a natural outgrowth of their fundamental concern for people.

Today, there are some twenty-five colleges, hospitals, museums, schools, and universities in Australia, Canada, England, and the United States which have benefited profoundly from the personal contributions of the Greens. These have included trusteeship, program development, endowments, facilities, scientific equipment, and program funds which have had a seminal role in advancing basic research and education on a global scale. Some fifteen major university or hospital facilities and twenty fully-endowed professorships, largely but not exclusively centered in fields of science and engineering, an endowed Master Teacher Chair held by a First Grade Teacher, fellowships to encourage women in science and engineering, life-saving medical equipment, an ocean-going geophysical and oceanographic research vessel, an educational television system, art objects, and a global system of earthquake detectors are involved in this staggering total.

The distinguishing features in the Greens' involvement with scientific endeavor are the degree to which they have personally immersed themselves in the institutions they serve, the range of scientific fields they have sought to foster, and the number of lives they have personally affected. They have been discriminating donors, seeking and fulfilling those opportunities where their support could make a qualitative difference—to people and to institutions.

The Greens are exemplary in so many additional, extraordinary ways. Unlike many prominent citizens who have concentrated their philanthropy and their attention on largely local institutions, Mr. and Mrs. Green have chosen to establish philanthropic roots on a global scale—in both public and private organizations. Their missionary zeal for inter-institutional cooperation and for cooperation between science and industry has altered the course of scientific education in the United States and accelerated the economic growth of the Southwest.

The idea of meeting jointly in Washington, D.C., to honor Cecil and Ida Green for their international example of private philanthropy led a group of college presidents and chancellors to come together and to contact close friends of Mr. and Mrs. Green to plan "An International Tribute to Cecil and Ida Green." During the summer of 1978, after several months of discussion, an ad hoc Planning Committee met in Chicago to map the outline and details of this Reception and Dinner. It became clear that the range of the Greens' generosity extended well beyond scientific and educational institutions to include civic and community development organizations, the arts, and innumerable charitable activities. Noting that Cecil and Ida had already been recognized in their receipt of top-level community awards for their munificence in the Dallas community, and by professional societies and industry associations, the Planning Committee determined that "An International Tribute to Cecil and Ida Green" should properly focus on their scientific and educational affiliations, without implying that their other activities were not also remarkable and important for their own sake. The desire to keep the size of the Dinner group within the number which could be accommodated by the National Academy of Sciences became the essential planning guideline. The Presidents and Chancellors of the Participating Institutions undertook the difficult task of determining who should represent their institutions among the legions of friends who would want to be with Mr. and Mrs. Green on this occasion.

Our purpose is to recognize the towering friendship which Cecil and Ida Green have given to students everywhere by dedicating their lives to advances in science, medicine, technology, and basic improvements in the quality of education. At a time when the Greens have already been recognized individually by the many institutions they have served, this combined Tribute has significant historic value in the annals of private giving to education, and it represents a unique, international expression of affection and appreciation by the grateful recipients of their support.

We who are privileged to claim the Greens as friends celebrate their example, and our own great good fortune in being part of their inter-institutional constituency. They inspire and ennoble the spirits of ordinary people. They are monumental proof that individuals can make a lasting difference as builders of scientific institutions in the twentieth century. More than 30,000 faculty and research staff, 200,000 students and 1,000,000 alumni are represented in the collective totals of the institutions they have fostered. No amount of praise can adequately convey the depth and intensity of our combined appreciation for Cecil and Ida Green, as individuals and as beloved colleagues.

During an active quarter-century career as an exploration geophysicist, co-owner of Geophysical Service Inc., and co-founder of Texas Instruments Inc., Cecil acquired—entirely by his and Ida's efforts—a substantial fortune. (They started married life with only a few hundred dollars apiece.) Beginning in 1950, and continuing to the present (1988), they have donated more than $150,000,000 of that fortune to educational, medical, and civic organizations and activities. Their extremely generous, creative, and far-reaching philanthropy has been recognized repeatedly not only by the 1978 International Tribute but also by a multitude of awards and honors from recipient institutions, many of which were represented at the Tribute.

Upon graduation from MIT as an electrical engineer, Cecil held a series of short-term jobs with three electronics companies—Raytheon in Cambridge, Wireless Specialty Apparatus in Boston, and Federal Telegraph in Palo Alto—before joining Geophysical Service Inc. in Dallas. For eleven years, from 1930 to 1941 (except for a brief return to Federal Telegraph in 1931–1932), he served GSI first as a party chief and then as regional supervisor. The responsibilities of these two positions took him and Ida tens of thousands of miles on an odyssey that necessi-

tated more than fifty stops along the way for temporary housing, and gave both of them a love of travel and a variety of
outdoor experiences that they still look back on with nostalgia.
In 1941, the Greens purchased a quarter ownership of GSI.

The purchase came when Cecil and Ida hesitantly agreed to
join three of their fellow GSI employees—J. Erik Jonsson, Eugene McDermott, and H. Bates Peacock—in buying the assets
of GSI for a total of $300,000, each to receive a quarter owner-
ship. The four partners closed the deal on Friday, December 5,
1941.

Two days later came the attack on Pearl Harbor.

On that fateful Sunday, Cecil and Ida could not help wondering if they had acted wisely in pledging everything that they
owned, in addition to arranging a large bank loan in order to
raise the $75,000 for the cost of their quarter ownership of GSI.

Undeterred by the advent of World War II, the four new partners of GSI took immediate action. Cecil, elected vice president
of operations, together with Peacock, was responsible for keeping the geophysical exploration program going and, if possible,
even expanding it. McDermott and Jonsson decided to seek contracts with the Armed Forces to make special war matériel. The
laboratory and manufacturing division created in Dallas for production of war equipment soon grew so rapidly that it outstripped the geophysical operations, and in 1951 it was made
the parent company—with GSI a wholly owned subsidiary—
and named Texas Instruments Inc. Green, Jonsson, McDermott,
and Patrick Haggerty (hired after discharge from the U.S. Navy)
were the co-founders.

On October 1, 1953, Texas Instruments common stock was
offered for sale to the public for the first time, on the New York
Stock Exchange. The first one hundred shares sold at $5.25
each. The electronics revolution was now in full swing. By early
May 1960, TI stock had skyrocketed to more than $210 a share,
and the Greens, as major stockholders, could now become philanthropists—what Ida had declared she would like to be at the
time they bought their quarter share of GSI, if she and Cecil
ever became rich. Now they were multimillionaires.

Actually, the Greens started their remarkable program of sharing as early as 1950, but did not initiate their major program until late in the decade. Since that time, however, they have given munificently of their means, at the same time maintaining a continuing interest in the purposes and projects they have supported. They have taken the time to acquaint themselves with their recipient institutions' current needs and future plans and have often helped those institutions by volunteering for fund-raising activities and by monitoring the success of their newly initiated projects. Following are the major results of their carefully planned philanthropy to date (1988), made possible by their donations totaling more than $150,000,000:

50 buildings, constructed, enlarged, renovated, or otherwise improved, together with furnishings, facilities, and special rooms, for 30 different institutions. Thirty of the larger buildings are named for Cecil and Ida.

Parcels of land, purchased for 7 different schools and research organizations.

28 endowed chairs, in 15 different institutions, ranging from appointments as master teachers in the primary grades, to distinguished professorships in leading universities. Most of the latter carry the Greens' names.

10 endowed awards to students, in the form of a gold medal awarded annually, fellowships, scholarships, internships and residencies—in a number of cases awarded to more than one person—involving as many as 50 recipients a year, in 10 different schools.

6 machines (computers, hospital facilities, geophysical instruments, etc.), a ship, and numerous other facilities for teaching and research in 8 different institutions, for research in science, medicine, earthquake detection, and oceanographic exploration.

Four additional projects not falling into the above categories were made possible by the Greens' personal efforts as well as funds:

Co-founding the Dallas branch of the University of Texas (UTD).

Co-founding TAGER (The Association for Graduate Education

and Research of North Texas), an educational audiovisual television facility serving schools and high-tech industries in the Dallas–Fort Worth metroplex.

Initiating a series of three international conferences on biological imaging.

Founding Cecil H. Green College (Oxford University).

Finally, Cecil, as chairman of numerous fund-raising campaigns has induced others to pledge millions for projects that he himself was also asked to support. This, and the time, thought, and effort that both he and Ida have given as part of their overall philanthropy have surely inspired others to emulate them in their own sharing.

1
THE GREENS' YOUTHFUL YEARS
(1900–1924)

CECIL'S BIRTH AND EDUCATION (1900–1924)

The first year of the twentieth century was less than eight months old when a son was born to Charles and Maggie Green in Whitefield, a northern suburb of England's industrial city of Manchester. The boy, born on August 6, 1900, was named Cecil Howard Green—Cecil, after Cecil Rhodes of South Africa fame, and Howard, his mother's maiden name.

The small cottage that was Cecil's birthplace (No. 23 Higher Lane) was located on a principal east–west street that follows the route of a military highway built by the Romans when they occupied Britain. No trace of the house remains, and the lot on which it once stood is today a paved parking area. Cecil remembers nothing of his birthplace, because he and his mother left Whitefield for Canada when he was not yet two years old. He has never returned to the actual site of his birth.

What Cecil does know about this earliest period is that conditions at the turn of the century were not very favorable for young couples starting life together in England. During the later years of Queen Victoria's reign, when the British Empire had extended its influence throughout the world, many young Britons, both men and women, single and married, left the land of their birth to seek their fortunes overseas, where they hoped opportunities were better. Among these was Charles Henry Green of Whitefield, who decided to quit his job in a local industry and try his luck in Canada. Leaving his wife and infant son behind, he crossed the Atlantic and found employment as a maintenance engineer in the coal mines of Nova Scotia near Sydney on Cape Breton Island.

When he had saved enough to pay for their passage, he arranged for Maggie and baby Cecil to join him in Sydney, but within a year, seeking better employment, he moved his small family—Cecil would be their only child—westward to Montreal. Life there was more difficult than they had expected, however, because most of the people spoke French, and Charles and Maggie knew only English. After less than a year, they relocated again, this time to Toronto, where English was the common language, conditions were more congenial, and Charles hoped the family could settle down for a while.

By this time Cecil was almost five, old enough to be aware of what was going on around him. His earliest memory is of an incident in Toronto that involved a pair of new shoes. As Cecil now recalls it, his mother took him along on a shopping trip to a local department store, where a salesman fitted him with a pair of new shoes, putting the old shoes in a bag for Cecil to carry home. While Maggie continued her shopping, Cecil strayed from her side and slipped out of the store. When she turned around to look for him, he was nowhere to be seen. After a frantic and unsuccessful search throughout the store, Maggie—greatly distraught—decided to return home. As she approached the house, Cecil saw her coming and ran down the porch steps crying out happily, "Mama, see my new pair of shoes!" Cecil doesn't remember just what happened after that, but he concludes—knowing his mother—that he probably received a severe reprimand. And to this day he cannot figure out how he safely crossed busy Yonge Street on his way home.

Shortly thereafter, several of Maggie's relatives who had emigrated from England to the west coast of North America wrote enthusiastically of the salubrious climate along the coast from northern California to southern British Columbia. Impressed by those letters, Charles and Maggie decided to move once more—to California. Upon arrival in San Francisco, Cecil's parents engaged lodging on Jackson Street near Van Ness Avenue, and his father soon found employment as a mechanic in the city's famous cable car railway system. Cecil has no memory of his fifth birthday, which must have been celebrated soon after the family's arrival. But what he does remember is the great earthquake of April 18, 1906, that shook San Francisco to its very foundations. Indeed, that event so profoundly impressed him, as a five-year-old boy, that he remembers it today, eighty-two years later, as if it had happened only yesterday, and whimsically credits it with introducing him to an aspect of earth science that he would refer back to as an exploration geophysicist.

Earlier that spring, Charles Green, dissatisfied with work on the cable car system, had decided to move his family to Vancouver, where Maggie had relatives. As when he first came to America, he traveled ahead to find work to his liking, planning to send for Maggie and Cecil as soon as possible. The earthquake struck before he could do so. Cecil remembers that he was awakened at about 5:30 in the morning, by ceiling plaster falling on his face and his bed shaking. He recalls that he jumped out of bed and tried to open the door but found it stuck: the door frame had been twisted by the shaking of the house. His mother had to call for help before they could get out. Barefoot, clad only in night clothes, with whatever belongings they could carry, Maggie and Cecil joined the crowd of stunned people milling about in the street. A soldier, seeing that Cecil had nothing on his feet, broke open a store window to get him a pair of shoes.

The skies were unnaturally brightened by fires everywhere in the city and surrounding lowlands, and Cecil has never forgotten the scene as dawn came to reveal the devastation caused by the earthquake and resulting fires.

Prohibited from returning to their living quarters because the house was to be demolished as part of a city firebreak, Cecil and his mother were directed to Golden Gate Park, where they found shelter in a tent, stood in food lines, and watched the city burn. He remembers vividly the pall of smoke and the flames that gave an eerie appearance to the midday sky.

While confined to the tent camp in Golden Gate Park, Mrs. Green had no way of communicating with her husband in Vancouver to tell him that she and Cecil were safe, since he had not yet been able to inform her of his whereabouts or prospects for employment. Under the circumstances it seemed advisable for Maggie and Cecil to go to Vancouver to try to find him there. So, after riding down Market Street in a horse-drawn wagon, from Golden Gate Park to the Ferry Building, they took the ferry to Oakland and then boarded a train for Vancouver. (Cecil remembers that he was much impressed en route by snow-capped Mount Shasta.)

In Vancouver, Cecil and his mother were incredibly fortunate.

On the second day after their arrival, while walking down Hastings Street, they were overjoyed to encounter Charles—one of the earliest examples of how good fortune has attended Cecil Green's activities for more than eighty-seven years.

Once again united, the Green family settled down in Vancouver. Cecil's father had found work to his liking as an electrician with the Portland Cement Company at Tod Inlet on Vancouver Island (now the site of the famous Butchard Gardens). This job did not last very long, however, and Charles soon obtained better-paying employment in the copper smelter of the Granby Consolidated Mining and Smelting Company at Anyox on Alice Arm just north of Prince Rupert, only about twenty-five miles from the Canada–Alaska border. Unfortunately for Maggie and Cecil, the smelter was some 450 miles northwest of Vancouver, which meant that Charles would have to be separated from his family all of the time except for a very few brief vacations. Anyox was too primitive for Maggie and Cecil to live, so they remained in Vancouver.

Cecil had a happy and healthy childhood, secure in the love and attentiveness of his mother and her steady training in ethical standards. Since she accepted his father's absence with courage, confidence, and serenity, so did Cecil, who maintained friendly and respectful feelings toward his father.

At six years of age, Cecil started his elementary education in Vancouver at the Fairview School, a one-room building at the corner of Broadway and Granville, then continued later in a school with four rooms, located at Oak Street and 25th Avenue. He subsequently entered King Edward High School in the spring of 1914 and graduated in 1918.

The Vancouver school system conducted a demanding program of study, starting with reading, composition, recitation, and arithmetic, followed by Latin, French, history and literature, mathematics, and the basic sciences. Regular oral and written testing was a standard part of instruction, and poor performance was not tolerated. A pupil had to pass special examinations to advance from the primary to the secondary level, and a rigorous provincial examination was a requirement for graduation.

Vancouver was a relatively small city at the time, and in the hours away from school a boy could wander into the large park or the countryside nearby to pick wild berries, follow a stream-let, or just enjoy the benign aspects of nature. Sometimes Cecil and his friends took a boat out into the bay to fish, and cooked and ate their catch on the shore. He did not engage in team sports to any degree, but did much walking.

Cecil studied the violin, too, as a boy and gained such skill that he played in the Vancouver Civic Orchestra. He took the violin to college and he played whenever possible, but the itin-erant nature of his years after college made transporting the in-strument impractical.

Cecil remembers little about his earliest school years, except that his mother always urged him to excel in whatever he did. Especially, she encouraged him to try to outdo his Vancouver cousins in their schoolwork—in friendly competition of course. He often comments, however, on the high quality of the edu-cation he received in his early years, and how well that training prepared him for college work. He has long regarded sound early instruction in the basic disciplines as the necessary foun-dation for future success. Indeed, it was this belief that later led him and Ida to support a number of preparatory schools in the Dallas–Fort Worth area, especially St. Mark's School of Texas in Dallas.

During Cecil's high school years, the countries of Europe were locked in the deadliest, most overwhelming struggle for sur-vival this world had yet seen. But Vancouver was on the western edge of a vast, thinly populated country, and communications were delayed and distant. The war never had sufficient re-ality for the boy to inflame his patriotism or arouse his moral indignation.

While in high school (1914–1918), Cecil delivered newspapers for a time; served as an evening usher at one of the theaters; worked one summer as an auto mechanic's helper, which proved valuable later on; helped some neighbors' boys to build a boat; and in the summer of 1918 worked as a caulker's helper in Vancouver's Coughlin Shipyards.

The caulker's job, which was inside the ship's bilge, was to caulk the seams between the steel plates that formed the bottom of the ship; this the caulker did with a pneumatic hammer. Cecil's job was to assist the caulker in any way he could, meanwhile enduring the high decibel level from the hammer's striking the metal plates. The constant din exacted its toll: Cecil recalls that at the end of each day's work he could not hear, and it took a night's sleep to restore his hearing. Although he did not realize it at the time, his hearing was being permanently impaired.

Cecil had learned from his different part-time jobs during high school that he liked to use his hands and enjoyed working with people. Furthermore he liked work that produced a practical result. So during the summer of 1918, after graduating from King Edward High School, Cecil decided to begin coursework at the University of British Columbia in the fall, and to prepare for a profession that would involve practical work with other people. Even at this early age he wanted to apply what he learned from his studies and from his numerous odd jobs.

After completion of the required first year of study in UBC's Liberal Arts Department, he was more than ever convinced of his desire to be an applied scientist—an engineer of some kind. So he transferred to UBC's School of Applied Science—it was not then called engineering—and during the next two school years (1919–1921) he concentrated on the courses in mathematics, physics, and chemistry that would prepare him to become an applied scientist.

At this early stage in his college education, however, Cecil was not yet certain which kind of applied scientist he wanted to be. It was natural for him to be interested in electricity because his father was a maintenance electrician. But Cecil had also gained some knowledge of machinery, as well as of electricity, in assisting that auto mechanic while in high school. In order to get more hands-on experience in the kind of work he most enjoyed, Cecil sought short-term jobs that could be done outside college during week ends, vacations, and summers.

One of Cecil's first outside jobs used his experience in auto mechanics. Having induced his mother to buy an automobile, and having learned quickly to drive and maintain it, he secured a part-time job as chauffeur for a wealthy retired farmer from the prairie provinces. His most important responsibility was to see that his employer was delivered home safely after a night of conviviality with his chums at their club in downtown Vancouver. It took patience on Cecil's part to wait the long hours before midnight outside the Hotel Vancouver so as to be ready to guide his unsteady passenger to the parked car and then later to his front door. Cecil's patience was well rewarded, however, by the generous wage that the farmer paid him for his services. The pay helped him to meet his college expenses.

One summer Cecil worked as an electrician's assistant at the same Anyox copper smelter where his father was employed. However, he deliberately elected not to work with his father, but with other electricians at the smelter instead. He wanted no favors!

The superintendent of the smelter, a Mr. Speight, soon took notice of the eager and personable young assistant and engaged him in conversation about his college work. What kind of professional career in engineering was he planning, and what school did he expect to attend if and when he left the University of British Columbia? Speight, a mining engineering graduate from the Colorado School of Mines (CSM), talked favorably about his alma mater and the challenges and rewards of the profession of mining engineering. Cecil liked the man, was much interested in what he had to say, and was strongly tempted to switch his interest from electrical to mining engineering. He also received a very good impression of CSM. Though Cecil did not finally attend CSM, he never forgot the discussions with Speight, and thirty years later became deeply involved in that school's affairs.

Back at college, Cecil contracted to help wire a neighbor's new house for electricity. During the same period he pursued amateur wireless as a hobby, and helped several of his schoolmates to organize a wireless club (figure 2). In recalling his student years at UBC (1918–1921), during a taping session, Cecil reminisced as follows about the Wireless Society:

. . . there was wireless telegraphy with a government transmitting station just a few miles away out on Grey Point, the present site of UBC. I decided to delve into this intriguing radio business by stringing a four-wire aerial between two tall trees on the vacant lot next to our house. With the help of descriptive matter from a commercial manual, I built a variable condenser, wound a tuning coil, and acquired a crystal detector with "cat whiskers." So—I was soon listening with real thrill to Morse Code emanating from the Point Grey station. A few of my classmates became similarly intrigued, and so was formed the first Radio Club [Wireless Society] of UBC. This bit of experience . . . steered my ambition in the direction of electrical engineering.

Clearly, Cecil was becoming more and more interested in the practical uses of electricity, but he knew that even with the completion of the program of study he was following, his degree would be in Applied Science: UBC did not yet award a degree in Electrical Engineering in the early 1920s.

Fortunately for Cecil, however, one of his chemistry professors was a graduate of the Massachusetts Institute of Technology (MIT), then still known as Boston Tech, and like Mr. Speight, he told Cecil about *his* alma mater, especially praising the Cooperative Course in Electrical Engineering. What Cecil heard about the course sounded exactly like what he wanted—a program in which classroom study alternated with practical work off-campus in a company such as General Electric (GE). The Co-op Course required three full years, including summers, starting after completion of the regular sophomore year—which meant prior preparation in basic mathematics, physics, and chemistry. After a total of five years schooling, students qualified for two degrees: a Bachelor of Science on completion of the senior year and a Master of Science in Electrical Engineering upon completion of the fifth year, including an acceptable thesis.

Wireless Society

J. W. Rebbeck   H. W. Gwyther   W. G. Walker (President)   C. H. Green

2   Cecil as a college student at the University of British Columbia and
at MIT. *Top:* Cecil H. Green, MIT, Course VI-A, '23. Courtesy of MIT
Class of 1923 Technique. Bottom: Amateur wireless was a hobby of Cecil's during his student years at the University of British Columbia
(1918–1921). He helped to organize the Wireless Society, and is shown
at the extreme right with three of his fellow members. Courtesy of
Bridgman Studio (Vancouver).

By the spring of 1921, when Cecil had completed three years
of work at UBC—the last two in applied science—he was ready
to move on. He had definitely decided on electrical engineering
as his field of specialization. Cecil's father favored his continuing his education at MIT—the new name replacing Boston Tech
when the school moved to Cambridge in 1916—and his mother
agreed, whereupon Cecil applied for admission to the Institute
as a transfer student to enter the three-year Cooperative Course
in Electrical Engineering in October 1921. His admission to MIT
was approved, and he confirmed his intention to attend—a decision that profoundly affected the lives of all three members of
the Charles Henry Green family, particularly because it called
for unusual sacrifices by both parents. It was the first of numerous turning points in Cecil's long life.

The plan that his parents had agreed on, in anticipation of
the Institute's favorable decision, was now put into action. The
Greens sold their home in Vancouver, together with its furnishings, and they also sold the automobile that Cecil had persuaded them to buy while he was attending UBC. His father
would continue to work as an electrician in the copper smelter
at far north Anyox. Cecil and his mother packed a limited selection of personal belongings and took the train to Boston. The
Green family would not be together again for almost three
years, until Charles came to Cecil's graduation in June 1924.

In Cambridge, Cecil and his mother rented rooms in a house
at 12 Farwell Place, a block west of Harvard Square. Here, for
three years, Cecil's mother prepared their meals, took care of
their lodgings, kept an eye on Cecil's health, encouraged him in
his studies, and did whatever else was possible to make their
temporary quarters comfortable. Few parents would have made
the sacrifices that Charles and Maggie Green made for Cecil,
even though he was their only child, and Cecil has never forgotten what they did. For many years after graduation, he faithfully wrote them a letter every week; he visited them as often
as possible, wherever they lived, and took them on sight-seeing
trips during visits. When Charles retired, Cecil and Ida bought
them a home in San Diego.

It was not until decades later, when he was freed from the pressing demands of his career and had time for reflection, that Cecil came to appreciate how much his parents had sacrificed to make it possible for him to attend MIT—especially his mother. His last appreciative act was to arrange that Maggie spend her final years in one of California's best rest homes, where he kept in touch with her by visits or telephone calls until she died on the last day of her ninety-fifth year.

When Cecil entered MIT in October 1921, he was classified as a member of the Class of 1923, but with a few extra subjects to make up. Though his preparation at UBC was excellent as far as it went, Cecil quickly found himself challenged to the limits of his ability and strength. That he met MIT standards, however, is attested by the excellent grades that he earned in every subject.

All of Cecil's work assignments away from the Institute were to plants of the General Electric Company. Of his five assignments the first three were to GE's plants in nearby Lynn, Massachusetts; the next, to the plant in Pittsfield, Massachusetts, where he tested high-voltage transformers; and the last, which came during his graduating year, was to the main plant in Schenectady, New York. During the last assignment he did the technical work of his Master's thesis under the supervision of Professor F. S. Dellebaugh, Jr.; its title,

"A Static Study of the No Load Flux Distribution in a Salient Pole Synchronous Alternator."

Like all MIT graduate theses, the original copy of Cecil's thesis is now preserved in the MIT Archives Special Collections.

The Green family was reunited in Cambridge to celebrate Cecil's receipt of two diplomas (SB and SM degrees) at Commencement on June 10, 1924. Immediately after the Commencement exercises, Cecil and his parents returned to the apartment near Harvard Square, and there packed the last of the furnishings in readiness for departure to Schenectady, where Cecil had accepted a full-time appointment in GE's main plant.

Upon their arrival in Schenectady, the family quickly secured suitable living quarters, and Charles inquired about employment at GE but found none, so he returned to his latest job, which was then in California. Maggie, as she had done so many times before, set about establishing a comfortable household for her husband and son.

At the GE plant, as a full-time employee, Cecil divided his time between research and design work in the Turbine Generator Design Department of the A.C. Engineering Division and work as an instructor in advanced engineering at the General Electric School. This experience—especially his teaching—made him aware of the critical importance of the academic/industrial cooperation and collaboration that he had enjoyed earlier as a Co-op student, when he had worked with and under many different GE scientists and engineers, in laboratories, shops, and classrooms. During the next sixty years, he would foster, promote, and financially support such opportunities for scores of other students.

Although the job opportunities and the beginning salary of $55 a week at GE were attractive, Cecil admits that there was an even greater attraction in Schenectady: a certain vivacious young lady in GE's statistics department with whom he had fallen in love during his student internship. Cecil quickly renewed his relationship with that vivacious young statistician—Ida Mabelle Flansburgh—who admits to having wondered, hopefully, if Cecil would return to GE after graduation from MIT. Now that Cecil had a full-time job, he felt that he could propose marriage to Ida. She quickly consented to engagement, with marriage at a not too distant date.

Charles and Maggie stayed on in Schenectady only until Cecil and Ida were married in 1926. They then left to establish residence in Tormey, near San Francisco and adjacent to the Selby copper smelter. Charles worked at the smelter until he retired in the early 1940s. He and Maggie then moved to a home in East San Diego that Cecil and Ida had purchased for them, and there they remained together until Charles's death in 1949.

Meeting Ida and becoming engaged to her, accepting a full-time job at GE, marriage, and, for the first time, experiencing life outside the family circle that his parents had created, together marked the second major turning point in Cecil's life. These events were also a turning point for Ida.

IDA'S YOUTHFUL YEARS IN UPSTATE NEW YORK (1903–1924)

Ida Mabelle Flansburgh was born in Pittsburgh, Pennsylvania, on January 19, 1903, to Louis Wilford Flansburgh and Laura (née Bolton) Flansburgh. Soon thereafter her parents moved back to the Schenectady area, the longtime home of her ancestors, where she grew up and received her early education. Her younger brother, Frederick Louis, was born in 1918. Ida and Frederick were the Flansburgh's only children.

The Flansburghs, whose forebears were early Dutch and Danish settlers in Upper New York State, were determined, hardworking country people who owned and cultivated large farms, raised cattle, made maple syrup in the spring, picked apples and dug their own potatoes in the fall, and carried on all of the other activities usual for a New York farmer in the early 1900s. The Boltons and their ancestors—Veeders, Vedders, and Furmans—were highly respected early seventeenth-century Dutch residents of the Schenectady area, many of whom were professional people—lawyers, commissioners, and merchants. Ida has always been proud of her ancestors on both sides and has often mentioned how she has tried to emulate them by being straightforward, respectful of learning, and considerate of the rights of others.

For several years during Ida's girlhood her family lived on the edge of the Adirondacks, and in that scenic region she developed a love for and appreciation of nature. This early exposure to country life contributed to the development of her forthright character and her good sense of values. It also gave her the practical knowledge and experience that would be valuable later on when, as a young wife, she would have to make a temporary home, often in some small town or remote village, for herself and Cecil while he and his geophysicist colleagues conducted fieldwork nearby.

Ida remembers her paternal grandfather, Henry Flansburgh, with special affection. He had wanted to be a lawyer but was forced to abandon that ambition because of family needs. Turning to another interest, biology, he set out to breed an all-white cow, and by selective crossbreeding finally succeeded. Likewise, being interested in developing new varieties of string beans—the kind that entwine corn stalks and poles—he ultimately developed thirty-two different varieties. Ida was always much interested in his different experiments.

Whenever she visited his farm, which was always a delightful experience, she would eagerly look in her room for the stack of books he had selected for her to read. Her love of reading, greatly stimulated by the thoughtful grandfather who knew just what books a little girl would enjoy, has persisted through the years. Ida remembers that early in her marriage, before she and Cecil had a radio, they often read books aloud with their guests for evening entertainment. And she often mentions how in later years her omnivorous reading helped her pass the lonely hours when Cecil had to be away on business.

Ida's pride in the family name, and the inspiration she gained from her grandfather Flansburgh, led her to request the use of her full name—Ida Flansburgh Green—when coupled with Cecil's in the naming of endowed buildings and scholarships.

Ida received her early education in the public schools of Scotia, a small town across the Mohawk River northwest of Schenectady. She had hoped to go to college after completing high school, but in 1920, in her junior year, she suffered a grave skiing accident and missed so much school that she finally withdrew in discouragement and started to look for work. Her inquiries at General Electric, where her father was employed as a mechanic in the assembly of steam turbines, brought her a position as file clerk in the Customer Statistics Department. In this role she monitored GE's complete list of customers; then she worked on requisitions. Finally she was advanced to the position of industrial statistician, in which she held her own against a dozen men at adjacent desks in her department. At this juncture, Ida decided to move into Schenectady from Ballston Lake, where she was living with her parents on their small farm. The move brought her closer to the GE plant and also made it much easier for her to enjoy evening activities in the city.

Pretty, with expressive blue eyes, red-gold hair, a slender figure, and a serene, melodious way of speaking, Ida was a popular companion for the numerous young men of her acquaintance who invited her to evening activities. She had won a prize in the plantwide health and personality contest, was a member of GE's Theatre of the Air, and gave expression to the more glamorous aspects of her nature in playing the star role in "Pollyana" and "Black Velvet," the first radio programs offered by GE's station, WGY. She was vivacious, friendly, and practical, and many of her admirers must have longed to inspire deeper interest on her part. But Ida was in no hurry to marry. The man whom she could love and admire, and share her life with, had not yet appeared on the scene; meanwhile, she was, in her words, "playing the field" and having a good time, to the frustration of her match-making women friends. Then she met Cecil.

It happened this way. Not long after Cecil's arrival in Schenectady in the fall of 1923 to work on his Master's thesis in one of the General Electric laboratories, one of Ida's office friends telephoned her, saying, as Ida reports it,

"Ida, I have found the right man for you at last. You know, the way you have been dating, you will never get to like anyone well enough to marry him. So you just come down to my floor. I want you to meet someone."

Always quick to accept such a challenge, Ida agreed to meet the someone. As she walked down the stairs, she saw a fast-stepping young man approaching her along the corridor below. She recalls that she looked him over quickly, then suddenly said to herself,

"*There* is the man I am going to marry, and I have never met him; I don't even know who he is."

When she reached her friend Ethel's desk, there was the same man, and he was introduced to her as Cecil Green. It was love at first sight. Cecil declared his affection after only three or four dates, but with her typical caution Ida wanted to know much more about this young man before she would finally make up her mind about him. During his first six months in Schenectady, Cecil became Ida's only suitor, and she corresponded with him after he returned to MIT to complete his Master's thesis and graduate. Nevertheless, she wondered if she would ever see him again. After all, the MIT Co-op students were only transients in Schenectady; they rarely returned to GE after graduation. Not so with Cecil: he returned to Schenectady immediately after graduation in June 1924 to take a full-time position with General Electric, and soon thereafter Ida happily accepted his proposal of marriage.

2
THE LEARNING-WHILE-EARNING YEARS (1924–1941)

CECIL AND IDA MARRY AND MOVE TO BOSTON (1926)

Cecil and Ida continued in their respective positions at General Electric until early in 1926, when Cecil received an attractive offer from the recently organized Raytheon Manufacturing Company,[1] in Kendall Square, Cambridge, near MIT.

The offer came from one of Cecil's MIT classmates, Miles W. Pennybacker, whom he had come to know quite well while they were working in several GE plants, including Schenectady, as Co-op students in electrical engineering. Pennybacker, who was sales manager of the young company, considered its future so promising that he made a special trip to Schenectady in order to tell Cecil about the company and urge him to accept the offer of a job. The company concentrated on making a thermionic tube, variously designated a "gaseous rectifier," "battery eliminator," or "S-tube," that had resulted from the research of C. G. Smith and Vannevar Bush at MIT. By converting AC from power lines to DC needed to operate radios, this tube could eliminate the 45-volt B batteries in the receivers of the 1920s. Pennybacker foresaw a tremendous market for the S-tube and was responsible for building up the company by successfully marketing the device. But the company also needed someone to direct and manage the production of the S-tubes, and that was the job for which Green was sought.

"You ought to leave this big outfit here," Pennybacker told Green during the visit to Schenectady. "You don't want to be merely a cog in a big machine! Why don't you come to Cambridge and get in on the ground floor of a new company!"

Raytheon's offer caught Cecil at a critical time. After a year and a half designing steam turbine generators, he was beginning to wonder about the future of his position: would he be

---

[1] Excellent discussions of the organization of Raytheon are given in the following books, which have been drawn on for some of the details of this section: Otto J. Scott's *The Creative Ordeal: The Story of Raytheon*, New York: Atheneum, 1974, 429 pp., and Vannevar Bush's *Pieces of the Action*, New York: Morrow and Co., Inc., 1970, 366 pp.

better off in a smaller company that was just getting started than in General Electric, which was large and rigidly organized, making advancement slower? Pennybacker's argument was persuasive.

Green heard an opposing point of view, however, when he conferred with his GE boss about Raytheon's offer. Cecil recalls that his boss's response was emphatic and to the point:

"See here, young man, don't conclude that you can't rise to the top in GE. Sure, it is a big company, but think of a bubble of gas at the bottom of a tank full of oil. The bubble rises through the oil regardless of the size of the tank and the amount of oil in it. You can do just what the gas bubble can do, right here at GE!"

That piece of advice also impressed Cecil, and created a dilemma for him: should he remain with GE, or should he accept Raytheon's offer? He discussed the Raytheon offer with Ida, and together they decided that it was attractive enough to accept, even though the salary was the same as GE's—$70 a week. That critical decision having been made, they decided to marry before they left Schenectady for the Boston area.

Their wedding took place on February 6, 1926, and the very next day Cecil departed for Cambridge to report for work at Raytheon and to find living quarters. Ida remained behind to pack their belongings and have them shipped to Boston. It was several weeks before the couple was reunited and their possessions installed in an apartment on Brainerd Road off Commonwealth Avenue in Allston, near the edge of Brookline.

Cecil immediately took up his new work in Raytheon's space in the Suffolk Building in Kendall Square, usually walking from the Allston apartment across the Charles River via the Cottage Farm Bridge (now called the Boston University Bridge) to Cambridge. It was a long walk but rich in the essence of Boston. He saw the river mirroring the changing seasons—blue in summer, gray and wind-tossed in stormy weather, white with ice and snow in winter. As he walked eastward along the Cambridge bank, and crossed Massachusetts Avenue, he saw the long row of townhouses, with their quaint chimneys and ventilators, against the background of the gold-domed State House and the

Custom House tower, Boston's only skyscraper at the time. And on his left, as he proceeded toward Kendall Square, was the massive complex of buildings constituting the Massachusetts Institute of Technology. He had been an habitual walker from his youth, and he remains so even today, though the distances have decreased to a couple of miles and his pace has slowed.

For Cecil, the Raytheon job meant a change from designing large steam generators to research and development on a variety of small electronic devices that were ushering in the Electronics Age. For Ida, it meant giving up the respected position that she had earned, with no certainty of finding comparable work in the Boston area. In fact, excepting brief jobs in Boston and later in Seattle, Ida never again sought or enjoyed paid employment, although later volunteer work with poor or handicapped people would occupy some of her time and energy. In 1926, married women were expected not to work for pay outside their homes unless necessity forced them to; a working wife reflected poorly on a husband's ability to support her. So Ida concentrated on the responsibilities of establishing and maintaining a household for herself and Cecil. Thus the move from Schenectady to Boston was another major turning point in both of their lives.

CECIL'S FIRST EMPLOYMENT BY RAYTHEON (1926–1927)

Cecil's work on the S-tube at Raytheon went well, and he was soon rewarded with an increase in salary. Ida, for her part, spent considerable time expanding her culinary repertoire, decorating their house, and especially reading. The two of them took long walks that commonly ended at the nearest library, where they borrowed a supply of books for the next week's reading. They became avid movie-goers, and found they greatly enjoyed making new friends. Hardly a week passed that they did not entertain in their small apartment or share an evening with friends.

Weeks added up to months, and soon they had been in the Boston area for almost a year. Ida was happy in her new life and felt no strong urge to seek other outside activities; she was content for the moment to spend as much time as possible with Cecil. But Cecil, although getting along well at Raytheon, began to feel what he calls "West Coast fever". That desire to return to British Columbia or California remains with him even today. After deliberating at length, he made the bold decision to move to California. Cecil had no certainty of finding work on the West Coast, and he realized that leaving Raytheon was a big risk, especially since the family savings amounted to only about $1,200 accumulated through thrifty living. Nevertheless, he was determined to go. As for Ida, she was not at all enthused about the move, particularly because it would take her much farther from her parents, who still lived in the Schenectady area. But never the one to turn down a chance to visit distant places and have new experiences, she finally agreed to break up their Boston household and move to San Francisco, near Cecil's parents.

Once the decision to go had been made, the question of logistics arose: how to get to San Francisco on such limited funds. After weighing the alternatives, they decided to go by automobile, camping along the way and preparing as many of their meals as possible. This decision meant buying a car and camping equipment. It would have to be a second-hand car, however, and the camping gear would have to be inexpensive yet suitable for a transcontinental trip.

THEIR FIRST AUTOMOBILE

The search for a used car began in January 1927. Fortunately for the Greens, the January 27, 1927, edition of the *Boston Globe* carried the following classified ad:

1926 Chevrolet Touring. Only been run a few thousand miles and will sell at a big sacrifice. Call Mr. Harson, 13 Mass. Ave., Cambridge—Porter 2405.

Cecil recorded the ad in his diary and then noted:

We both went over to the garage to look at it and made a date for a road exam tomorrow at 1 p.m.

The *We* in the entry is underlined to show the nature of their marriage partnership: Cecil and Ida were doing things together and making decisions together. Ida was even going to help decide whether or not they bought that car! Cecil's entry for the next day (January 28, 1927) reads:

Took spin in our prospective car at 1 p.m. today. Performed great on the boulevard so made $10 deposit arranging to pay balance tomorrow.

The automobile that Cecil and Ida bought for some $200— they don't now remember the exact amount—was a 490 Model Chevrolet touring car with four cylinders, equipped with side curtains having small windows of celluloid to admit light when the curtains were buttoned to keep out the rain and wind. There was no spare tire and no place for one. If a tire went flat, one had to jack up the car, pry the small-diameter casing off the wheel (a portion of a broken spring made an excellent lever), remove the inner tube, repair the puncture by cold patching (by sticking a rubberoid patch to the surface of the inner tube with a special adhesive), replace the tube in the casing and the casing onto the rim of the wheel (being certain to position the valve stem properly in the hole in the rim), and inflate the tube with a hand pump until the casing showed no flattening when the jack was removed.

Cecil truly loved their first automobile, and he soon personalized it by naming it "Tillie." Of course, he had driven cars before and had even been in charge of one as a chauffeur, but Tillie was different! He has often remarked that it was the best and most important car that he has ever owned because it was the first one. Owning an automobile set him and Ida free to go where they wished, and their elation at the prospect of seeing every part of North America was unbounded. They looked forward eagerly to the time they could show off Tillie to Ida's folks near Schenectady, and later on to Cecil's parents and other relatives on the West Coast.

Their first trip was to Ballston Lake, a few miles north of Schenectady, to visit Ida's parents. With characteristic meticulousness, Cecil recorded the following in his diary for Saturday, April 16, 1927:

Started on first long trip in car. Left Allston 2 a.m. via Mohawk Trail to Ballston Lake arriving at 11 a.m. Trip 240 miles. Ate breakfast in Greenfield [Massachusetts].

By mid-July they had made several round trips to Ballston Lake. It mattered little to them when they started or arrived, either going or returning—such was their enthusiasm and exhilaration over the freedom to make the 240-mile drive. A warm welcome and good food always awaited them, regardless of the time of their arrival. In the summer of 1927 there was also a one-day trip to Newport and Providence, Rhode Island, a 2,030-mile tour of upstate New York and adjacent southern Canada (Ontario and Quebec), and a short trip to Concord and Lake Winnipesaukee in New Hampshire. They clearly were making trial runs for their long drive to the Pacific Coast.

THE FIRST TRANSCONTINENTAL AUTOMOBILE TRIP (SEPTEMBER–OCTOBER 1927)

Meanwhile, between work periods at Raytheon, Cecil was carefully preparing for their cross-country trip. He listed and was accumulating the tools, equipment, and supplies that he assumed he and Ida would need for camping out and for keeping Tillie in good running condition. His two lists follow:

For Camping	*For Tillie*
2 Folding Camp Chairs	Curtain Fasteners
1 Double Cot	Blowout Repair Kit
1 3-burner Coleman Stove	Wheel Pull & Block
1 Umbrella Tent 7′ × 9 ½′	Automatic Windshield Wiper
1 Canvas Carrying Case	Sun Visor & Maps
1 Gasoline Can	Brake Lining
1 Tent-light Extension	Valve Compound
Bed Clothing	Piston Rings
Cooking Utensils	Adjusting Wrenches
Tinned Food	Hand Pump

It is evident from Tillie's list that Cecil anticipated the need to repair punctures and blowouts, reline brakes, grind valves, tighten bearings, and put in piston rings, as well as to change oil and grease the chassis, in order to keep the car in excellent running condition on such a trip.

Cecil also constructed a large storage box, which he attached to the running board on the driver's side, to hold cooking equipment and staples. Fresh food would be purchased along the way, and Ida would cook their meals on the little Coleman stove.

Nothing in their experience had prepared the Greens for this first transcontinental trip. As the day of their departure approached, Ida, especially, became increasingly uneasy about the whole affair—leaving behind family, friends, and familiar settings, by a mode of travel that was by no means conventional in 1927. But she knew that making the trip was what Cecil most wanted to do. Interesting and successful as his work at Raytheon had been, Cecil could only satisfy his yearning for the land of his youth by returning to British Columbia. It would not be the only time that Ida would loyally follow her husband on a new and uncertain adventure.

Their first stop was with Ida's parents, as recorded by Cecil in his diary entry for August 31, 1927:

Leaving Boston (Allston) 3 p.m., odometer reading 5,083; arriving Ballston, N.Y. 12:15 a.m. [Sept. 1], via Albany and Schenectady over Berkshire Trail, 222 miles.

In the 1920s there were no great expressways. Few main highways were even paved; most had only a thin layer of gravel or crushed stone. They commonly developed a ribbed or washboardlike surface, not to mention innumerable potholes. In addition there was dust to torment the motorist and, after rain, sprays of water from the potholes. And woe to the unfortunate driver who slipped off the crown of the road into the side ditch!

Secondary roads not only offered all the obstacles just mentioned but also challenged the automobile driver with loose sand or treacherous, sticky gumbo (a fine silty soil). And unless they were graded regularly, these roads were likely to provide single and double pairs of ruts to bedevil the driver. Tire trouble was endemic, the result of the punishing road surface, with nails, loosened from vehicles and shod horses, and sharp-edged stones.

As to camping areas, general stores, garages, and filling stations along the highways of the late 1920s, they were problematical—certain to be found if one drove far enough but seldom found between towns. For lodgings, there were private homes that opened their doors to the tourists; clusters of small single and double cabins with inviting names; and open camping areas, with or without toilet facilities, where more venturesome tourists could pitch their tents, dip water from a nearby stream, and cook their food on a Coleman stove or over an open fire. Road maps and signs often left much to be desired.

These were the conditions that Cecil and Ida faced on their first transcontinental trip, particularly west of Cleveland. Ida recalls that she was "scared stiff" at the start because she had never camped out before, didn't know what to expect, and feared the West. It took more than a little courage for her to start on what turned out to be a roundabout journey of 4,388 miles through sixteen states plus the District of Columbia (see map, figure 3).

Reading the record of the Greens' early travels by car, one has the definite impression that they were totally unmindful of the theorem that the shortest distance—and presumably the fastest route—between two points is a straight line. Almost invariably they took a longer way, detouring to right and left to see as much of the country as possible.

After a few days of visiting with Ida's parents in Ballston Lake, Cecil and Ida set forth on September 7, 1927. Before heading west, however, they first drove south, stopping for a few days in New York City and Washington, D.C., then proceeded on a westward route that took them through Pennsylvania, Ohio, Indiana, Illinois, Missouri, and Kansas to Denver, with nightly stops along the way where they pitched their tent.

Not until the end of the fourteenth day, in the vicinity of

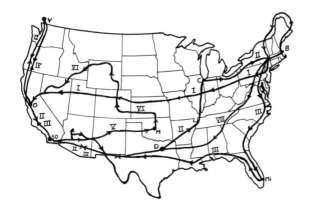

3   The Greens and their longer auto trips (1927–1931). *Top:* Map
showing the routes taken by the Greens during the first few years of
their married life. I. Boston to Oakland. II. Oakland–Vancouver–Oakland–San Diego–Chicago–Boston. III. Boston–Oakland via Miami and
the Gulf states. IV. Oakland–Vancouver–Oakland. V. Oakland–San
Diego–Maud, Oklahoma. VI. Guthrie–Denver–Yellowstone–Palo Alto.
VII. Palo Alto–New York: Cecil by train; Ida with auto by ship via Panama Canal. VIII. New York via Gulf states to Dallas. *Bottom:* Cecil and
Ida Green, the young couple in 1932. Photo from the Green collection.

Granby, Colorado, did they have to seek more substantial shelter because of rain and snow. But then their strength was tested; for the next three days they encountered highway construction, with a road surface of gumbo soil made sticky and slippery by recurrent rainstorms. Salt Lake City offered a municipal auto camp where they stayed a day to clean the mud off the car, repair a battery lead, and grease the chassis, and, of course, to visit the State House, the Mormon Tabernacle, and Solitaire Beach.

Resuming travel on September 29, Cecil wrote,

Left S. L. City municipal camp at 8:45 travelling over gravel roads   route 40–Victory Highway to Elko, Nev. arriving at municipal camp where we took two-bed cabin with coal stove, electric lights, screens, for $1.00. 8889–8619 = 270 miles. Intermittent clouds, rain, hail, and strong, cold wind. 45 miles straight away across Great Salt Lake Desert.

It took a day and a half over washboardlike gravel roads to reach Reno, and more time to reach Truckee, California, following a dirt road through Emigrant Pass in the Sierra Nevada. Then to Auburn, where they made camp for the last time. At one point along the road an auto had carried eight men over the outer edge, killing one.

The last day of their trip, October 2, 1927, took them down the Sacramento Valley via Martinez to Tormey—the small town beside the Selby copper smelter—where they arrived at 4:00 p.m., after repairing three punctures en route that day, to be warmly greeted by Cecil's parents.

Cecil summarized the trip as follows:

Time elapsed ....33 days

States visited (including Washington, D.C.) 15

Miles travelled 4,388

JOB HUNTING ON THE WEST COAST (1927–1928)

Having arrived in Tormey (Selby)—the first goal of their long trip across the continent—Cecil and Ida stopped only long enough for a short visit with the elder Greens, then pressed on toward Vancouver, the more important goal of their trip, where Cecil hoped to find a job in the city that he still considered home. On the way through Seattle, however, Cecil decided to look up an MIT classmate and fellow Canadian, Thomas M. Rowlands, who had grown up in Victoria, B.C., but had moved to Seattle when he could not find employment in the Vancouver–Victoria area. A naval architect, Rowlands had designed and built his own boat, the *Bettybelle*, on which he and his wife were living on Lake Washington.

After a brief exchange of greetings, Tom said to Cecil,

Before you go up to Vancouver, why don't you try to get a job here in Seattle? You and Ida can sleep here on the boat with us while you are looking.

With funds running low, Cecil and Ida happily accepted the generous invitation, and Cecil immediately started his search.

First, he found an ad in a Seattle newspaper that asked, "Would you like to earn $10,000?" He answered the ad, made an appointment, and received his first lesson in "come-on" advertising. Cecil learned that if in a year he could sell a simple little turn-indicating device to the owner of every auto registered in the State of Washington, his commission would add up to $10,000 for the year.

Another ad invited, "How would you like to be a salesman and sell insurance?" Again he arranged an interview, at the conclusion of which the interviewer gave him some advice that he has never forgotten:

"Mr. Green, you don't want to be an insurance agent. You'll be throwing your education away. Furthermore I don't think you are cut out to be one . . . Don't accept my offer or anybody else's offer as an insurance salesman. Try to stay nearer your own field".

Then he added, to Cecil's surprise:

"But I'd like to see you stay here, and I would be willing to lend you $5,000 because you are going to make it all right, whatever you ultimately do".

To which, as Cecil remembers, he replied that he wouldn't want to start work on such a basis.

Cecil's third effort to start on a new career seemed to hold more promise. Neon signs were just coming onto the market in Seattle, and he decided he would set up his own business in Vancouver making and selling neon signs where they had not yet been introduced. He already knew glassblowing and how to handle neon gas from his work at Raytheon, but his responsibilities there had been in production, not in sales. To be successful in the neon sign business, Cecil knew he would have to learn the techniques of selling. So, in the last two weeks of October, he took a temporary job as a salesman with a company called Electrical Products Corporation in Seattle. There he quickly saw that the successful selling of neon signs was simple. The key was to make a first installation. Then other stores in the same neighborhood would quickly place orders to keep pace.

When he was satisfied that he understood enough about selling, Cecil approached the management of Electrical Products Corporation with a proposition: the corporation should set up a branch operation in Vancouver, which he, Cecil, would manage.

"No go," was the Corporation's reply. "It's out of the country and we don't know the Canadian laws, and you still don't know enough about the business. But we've shown you the tricks of selling the signs, and we'll sell you a license under the Claude patents if you want to go up there."

This seemed to Cecil an excellent proposition; furthermore the prospect of a shop of his own greatly raised his spirits.

So on October 29, 1927, having reassembled their camping equipment, Cecil and Ida departed Seattle for Vancouver, with high expectations. Having reached their last few dollars, they welcomed an offer of housing from Cecil's aunt and uncle, Polly and Jack Fox.

THE NEON SIGN EXPERIENCE IN VANCOUVER

Once in Vancouver, Cecil immediately pursued his hopes of selling neon signs. Only, before leaving Seattle, he had decided that he would set up his own shop—not only to sell the signs but also to make them—rather than working as a subsidiary of Electrical Products in Seattle. This meant that somehow, from some source, he would have to obtain the funds to set up such a shop, since his and Ida's resources were at a low ebb.

Remembering the wealthy prairie farmer for whom he had worked as chauffeur while a student at UBC, Cecil decided to pay Mr. Nealy a visit. He enthusiastically described the characteristics of the neon sign that made it so superior to an array of incandescent bulbs. The glass tube could be twisted to form words, symbols, and the like; it required only one-tenth the electricity of a similar sign made of ordinary bulbs; and there were no bulbs to replace. Would Mr. Nealy be willing to lend him the funds to start up his business? Much impressed by Cecil's glowing description of this new type of electrical sign, which had not yet been seen in Vancouver, Mr. Nealy replied:

"Well, I tell you what—I'll lend you the money to set up your shop provided you can show me some evidence of customers—some orders!"

Cecil set out to meet that challenge. He visited Vancouver's largest department stores—Hudson's Bay Company, Woodwards, Spencers, Birks, and others. Several of these potential buyers showed interest. However, when they wondered about maintenance and replacement problems, and asked Cecil where his shop was located, he had to admit that he did not yet have a place of business. "Oh! Well, come back when you have one!" they replied.

Furthermore one of his potential customers asked Cecil if he had a city license to make and sell the new electrical sign systems. Cecil hadn't known about this requirement, so he visited the official in charge of licenses, and described the neon sign. He made the mistake, however, of emphasizing that the sign required only one-tenth the electrical power of a similar sign of conventional incandescent bulbs. Whereupon the official exploded,

"Now see here, young man, you mean to tell me that this new sign that you want to make and sell is a small glass tube that can be bent into all sorts of geometric forms or figures, and particularly that it uses only about one-tenth of the current required by a string of ordinary light bulbs? Do you know the main purpose of an electric sign, Mr. Green? It is to help illuminate the street. I'll tell you what I'll do, though—if you will surround your sign with a border of standard incandescent light bulbs, I will approve your license."

Then and there Cecil's first effort at entrepreneurship came to an abrupt end. But the neon sign experience had been invaluable. Cecil had learned that technical knowledge and experience alone were not enough to establish a business; venture capital was also required—and, of course, it helped to have friends in the right places. Although Cecil continued a vigorous search for employment in the Vancouver area for a time, he and Ida finally concluded that further efforts would probably be fruitless. So on November 29, they bade the Foxes goodbye and drove back to Seattle.

RESUMPTION OF JOB HUNTING IN SEATTLE
This time in Seattle the Greens rented a cabin, and Cecil again started looking for work, but to no avail. However, by early December, Ida had found temporary work as a lingerie saleswoman at Frederick and Nelson's Department Store in Seattle. She performed so well that the manager offered her a full-time job after Christmas—an offer that left Cecil rather chagrined in view of his own jobless situation. But alas, another chance for a career for Ida was nipped in the bud. Discouraged by his futile job search, Cecil had been working on Tillie: he bought a new 6-volt battery from Sears Roebuck Company for $7.35, changed the oil in the engine and transmission, greased the chassis, and otherwise put the car in good condition for the next long trip, because he had decided to return to San Francisco at the beginning of 1928 to renew his search for employment there. First, however, there was a hurried all-night auto trip back to Vancouver to spend Christmas day with Uncle John and Aunt Polly Fox. On the day after Christmas, they started for San Francisco. Once again during the following month (January 1928), Ida

had to wait patiently and hopefully while Cecil trod the streets of the San Francisco area looking for a job. But again, luck did not attend his efforts, and by mid-January he had decided to return East—at least as far as Chicago—to look for work. It was a bitter disappointment. The Greens had come West with high hopes that Cecil would find a satisfying job in his field, and Ida had come to prefer California over Massachusetts as a place to live. Even though Cecil was unsuccessful in his search for employment, he and Ida did, during this period, seize every opportunity to see as much new country as possible, by taking trips along the scenic Pacific Coast as well as through the mountainous country inland.

THE SECOND TRANSCONTINENTAL AUTOMOBILE TRIP: SAN FRANCISCO TO BOSTON VIA CHICAGO AND SCHENECTADY (FEBRUARY AND MARCH 1928)

Having decided to return East to continue his job search, Cecil began to plan their second transcontinental trip. (See accompanying map, figure 3.) By now, Tillie had gone more than 15,000 miles and needed to be thoroughly overhauled. The paraphernalia for camping had to be reassembled. And finally, the route they would take to Chicago, their first goal, had to be determined. Cecil decided that the most interesting, as well as the most practical, route to follow, given the time of the year, was through the southernmost states. So it was that Cecil and Ida bade the elder Greens goodbye and headed southward from San Francisco on February 6, 1928. Cecil noted in his diary that this was their second wedding anniversary and that they took a twelve-mile drive along the ocean in the moonlight on their way into Monterey.

Proceeding southward they passed through San Miguel, San Luis Obispo, Santa Barbara, Beverly Hills, and Hollywood, then on to Long Beach, Pasadena, and San Bernardino. They spent several days in the Los Angeles area; then they moved on to San Diego to visit some of Cecil's relatives, who drove them down to Tijuana, Mexico.

Finally headed east on February 16, they followed Route 80 through Phoenix to Tucson, thence back north to Prescott and Ash Fork, Arizona. On Washington's Birthday, they thought it would be exciting to drive north to see the Grand Canyon. Cecil's log for the day reads:

Leaving Ash Fork via Williams en route to Grand Canyon—but stopped by frozen ruts and snow. Put on chains and started back to Ash Fork. Thawed mud clogged wheel against fender. Had to stop and dig out sticky mud. Road again froze after sunset, and so finally got back to Ash Fork after 11 p.m. 76 miles; Cabin $1.25; gas 29¢/gal.

The next day they drove east through Flagstaff to Holbrook, visiting Meteor Crater on the way. They stopped to see the Petrified Forest and the Painted Desert, then proceeded southward to return to Route 80 at Springerville near the Arizona–New Mexico line, thence eastward over the Great Divide at 8,300 feet to Socorro, and finally south to El Paso, Texas, and the Mexican border. Of course they had to walk across the Rio Grande into Mexico and Ciudad Juarez (bridge toll—2 cents) before going on to Van Horn, Texas, where they paid $2.00 for a concrete cottage with shower, bathroom, and garage, and only 15 cents for a gallon of gasoline. From here on, their progress accelerated as they traveled through Pecos, Midland, Abilene, Fort Worth, and Dallas to Texarkana, one day making the impressive total of 319 miles. Although we know that the Greens were interested in seeing as much of the country as possible, they did not record their opinions of this East Texas country, which they would soon enough come to know intimately when Cecil left electrical engineering to become an exploration geophysicist in the petroleum industry.

Dirt roads lay ahead when they had passed through Texarkana into Arkansas, and they had to reach Little Rock before they again found gravel roads. Their course took them across the Mississippi River at Memphis, thence northward on gravel and paved roads to the junction of the Ohio and Mississippi rivers at Cairo, Illinois. From here it was an easy run over paved roads to Chicago, where they arrived on March 6, 1928, exactly a month after leaving San Francisco. More than 5,000 miles had been added to Tillie's odometer, and there were memories galore of magnificent scenery and remarkable adventures. It was indeed "the springtime of youth" for the young couple, and they now look back on this particular trip with a strong feeling of nostalgia. Many times in the future they would drive the highways between California and Texas, but no other trip would offer the excitement of that first venture in their cherished Tillie.

That Cecil and Ida could make such a long, strenuous journey across unfamiliar country, confined day after day to their car, jolted and shaken over so many miles, with a breakdown always threatening—that they could accomplish this and come out cheerful says a great deal about the young Greens. With those qualities they could go far and achieve much—as indeed they did.

Upon arrival in Chicago, they rented two furnished rooms in the northern section of the city, on Sheridan Road, and Cecil immediately began making appointments for interviews. He also called at the office of the American Institute of Electrical Engineers (AIEE) to describe his training and seek help in his search for a suitable position as a research engineer.

In four days of intense effort, Cecil received only one job offer which he refused because it did not interest him. He then decided to go to Boston, feeling sure that prospects were better there; indeed, he hoped to return to his former employer, Raytheon. As he frequently remarks, in the 1920s a successful job search relied far more on personal interviews and contacts than on letters of inquiry and scrutiny of resumés.

Three days later, on March 14, Cecil and Ida reached Ballston Lake, delighted to be back home with the elder Flansburghs and Ida's brother, Frederick Louis. Cecil's log of the trip from Tormey to Ballston Lake includes the following summary of expenses, which shows how inexpensively people could travel long distances by car across the United States in 1928:

Summary of trip costs	Gas and oil	Cabins, food, etc.	February 6 to March 14, 1928 (38 days)
Selby, Cal.–Chicago, Ill.	$75.00	$30.85	
Chicago–Ballston Lake, N.Y.	11.32	5.50	
Totals	$86.32	$36.35	
Total for trip		$122.67	

Mileage: 21.203—15,208
(ca. 6,000 miles)

JOB HUNTING IN NEW YORK AND BOSTON

During the next three months, from March 14 to June 23, 1928, Cecil and Ida used the Flansburgh home in Ballston Lake as a base of operations. Cecil made two trips to New York, where he had interviews with more than a dozen potential employers and checked repeatedly with the AIEE office. But for all his efforts he received only one offer, and after accepting it he quickly changed his mind because of the nature of the work and the low salary.

Finally, on Sunday, June 24, Cecil and Ida packed Tillie for camping and headed east, reaching the Boston Auto Camp at 1515 Commonwealth Avenue in late afternoon. They pitched their tent, and Cecil planned his program for the remainder of the month.

A week or so after the Greens' arrival at the Boston Auto Camp, the *Boston Post* heard about the young couple and in the July 4, 1928, edition published a feature story about them head-lined, "Find Boston Best Place for a Home." With the story was a picture of Cecil and Ida standing in front of their tent, with one of Tillie's front lights to their left (figure 4).

The brief article, which was kindly supplied by the Boston Public Library, ran as follows:

BOSTON POST, WEDNESDAY, JULY 4, 1928

Find Boston Best Place for a Home

MR. AND MRS. CECIL GREENE
Camping in Boston after traveling 21,000 miles in 10 months.

Others may rave about California, Florida, and other sections of the country, but Mr. and Mrs. Cecil Greene, after a 10-month auto tour planned primarily to visit every part of the country with a view of selecting a permanent home, have finally arrived back in Boston, and here they intend to settle down.

"We've covered 21,000 miles in our 10 months' tour," said Mr. Greene out at the Boston Auto Camp, where they are tenting out until they find an apartment here, "but its great to get back again to old Boston.

"After I graduated from Tech I remained here in Boston, but last year Mrs. Greene and I talked the matter over and we decided to go out and see if there was a place that appealed to us more than Boston, and where we felt there might be greater opportunities.

"We started out in September of 1927 and in our travels, we have not only been all over the United States, but up through Canada and down into Mexico. And, after seeing it all, we are both satisfied now that Boston is the best place after all."

Mr. and Mrs. Greene have lived out in the open practically the entire time during the last 10 months, and they both say it will seem rather strange at first when they get back into a real house again.

4 Cecil and Ida in front of their tent and table, in the campground on Commonwealth Avenue (Boston), after arrival from their second transcontinental auto trip from San Francisco, as pictured in the 1928 *Boston Post* story. Tillie's right front light and fender are visible on Ida's left. Courtesy of the *Boston Post* and Boston Public Library.

Others may rave about California, Florida, and other sections of the country, but Mr. and Mrs. Cecil Greene [sic], after a 10-month auto tour planned primarily to visit every part of the country with a view of selecting a permanent home, have finally arrived back in Boston, and here they intend to settle down.

"We've covered 21,000 miles in our 10 months' tour," said Mr. Greene out at the Boston Auto Camp, where they are tenting out until they find an apartment here, "but its [sic] great to get back again to old Boston.

"After I graduated from Tech I remained here in Boston, but last year Mrs. Greene and I talked the matter over and we decided to go out and see if there was a place that appealed to us more than Boston, and where we felt there might be greater opportunities.

"We started out in September of 1927 and in our travels, we have not only been all over the United States, but up through Canada and down into Mexico. And, after seeing it all, we are both satisfied now that Boston is the best place after all."

Mr. and Mrs. Greene have lived out in the open practically the entire time during the last 10 months, and they both say it will seem rather strange at first when they get back into a real house again."

Soon after their arrival in Boston, Cecil and Ida got in touch with their good friends Roland and Helen Beers, and Cecil immediately inquired about the possibility of employment in Raytheon's neon sign division where he had worked previously. Roland encouraged him to contact Raytheon, so on July 1, Cecil made an appointment to visit his former employer. His diary entry for the next day, July 2, 1928, records the results of that appointment:

Saw Shultz, [David T. Schultz] . . . at 9–10 A.M., of neon sign division of Raytheon Mfg. Co. who both wished for my return. Saw Shultz [sic] again in afternoon who offered me production work on neon signs with consideration [for salary increase] in 6 months. I accepted. We found Apt. #32 at 1615 Commonwealth Ave., Allston acceptable. Paid first month's rent—$45.00.

Upon his acceptance of Raytheon's offer, Cecil was instructed to report for work at the company's building in Cambridge on July 9.

Then on July 4, the Greens left their camping equipment in their newly rented apartment, and following their customary practice of sharing good news with Ida's parents, they drove to Ballston Lake to spend a week celebrating their good fortune.

It was a time for rejoicing. The previous nine months had been most discouraging, with Cecil meeting frustration at every turn and their total savings going down as inexorably as the sand in an hour glass. Indeed, he had found it necessary to cash his last traveler's checks during their spring stay with the Flansburghs. The Greens' financial reserve had sunk to less than $150 by the time they reached Boston, and it was drastically reduced by payment of the first month's rent and purchase of some basic equipment for the apartment. Indeed, Ida had only about $15 to cover food expenses until Cecil's first payday. Never again would they come so close to being broke.

THE SECOND TIME AROUND AT RAYTHEON (1928)

When Cecil reported to the Raytheon plant in Cambridge on July 9, 1928, he was assigned to the new neon sign department and put in charge of pumping the gas into the tubes. Two months later, when manufacture of neon signs was terminated, he was transferred to research on photocells and coated filaments with Percy Spencer, inventor of the bi-metal thermostat. Cecil did not care much for this kind of work, so he welcomed Spencer's suggestion that he interview J. A. Proctor, General Manager of the Wireless Specialty Apparatus Company (WSAC) in Jamaica Plain, a southern section of Boston, with the hope of obtaining a foreman's position in their Canadian Department, also located in Jamaica Plain.

Cecil promptly arranged an interview with Proctor and then with W. L. Hodges, Production Manager of WSAC's Faradon Condenser Department. Hodges first took Cecil on an inspection tour of the department, then at lunch offered him the position of Production Engineer of the Canadian Department. The contract specified a salary of $3,600 a year, starting on September 15, 1928, with an increase to $4,000 at the end of six months, if mutually agreeable. Cecil signed the contract with considerable relief and immediately turned to completing reports on his latest research at Raytheon: the status of the experimental neon sign electrode, setup for coating the filament wire of the radio tube, and work on a portable thermocouple.

A YEAR AT WIRELESS SPECIALTY APPARATUS COMPANY

Cecil reported for his new job at the Wireless Specialty Apparatus Company on Wednesday, September 26, 1928. As Production Engineer in the Faradon Condenser Division, he was first assigned to oversee the work of several dozen men and women—the number varied with the orders in hand—who made the paper condensers.

Problems with impregnation, winding, and testing of the condensers immediately demanded his attention because they were adversely affecting the quality and production cost of the product, and hence profits. He was soon involved in revising production schedules and wage scales, developing a program to reduce paper and foil wastage in the winding room, organizing an inventory control system to monitor production, and attempting to resolve the "niggling and corrosive" personnel conflicts that arose unexpectedly. He quickly discovered that he enjoyed working with the people under his supervision, even though their problems sometimes drove him to distraction and required many extra hours of his evenings and weekends. Nevertheless, it was during his year at WSAC that Cecil definitely decided that he preferred to be involved with personnel management rather than with the design and development of instruments, machines, and the like.

Cecil's work at WSAC at first seemed to go well. At the review on December 15, as provided in his contract, his salary was increased to $4,000 a year effective the following March 15. Several days before the latter date, Hodges even went so far as to compliment him on his performance, saying his "good work had been very instrumental in lowering and regulating costs."

Hodges did not know, however, that on that very day (March 11, 1929) Cecil had sent a job application to a company in Oregon. Actually, the idea of going back to the West Coast had been recorded earlier in Cecil's diary on the first day of 1929:

This volume [Cecil's diary] opens with a New Year with everything progressing favorably in matters of health and financial income. It is our earnest hope on this day that these items may continue, so that we be successful in the purchase of a new Chevrolet next spring as well as [in saving] enough funds to make our return to the West Coast a possibility next year at least.

Although trips through New England and Eastern Canada, and the frequent weekend drives to Ballston Lake to visit Ida's parents, were quite adequate to satisfy Cecil's desire to travel, he really wanted to return to California or Vancouver. By October he and Ida had saved $1,000, enough to meet the expenses of the return to the West Coast and of living there while Cecil searched for a new job.

Two other factors contributed to the decision to leave New England. The Greens' close friends of Raytheon days, Roland and Helen Beers, had gone to Texas where Roland had taken a job as a field-party chief with a geophysical exploration company called Geophysical Research Corporation (GRC)—a subsidiary of Amerada Petroleum Company. But, more important, a series of difficulties arose at Wireless in the summer and fall of 1929. Orders for condensers began to fluctuate in size and number. Firing and then rehiring personnel in the winding and impregnating room in the Faradon Condenser Division became a recurring necessity that undermined morale. Then when the quality of the condensers fell, causing a large increase in rejects, Cecil was required to spend many hours of overtime attempting to resolve the problems of production and personnel. At the end of September, when operations had deteriorated to a critical stage, a new management took charge. Cecil was given new responsibilities for scheduling jobs in the factory. There was more overtime, and conference after conference. Cecil finally noted in his diary for September 23, 1929, ". . . my job seems to have degenerated into plain estimating."

50

Chapter 2

During those troubling months Cecil applied through several employment agencies for jobs that might be more satisfying, hoping, especially, that something would turn up in California. Then on Friday, October 25, 1929, came an exciting inquiry from the Engineering Societies' Employment Service in California: Would Cecil wire his address to the Federal Telegraph Company (a subsidiary of International Telephone and Telegraph Company—IT&T) in Palo Alto?

The next day came a 200-word night letter from A. S. Brown of Federal Telegraph asking Cecil to send him details of his supervisory work at Raytheon and Wireless Specialty and of his knowledge of radio and vacuum tube construction. Before the afternoon was over, Cecil had wired back his own 150-word night letter. Then he began the anxious wait for Brown's next communication.

When it came, on November 2, exactly a week after Brown's night letter, Cecil wrote,

Motored to work and came home promptly when I found that the expected telegram from Palo Alto had arrived stating that they were offering me the position and that they wanted to know when I could report.

Cecil replied that he could be on the job on December 2 and quickly gave his week's notice at WSAC. He now recalls that the most important thing he learned during his year and forty-four days at Wireless was the tremendous challenge of trying to get different types of people to cooperate and collaborate in a single effort.

Moreover, his hard-working immediate superiors, W. L. Hodges and J. A. Proctor, had taught him much about managing a business enterprise. Tough-minded and exacting, they had demanded extreme—sometimes unreasonable—effort from him, yet he left Wireless Specialty without any resentment toward them.

The Greens remember their sixteen months in Boston, from the day Cecil started his second period of employment at Raytheon until he and Ida departed for California, as exciting and rewarding for both of them—one of the happier times of their long married life. They made many new acquaintances, renewed former friendships, and week after week enjoyed entertaining these friends at teas, dinner, and weekend picnics. Evening entertainment often consisted of reading popular novels of the day aloud to one another, playing card games, or listening to the radio that Cecil assembled from parts procured from his companies' supply rooms. They typically ended such an evening with hot chocolate and generous servings of Ida's tasty cookies. From these early beginnings developed their pleasure in entertaining and the generous hospitality that soon became their hallmark.

During their private evenings, Cecil and Ida both read omnivorously, and Cecil's diary mentions many classic works: *The Three Musketeers, Anna Karenina, The Bridge of San Luis Rey,* and J. H. Robinson's *The Mind in the Making.* Frequently, when alone, they went to the movies, and seldom did they miss seeing at least two shows a week. Among the highlights were: Lillian Gish in *Wind,* Wallace Beery in *Beggars of Life,* Pola Negri in *Loves of an Actress,* Emil Jannings in *The Patriot,* Al Jolson in the "talkie" *The Singing Fool,* and John Gilbert in *Masks of the Devil.* At Boston's Repertory Theatre they saw *The Tempest* and *Charlie's Aunt,* among other plays.

During their first fall in Boston Ida attended Brighton High School to resume the education that she had interrupted when she started work at General Electric. She registered for a full program of five major subjects, and, in spite of the fact that she had been out of school since 1920, she made a superior record, carrying five hours of credit in English, French, and U.S. history, four in chemistry, three in drawing, and one each in gym and glee club. Brighton High School records reveal that Ida completed the eleventh grade in June 1929. Cecil's resignation from WSAC and their departure from Boston for California in early November, however, forced Ida to withdraw from twelfth grade and thus frustrated again her hopes of completing her secondary school education. On this occasion, however, as throughout her married life, Ida's loyalty to Cecil came first.

THE THIRD TRANSCONTINENTAL AUTOMOBILE TRIP: BOSTON TO CALIFORNIA (NOVEMBER 9–26, 1929)

Never ones to tarry once they had decided to do something, and spurred additionally by Cecil's agreement to report for work in Palo Alto on December 2, the Greens departed for California almost immediately after Cecil had bade farewell to his fellow workers at Wireless Specialty, on November 9, 1929. At last they were to realize their hope to settle in California. On this, their third transcontinental auto trip, they chose to follow a southern route. Touring as far south as Miami, they then headed west along the Gulf Coast, ignoring the many enticing tourist attractions across Georgia, Alabama, and Mississippi. They crossed the Mississippi River at New Orleans, pushed through the Acadian country around Jennings, and drove across Texas, making more than 400 miles a day and denying themselves their customary detours to places of interest off the main highway. Across Arizona the traveling was harder; the roads, dusty with stretches of "washboard," slowed them to a day's run of only 329 miles. In San Diego they stopped only long enough to permit a brief visit with two of Cecil's relatives before they drove the last lap of their trip to Tormey, where the elder Greens were living.

TEN MONTHS AS PRODUCTION ENGINEER AT THE FEDERAL TELEGRAPH COMPANY IN PALO ALTO (DECEMBER 1929–OCTOBER 1930)

Cecil reported for work at the Federal Telegraph Company in Palo Alto on Monday, December 2, 1929. A. S. Brown, who had made him the original offer of employment, immediately took him to the Vacuum Tube Laboratory and introduced him to Charles V. Litton, who was in charge. Brown and Litton then described the important project in which Cecil was to be involved—completing the development of a high-voltage vacuum switch.

On his second day at work, Cecil was advised by Brown that he should spend all of his time in the tube laboratory with Litton. And the first thing that Litton did was to introduce Cecil to the machinists, to the glassblowers and their assistants, and to a Mr. Wagner, who was studying the mercury relay problem. Cecil would soon be working closely with all of these young engineers and technicians, and he plunged into his first assignment with enthusiasm. Indeed, before the day was over, Cecil had started

a layout drawing of a mercury discharge tube with shield and aperture, using hot cathode, for use at 150 words per minute (60 cycles), 15,000 volts at 15 amps.

When Cecil's completed drawing of the tentative high-power mercury switch was presented to Litton on the third day, Litton immediately turned the drawing over to a draftsman, started a machinist on fabricating the molybdenum shield, and himself prepared to do some glasswork during the evening. While one of the glassblowers began construction of the mercury vapor tube, Cecil completed designing the magnetic circuit and then began thinking about how the mercury could be introduced most effectively into the tube. He finished the busy third day welding a 9-inch strip of very thin nickel to a large three-wire stem. It was a heady atmosphere, and Cecil was delighted, because the tube laboratory under Litton's direction was a fast-moving, efficient, and cooperative group of imaginative engineers and skilled technicians who were pushed to the limit of their special abilities by an even more ingenious and skilled director.

Cecil was deeply impressed by Litton's ingenuity and his unusual skills in the shop and in the conference room. According to Cecil, Litton could go into the company's machine shop with a piece of new equipment in mind, take over the lathe or milling machine, and equal the best machinist in producing the desired component.

Litton was also reputed to know more about glass technology than any of his peers in the United States and was even considered the equal of the famed glassblowers of Europe. His expertise in glassblowing included a special ability to connect glass to metal—a connection that was absolutely necessary for the water-cooled 20 kw power transmission tubes. As with the machinists, Litton could equal or even surpass his glassblowers at their own trade, and often did so in his own research and development.

Then, as Cecil remembers, if the occasion called for it, Litton could walk from the shop or laboratory into a conference room full of mathematicians, physical scientists, and engineers and cover the blackboard with the theory that was behind some ingenious idea or successful invention.

According to Cecil, many of Litton's associates considered him a true genius. Their judgment is supported by the record of patents and inventions that laid the foundation of Litton Industries, one of the giants of today's electronics age. Cecil often mentions how much he learned by just watching Litton in action.

A second aspect of Litton's nature that affected Cecil and Ida was his unconventional work schedule. Being unmarried, Litton gave little thought to anything but his job. Developing new and better power vacuum tubes for IT&T's transmitting stations was his major responsibility, and to that he devoted his full time and energy. A workaholic, Litton customarily arrived at work late in the forenoon or early in the afternoon, and once started on a project he was likely to stay with it until he finished, even if that took part or all of the night. He defended his after-hours work by pointing out that temperature conditions for glassblowing were better at night than during the day. Furthermore, at night it was quiet—no one walking around, no telephone ringing, no disturbances, to break his concentration.

As a conscientious and ambitious production engineer in general charge of the tube laboratory, Cecil had to be on the job at eight o'clock in the morning. And, of course, he had to be around at the end of the work day to secure the laboratory for the night. But that was not all. If Litton was engaged in some exciting project when quitting time came, he would suggest that Cecil stay on for a few hours to see how his idea worked out. Cecil not only felt that Litton's suggestion had the force of a demand; he also realized that watching this genius in action was undoubtedly a unique opportunity to learn new techniques.

Unfortunately, however, that opportunity didn't greatly impress Ida, who ate dinner by herself and spent the evening alone in the apartment. Nor was she particularly pleased on those occasions when, if the hour was not too late, Cecil invited Charlie to stop at the apartment on his way home for a sandwich and some tea—which Ida was expected to produce on short notice, or without any notice at all. In due course Ida began to wonder just what kind of future she could look forward to with a husband who seemed to have to work day *and* night to satisfy his boss. So she took action.

Close to midnight one evening, as Cecil and Charlie were leaving the laboratory after one of their typical "extended" work days, they saw a determined female figure rapidly approaching them in the rain down faintly lighted University Avenue.

According to Cecil,

Ida was madder than a wet hen! When she reached us, she really bawled out both of us [using some words not normally expected from young women in the 1930s]. Litton was at a loss to say much, but brought us home in his Chrysler roadster. Although Charlie got the point of Ida's remarks, it didn't sink in very deeply.

Charlie did send Ida a box of chocolates the next day, and later he invited her and Cecil to be his guests on a weekend trip to the mountains. His effort to make up for the unpleasant incident, however, did not at all change Charlie's habits or Ida's feelings about his self-centered monopolization of Cecil's time.

There were other frustrations, too. Being single, Litton had more freedom than most to do whatever he wished when not at work. It was not uncommon for him to take off a few days now and then for fishing, mountain climbing or seashore activities— confident that Cecil was at the laboratory seeing that everything was going along smoothly.

As time passed, frustrations of overtime and extra responsibilities began to make Cecil think about looking for another job. His discontent was increased when reduced orders for the products of the laboratory necessitated many layoffs, and increased the work load of those employees who were asked to work over-

time. The resulting fatigue led to a lower standard of performance; more components were rejected; and Cecil found himself
becoming responsible for every activity of the tube laboratory,
from maintenance and inventory control to product design,
product testing, and final inspection. By early May he was
spending four nights a week, on the average, at the laboratory,
and conditions worsened during June, when more layoffs led to
the departure of some of the most skilled laboratory personnel.
As troubles proliferated in August and early September, Cecil
started looking for another job, although Litton assured him
that his job was secure at least until the end of 1930—and despite the worsening of the Depression. By late September the
number of personnel in the laboratory had decreased from
thirty-two to nineteen. Fatigue, anger, and jealousy were making work more difficult and mishaps more common. So Cecil
was more than ready for an offer that came as a result of Ida's
efforts behind the scenes, even though he gained valuable experience working under Litton.

THE GREENS' LIFE-STYLE IN PALO ALTO (1930)

Despite Cecil's stressful work schedule at Federal Telegraph,
when he was not at the tube laboratory, he and Ida were able to
develop for themselves a life-style that was enjoyable and relaxing. Of greatest importance to both was the satisfaction of living
in Palo Alto and having that Chevrolet mentioned in Cecil's
diary entry for New Year's Day, 1929. After having bought the
new Chevrolet coach[2] that brought them to California and pay-

[2]The Greens turned Tillie in for a new 1929 car on October 13, 1928, as
recorded in Cecil's diary for that day:

This was our last ride in our faithful old car Tillie—took her down to
[Boston's] Commonwealth Chevrolet Co. when we received bill of $140
receipt toward next year's car. Her mileage 24,441.

Tillie had had only about 2,400 miles on the odometer when the Greens
purchased her in January 1927 for approximately $200. The "490
Chevie" had carried the Greens safely across the continent twice: first
from Boston to San Francisco and Vancouver; then up and down the

ing the expenses of the trip from Boston, they still considered their financial resources adequate for a $55-per-month apartment at 159 Emerson Street that fulfilled all the hopes of which Cecil had written in his 1929 diary. They soon began to make new friends and to resume the life-style that they had enjoyed in Boston. They immediately drew out books from the nearest library and resumed their customary practice of reading during evenings when they did not attend a movie (and when Cecil was not at the laboratory)! Likewise, they resumed their long-time habit of seeing every movie they could, especially since the new "talkies" were rapidly replacing the silent pictures of the earlier 1920s.

On Sundays they took long walks, as had been their wont from their earliest Boston days, and they seldom missed a short evening walk during the week—something they have still done until quite recently.

At this time Cecil discovered that their landlord, a Mr. Haskell, was—like himself—an avid stamp collector, and at the same time was president of the Palo Alto Philatelists' Club. The two men greatly enjoyed comparing their latest acquisitions and exchanges, and Cecil's present collection owes many of its choice stamps to that year of residence in Palo Alto.

Palo Alto is within approximately 35 miles by road from Tormey, near Oakland, where the elder Greens lived, so Cecil and Ida were able to visit his parents frequently. Sometimes they stayed overnight in Tormey, and sometimes the elder Greens were brought to Palo Alto, as happened at Christmas time in 1929. As the year ended, Cecil recorded in his diary their joy and their future hopes:

West Coast on many short trips; and finally back across the United States from Tormey to Boston, to finish with several visits to Ballston Lake. They turned in their cherished vehicle after twenty-two months and 20,000 miles, receiving a receipt worth $140, more than half what they paid for the car. They would use that receipt on March 10, 1929 as partial payment for that second Chevrolet Cecil so much wanted, a new Chevrolet coach—list price of $595—which they promptly dubbed "Tillie II."

The first day in this New Year [1930] finds us both with thankfulness that our two wishes, to get a new car and reach California again, have been granted. And now for the coming year our ambition is to become definitely located and settled with health above all.

But as it turned out, for the reasons noted previously, this wish for a sense of permanence was not to be realized, for on October 4, 1930, Cecil was saying goodbye to his fellow workers at the Federal Telegraph Company, having accepted a job as a field-party chief with Geophysical Service Inc. (GSI) of Dallas, Texas. This change in employment was the direct result of Ida's correspondence with Helen and Roland Beers.

THE GREENS AND THE BEERSES
The Greens initially met Roland and Helen Beers in Boston soon after Cecil took his first job with Raytheon, where Beers was working in 1926. They became close friends, and even though the Beerses left Boston for the Southwest in late 1928—Roland had accepted a position with a geophysical company in Tulsa—Ida and Helen kept in touch by letter. It was early in June 1930 that Roland became chief of the first field party of a newly organized company called Geophysical Service Incorporated (GSI) of Dallas.

At the same time, in the early months of 1930, as the Great Depression began, Ida was writing to Helen Beers about Cecil's intolerable work load at Federal Telegraph and her unhappiness with his work schedule, and asking if there might be an opening in the geophysical company that Roland had recently joined.

Ida's letters were just what Roland needed to refresh his memory of the many good times he and Helen had had with Cecil and Ida during the Raytheon days in Boston. Furthermore Geophysical Service was doing a booming business and badly needed additional personnel, including more party chiefs. Roland and Cecil had worked in adjacent laboratories at Raytheon in Cambridge; both were trained as electrical engineers, and Roland was certain that Cecil could fill the job of party chief at GSI.

So in late September of 1930, Roland wrote an enthusiastic

letter, reproduced in appendix B, that described his field work as a GSI party chief, and urged Cecil to consider joining GSI immediately. The company could offer Cecil a much better salary than he was getting from Federal Telegraph. Cecil recorded his response as follows on September 30, 1930:

Wrote 2-page typed letter to R. F. Beers telling him I was quite open to his proposition.

The next day a telegram from Beers informed Cecil that the president of GSI wanted him to join the company. Not only was the salary attractive; even more attractive was the prospect of traveling around the country from town to town and working with a crew in the wide-open country of the Southwest. No job could better satisfy Cecil's wanderlust. Cecil accepted GSI's offer and the next day, October 2, informed Litton that he intended to resign.

Work had slackened at Federal Telegraph, and there were other company troubles brewing. Nevertheless, both Brown and Litton urged Cecil to stay. Certain that business would soon pick up, they suggested that Cecil propose whatever salary increase would be satisfactory.

Litton's attempts at persuasion were of no avail, however, because Cecil and Ida had made their decision—a decision that would change their life-style drastically, present a new kind of challenge to Cecil, and lead them to a golden future. But good fortune would not come until after a decade of "stop-here-go-there" living, as Cecil pursued his geophysical exploration work. Litton expressed keen disappointment, saying that he had wanted Cecil to succeed him at Federal Telegraph when he, Litton, was promoted.

CECIL LEAVES FEDERAL TELEGRAPH AND JOINS GEOPHYSICAL SERVICE INC. (1930)

Ida was clearly responsible for GSI's offer to Cecil, in that the offer came as a result of her correspondence with Helen Beers. Although it was Roland's enthusiastic letter that drew a positive response from Cecil, the official offer of a job came from GSI's vice president, Eugene McDermott, who later became one of the Greens' closest friends. Indeed, Ida has remarked often that Cecil and Gene were almost like brothers.

Having anticipated an acceptable offer to Cecil from GSI, Ida had already begun the the now-familiar task of sorting, discarding, packing, and storing their household effects. After definitely accepting the GSI offer, Cecil turned to preparation for a quick trip to Vancouver to say farewell to his relatives there. That trip began on October 7, when Cecil, Ida, and Cecil's parents all crowded into the car with suitcases and cartons. Riding on wet roads through fog, rain, and snow, they reached Vancouver, some 1,030 miles from Palo Alto, in four days, stopping at tourist cabins only long enough for short nights of rest. They had to hurry because their time was limited. Indeed, it was more limited than they knew, for upon returning to Palo Alto, Cecil found a telegram from Beers urging him to come to Seminole, Oklahoma, as soon as possible. Cecil immediately wired that he would appear on Sunday, October 26. Accordingly, Tuesday, October 21 found Cecil and Ida hurriedly packing and making final preparations for an early start from Tormey the next morning—but not before they had enjoyed at least one more of Maggie Green's English breakfasts.

Five days and 1,835 miles later, the Greens reached Seminole on schedule, to be greeted by Roland and Helen Beers, whose house guests they were to be while in Seminole. Cecil and Roland stayed up well after midnight discussing GSI matters, while Ida and Helen brought one another up to date on happenings since their last meeting in Boston in late 1928.

CECIL'S FIRST PERIOD OF EMPLOYMENT WITH GSI: LEARNING THE ART OF THE "DOODLEBUGGER" (OCTOBER 1930–JULY 1931)

When Cecil and Ida arrived in Seminole, they had little idea of what exploration geophysicists—*doodlebuggers*[3] they were

[3]Two definitions of the noun *doodlebug* are given in *Webster's Second New International Dictionary:*

dubbed in the early 1930s—actually did. Of course both had read and reread Roland's letter and had discussed it at length before they departed for Seminole. But they had many questions and were eager to learn more about the life they would be leading in the new job. So during the week after their arrival, while the personnel and equipment of Field Party 310 were being assembled in Dallas, Cecil discussed a wide range of technical and business matters with Roland while Ida quizzed Helen on the responsibilities of the party chief's wife.

An unscientific device, with which it is claimed that minerals, including petroleum, water, etc., may be located. Cf. divining rod. The larvae of an ant lion; loosely, any of several other insects. U.S.

Just when and how geophysical prospecters were first dubbed "doodlebuggers" is uncertain. However, throughout historic time, certain individuals (e.g., "dowsers") have claimed that they could determine the presence of water, oil, minerals, etc., below the surface by means of a forked stick or divining rod. These individuals have been widely discredited because they have no scientific explanation for the occasional successes of their nongeological prospecting methods beyond what is possible under the laws of chance.

So when geophysicists began using their special devices ("black boxes") to measure gravity and the different physical properties of subsurface rocks, the nonscientific public became suspicious of the instruments and the way they were used. Doubts arose especially when the geophysicists explained that they were prospecting for oil reservoirs by exploding dynamite in the bottom of shallow holes they bored in the soil. Little wonder that the nonscientific public was suspicious, when the physicists themselves had only recently discovered how they could apply the laws of physics to determine the structure of rocks below the surface without drilling into them.

My own guess is that the term *doodlebugger* was coined by an imaginative person who likened the methods of the earliest geophysical prospector to the doodlebug ant lion. The larva of the ant lion, a neuropterous insect, digs a conical pit in sand, then lies just below the surface at the bottom of the pit, waiting to seize ants and other insects that have slid into the pit and are struggling to climb up the sandy slope.

The interested reader will find much more about the art of doodlebugging in L. W. Lau's article "Black Magic in Geophysical Prospecting," *Geophysics* 1/1, pp. 1–8, January 1936.

Being a rapid learner and possessing a remarkably retentive memory for technical details, dates, and the names of individuals and places, Cecil made the most of this opportunity. He made several visits to Roland's field party to become familiar with the field equipment and the procedures for its use to acquire field data. Then back in his office Roland showed Cecil how the data were processed, interpreted, and prepared for presentation to the client. Cecil remembers Roland during this experience as an excellent teacher, but he soon realized that, extensive as the field observations and office instructions were, there was much more he had to learn about both people and equipment. He recognized that only by successfully managing his own field party could he become a true doodlebugger. He could hardly wait for the week to end so that he could start his own contract.

Helen Beers, who had insisted that Ida and Cecil stay as house guests for the first few days, spent much of this time describing to Ida her own experiences as the wife of a GSI party chief. Inasmuch as Ida fully intended to accompany Cecil wherever he took his field party, she welcomed the account of the Beerses' varied experiences as doodlebuggers. By the end of the week, Ida was as eager as Cecil to get on with their new life, even though she admits to some uneasiness about whatever adventures might lie ahead, especially when she recalled that she was really responsible for this new turn in Cecil's career.

GSI PARTY 310'S FIRST CONTRACT—
MAUD, OKLAHOMA

An entirely new life-style began for the Greens when they arrived in the small Oklahoma town of Maud (population 265) on Sunday, November 2, 1930—a drastic change from the comfortable environment and predictable kind of life they had enjoyed in Palo Alto. For the next decade there would always be uncertainty about their housing and surroundings as they moved again and again, from one small town to another in Kansas, Oklahoma, Texas, and Louisiana. It was also an entirely new aspect of applied electrical engineering for Cecil, wholly different from the controlled laboratory and production environments of his previous work.

Yet both Cecil and Ida remember the years from 1930 to 1936 as one of the happiest and most exciting periods of their long married life together. They were young in spirit, enjoyed good health, and liked to travel and to be outdoors. So they looked forward eagerly to their adventures as doodlebuggers.

Cecil and Ida were now in on the ground floor of the company which for the next forty-five years would absorb the full measure of their effort, time, loyalty, and creative achievement. As a co-owner of the company, beginning in 1941, Cecil would be directly involved in the administration and restructuring of the organization during World War II which led to the founding of Texas Instruments Incorporated (TI).

Even while these developments were taking place, actual fieldwork began in May 1930 after Karcher had hired several party chiefs who had worked for him when he was president of Geophysical Research Corporation (GRC). Included were Roland F. Beers, chief of the first party, Party 301, working around Shawnee and Seminole, Oklahoma; Kenneth E. Burg, chief of Party 302, working out of Oklahoma City; Henry Salvatori, chief of Party 303, based at Enid, Oklahoma; and H. B. Peacock, chief of Party 304, based at Palestine, Texas. By the end of 1930, eleven parties were conducting reflection seismic exploration in Oklahoma and Texas. Of these, Cecil H. Green was chief of Party 310, first based at Maud, Oklahoma.

This rapid expansion of the recently organized GSI was possible, despite the economic depression that began in 1929, because of the effectiveness of its new exploration method. Indeed, the reports of the initial eleven parties' activities in 1930 contained no hint of concern about the deepening depression.

Would the years immediately ahead be as full of exciting activities and memorable events as 1930, which was coming to an end as Cecil and Ida started their new life as doodlebuggers? Without any doubt, when 1930 ended, historians could record it as a fantastic year during which the United States took its last faddish flings before sliding into the Great Depression.

GEOPHYSICAL SERVICE INC.[4]

Geophysical Service Inc. (GSI) was launched at Dallas, Texas, on May 16, 1930, by J. Clarence Karcher and Eugene McDermott as incorporators. It was one of the very first independent prospecting companies established to do reflection seismic exploration for petroleum-producing companies that wanted to increase their resources by locating and acquiring rights to subsurface structures likely to contain oil and gas. GSI's responsibility was to locate such favorable subsurface structures in areas leased or owned by its clients, generally major oil companies. If promising sites were found by GSI, it was then up to the client to determine the presence or absence of petroleum by drilling a wildcat well to the reflecting zone.

Karcher, who originated the reflection seismic method for petroleum exploration, became president of the new company and McDermott vice president, with company headquarters established in Dallas. By June, a month after the charter was signed, Karcher had a dozen contracts in his pocket. He and McDermott established a laboratory in Newark, New Jersey, to assemble the necessary instruments and hired Alfred Morel, draftsman; Tony Case, electrical assembler; and Henry Stoll, toolmaker. In the same period they added three people to the office staff in Dallas: Keating Ransone, manager; William C. Edwards, Jr., bookkeeper; and Bonnie Scudder, secretary. By midsummer, J. Erik

[4]The information in this and the following section is based on the following sources: 1) taped interviews with Cecil and Ida, together with their diaries; 2) items from GSI's monthly *The Grapevine*, especially the 25th Anniversary number (Vol. 11, No. 8, September 1955); and a special report that Cecil prepared at my request, entitled "The Earliest Days of Reflection Seismic Prospecting for Petroleum," which is included as appendix C; and an autobiographical article by J. C. Karcher entitled "The Reflection Seismograph: Its Invention and Use in the Discovery of Oil and Gas Fields," published in SEG's *The Leading Edge*, November 1987, Vol. 6, No. 11, pp. 10–19.

Jonsson, a recent graduate of Rensselaer Polytechnic Institute (RPI), had been added as laboratory superintendent in Newark.

Elder citizens of today (1988) can hardly have forgotten all the events, activities, personalities, and general, public exuberance of 1930—the year Geophysical Service Incorporated was launched and Cecil joined the company as its tenth party chief.

To remind the reader of 1955 what was going on in 1930, when GSI started, the editor of GSI's monthly the *Grapevine* (Vol. 11, September 1955, p. 2), in celebrating the company's 25th Anniversary, listed the following items:

IT WAS 1930 . . .

. Herbert Hoover was in the second year of his term as president of the United States and Huey Long, the Kingfish, was riding high as governor of Louisiana.

· Construction had begun in New York City on the 100-story Empire State Building, the world's tallest. Also Rockefeller Center, soon dubbed Radio City.

· Bobby Jones was golf champion, Bill Tilden and Helen Wills Moody led in tennis, Babe Ruth was Sultan of Swat, and Gallant Fox consistently ran in the money.

· Prohibition popularized bootlegging, rumrunning, and the speakeasy. Legs Diamond and Al Capone were No. 1 public enemies. Dance marathons, miniature golf were fads.

· Vaudeville was on the way out and talking pictures began to appear: "Grand Hotel," "Little Caesar," and "The Green Pastures." Movie stars included Wallace Beery, Marie Dressler, and Garbo. Claire Booth Luce was a Broadway hit and a young French soprano made her debut at the Met—Lily Pons. The jazz boys were Duke Ellington, Count Basie, Louis Armstrong. Radio favorites were Kate Smith, Paul Whiteman, Amos and Andy. The snappy comeback was "So's your old man!"

· Evening gowns were floor length again. Colorless nail polish was in vogue. "Harper's Bazaar" said: "In designs for bathing suits, attempts at Art, as well as all imaginative designing, should be avoided. Swimming is a sport and was never meant as an opportunity for sartorial display. There should be hardly any difference between a man's and a woman's swimming suit."

· Automobile notes: Chevrolet-6 sports roadster, $555 FOB, wire wheels, large ornamental hub caps, and spacious rumble seat . . . Chryslers, from $795 up. The Eight and Imperial Eight closed models were factory-wired for Transitone, the pioneer automobile radio . . . Cadillac V-12 and V-16, $3795 to $4895 FOB . . . Lincoln as low as $4200 FOB . . . Pierce Arrow, custom built, up to $10,000 . . . Studebaker featured free wheeling (you need the clutch only to start or back up). President 8, convertible, 122 h.p., $1950. Six wire wheels. Bumpers and spare tires extra.

· There had been a nose dive after the stock market crash October 24–29, 1929. By 1930 the Great Depression was under way. Apple selling, unemployment, purchasing power down, relief necessary, breadlines in the East. Fear set the mood—fear of the unknown.

A GSI FIELD PARTY OF THE 1930s
A typical GSI 1930 field party, according to Cecil and *The Grapevine,* consisted of three components: the crew members, the motor equipment and instruments, and the families of crew members. The usual crew, ten to twelve in number, consisted of the following key personnel: *party chief, computer* (a person, not a machine, at that time), *observer, observer's assistant, shooter* and *helper, labor foreman, shot-hole diggers, surveyor* and *rodman,* and several *jug hustlers.* Key personnel were hired by the company's Dallas headquarters. A typical crew's equipment consisted of four or five trucks: one to carry only explosives; another to carry caps along with hand augers, seismometers ("jugs"), and cables; a recording truck containing the sensitive electrical recording instruments; a truck with a water tank; and a pickup truck to carry surveying instruments, spare parts, and the like. A drilling rig would soon be added as hand-operated augers gave way to a power-driven drill. Each crew negotiated for office space for the party chief and his computers, and garage space for repair work and storage, in a town near the site of the work.

Men naturally preferred to have their families with them, so the families typically accompanied each crew. The wives provided a homelike living environment as best they could in their rented quarters, and they prepared better food than the small-town restaurants generally offered. In addition, the wives often arranged a variety of social gatherings in which all members of the party could participate. The sense of a happy family that members of the party displayed could be very helpful in the crew's relations with the local public. It was a special advantage when a GSI party returned to an area on a second contract. Local arrangements were then much more easily and agreeably made with the townspeople.

Ida's account of her activities as the wife of the chief of Party 310, reproduced on a following page, gives an idea of the activities through which she and other wives supported the crew and enjoyed one another when their husbands were at work.

Actual operations in the field demanded the highest possible degree of cooperation and collaboration. This was especially important because explosives (dynamite) were involved in those early days, and accidental detonation could be disastrous. In addition, handling of vehicles was always a source of danger; trucks were frequently stuck in the mud and occasionally went out of control down a slope during a rainstorm or as the result of hitting a deep pothole or a rut full of water.

An exploration program began when the GSI staff in Dallas negotiated a contract with a client to conduct a reflection seismic survey of a specific area of land. When the project was assigned to one of the party chiefs, his first task was to visit the area to be surveyed, determine its boundaries, and note its topography and other features. He then had to find office space and garage facilities as near the area as possible. With these arrangements made he could summon from its previous base the procession of GSI vehicles—trucks and company cars carrying crew members and their equipment. They were followed by the families of party members in their own cars laden with household necessities. The crew members commenced work as soon as possible while their wives canvassed the town in search of living quarters.

First the surveyor and his rodman laid out the profile lines (see accompanying diagram) marking the location for shot

PARTY CHIEF AT WORK AND PLAY. Above, Cecil Green starts out for a busy day in the field. Below, he and wife, Ida, pose in holiday mood on Sunday at Turner Falls, Okla.

EXPLORER'S WIFE WRITES OF PIONEER CREW'S GYPSY LIFE

By Ida M. Green
Party 310
Maud, Oklahoma

LIFE MOVES swiftly on our party since moving about every few weeks seems to be our fate. After dashing around in some small town to find a place to live, we may end up with two shabbily furnished rooms, or in rare cases a nice private home which happens to be for rent. First there, first served is the policy which equalizes living conditions for everyone at some time or other, especially in these Oklahoma boom-times. If you end up living in a place where you share the bath with three other families, maybe the next time you'll hit the jackpot and get an attractive home. One soon becomes philosophical. The business is so new that small towns haven't adjusted to oil people and regard them with suspicion and concern. We try to keep a good reputation, but perhaps the last crew forgot themselves and left the barn doors open and so we find ourselves trying to close them.

First thing the wives do after getting located is to rush around to see what the grocery stores (not more than two) have to offer in foodstuffs. Vegetables are scarce other than turnips and mustard greens, with a few other specialties thrown in. Fresh killed yearling beef is sometimes good and then again tough. Chicken for Sunday dinner is found by going down to the local feedstore on Saturday. Sacks of Purina Chow and other feeds fill the place, but out back is a pen or yard filled with fast-stepping chickens. The man snares one with a wire and I ask him if he would mind cutting its head off. That accomplished, I return home, proceed to remove the feathers, and then do some major surgery. Come Sunday, golden brown fried chicken with cream gravy graces the festive board.

We crew wives drop in on each other for a cup of coffee and to catch up on current news. Occasionally a bridge luncheon takes place, or a get-together of the crew with all the wives showing off their culinary art, with fried chicken and all the fixin's.

Dr. Karcher and Mr. Mc drop in and the conversation is mostly about Viola —whether high, low, or a closure. (Viola is mighty important! For a long time I thought Viola must be quite a gal until after a few sessions I learned differently.) As long as we have a table, chairs, water tap, and some kind of flame they are always invited to "set up" and eat with us. Our geologist, Mr. Ronald Cullen of Twin States Oil, comes to see how things are going and he, too, shares our meals. We are happy to have him from the big city.

Life is settling down, or so it seems. Suddenly, it is whispered around — it is rumored — that we are going to move. The women begin to lament — they have just stocked up on a lot of perishable food or groceries, or someone has been silly enough to have done some extensive housecleaning. Such things always bring on a move. Plans are made to move as quietly and as secretly as possible since the general idea is to outwit the scout and leave him behind. Locations for leasing are kept secret by the oil companies. Sooner or later the scout shows up and follows the crew to the field, and like Sherlock Holmes minus the magnifying glass makes collection of evidence — even to small scraps of records that somehow get scattered around.

Some of the single men are marrying, and the married folks are beginning to have families. Life isn't too monotonous and even has an element of adventure in it. The crew is like a family —

Cecil H. Green . . . . Party Chief
Chester Donnally . . . . Computer
Bill McDermott . . . . . Observer
J. D. Strange . . . . . . Surveyor
Eldridge Tabor . . . . . . Helper
Finis McCluney . . Labor Foreman
Joe Arnold, Joe Rowland,
  W. L. Horne, Perry
  Horne . . . . Helpers

our joys and sorrows are shared, and we make an effort to get along together.

The advantages are: time for hobbies, family ties are closer, and appreciation for any kind of entertainment. Children seem precocious and extroverted, meeting people easily. There is less expense in keeping up appearances. Last, but not least, we always see and learn about something new or different, and can put away a nice nest egg for the future. Such is the doodlebugging life.   ●

Note: The Viola mentioned in the third paragraph in the column to the left is a hard limestone in the subsurface in Oklahoma and surrounding states that reflects sound waves sent downward from the explosion of dynamite detonated in shallow drill holes (see figure 5). Its structure is important in determining if a possible petroleum reservoir is present. Photo of Cecil and Ida by J. W. Thomas 1932. Article from the *GSI Grapevine,* 1955 holes, and the drillers and their helpers followed, boring holes 19 to 20 feet deep. In the very earliest years (1930–1932), three or four holes had to be bored at a fixed distance of a few feet around each shot point because it was usually necessary to detonate three or four different sized charges before an optimum record was obtained.

These early-day shot holes were hand-drilled by three-man crews, using a T-shaped auger, the handle forming the crossbar. With one man sitting on the handle for weight, and the other two at opposite ends of the handle, facing in opposite directions, the driller's chant was "Bear down, turn right." Fortunately for the drilling crews, a motorized drill rig was soon developed.

A typical recording setup in the earliest 1930s comprised five seismometers (jugs)—each cylindrical in shape and weighing about 30 pounds. The jug hustlers placed these seismometers about 100 feet apart—all in a straight line from the shot point—the nearest generally being between 1,000 and 2,000 feet from the shot point. This lineup of shot point and seismometers was referred to as a "profile." Each jug was connected via a multiconductor cable to the recording truck, and each was buried in a shallow hand-dug hole to avoid wind disturbance (see figure 5).

Meanwhile, the shooter was assembling the explosive charge, one or more sticks of dynamite into one of which he pressed a cap attached to a wire that led to his hand-operated "blaster." This dynamite charge was pushed to the bottom of the shot hole with jointed tamp rods; then the hole was filled with water, so that most of the energy was directed downward.

The recording truck was connected by a telephone line to the shot point so that the observer in the truck could tell the shooter when to fire the dynamite charge. Just before the firing, the observer started hand-cranking the strip of sensitive recording paper through the camera and on into the developer box.

The principle of the system, as indicated on the accompanying diagram (figure 5), was that the downward energy released from the explosion was reflected back to the five seismometers

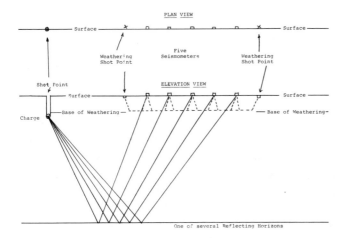

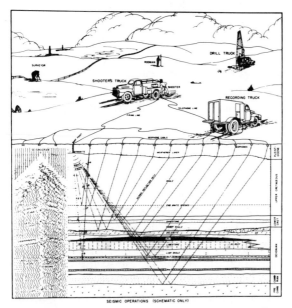

5   A GSI field party in action. *Top:* Schematic diagram prepared by
Cecil H. Green to illustrate reflection seismology. *Bottom:* Sketch of a
reflection seismic party, with the components deployed for operations. The record (seismogram) obtained from a shot is shown in the
lower left corner of the sketch. (See also figure 6.) Courtesy of Texas
Instruments Inc.

on the surface from subsurface rock layers. This reflected energy in the form of sound waves was only a tiny fraction of the total energy released by the explosion—actually far less than that produced by a human footstep, which meant that all crew members and equipment must remain "frozen" during the shot in order not to introduce spurious signals. The sound waves impacting the seismometers were strong enough to shake them ever so gently, causing them to send an electric current along the cables to a camera in the recording truck. There the reflections were highly amplified and recorded, along with the ground roll, on the sensitized paper strip.

When developed, the exposed paper strip, a "seismogram" (figure 6), shows the arrival time and amplitude of the individual reflections that struck each of the five seismometers. The process was continued with charges of varying sizes detonated in different holes until the observer obtained the best possible record.

The time that the reflected waves took to traverse the weathered layer of surface material—soil, gravel, and perhaps vegetation—under the seismometers was next determined from refraction information obtained by firing small dynamite charges in shallow holes in line with the seismometer array and 100 to 200 feet from the nearest seismometer.

Finally, the surveyor's measured distance from shot point to each jug in the seismometer line was checked by the shooter who detonated a very small charge in the air above the shot point. (See Cecil's discussion, in appendix C, for further details of the above procedures.)

Once this shooting sequence was completed, the observer, who was the senior man next to the party chief and was in direct charge of field operations, delivered the day's seismograms to the crew chief in the party's temporary office in town. The latter's responsibility was to identify the reflections. The computer next determined the arrival times of all reflections on each profile. Then, using a "time-depth chart," he calculated the depth of each reflecting layer, and gave this information for each jug point to the party chief.

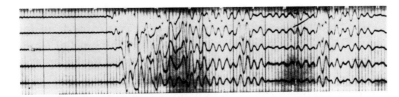

6   A seismogram. Karcher's new method of reflection seismology
produced records (seismograms) that showed structures in greater detail and of better quality than had ever been obtained before. The
above seismogram shows five channels—the kind of record that
would be obtained from the layout illustrated in the lower diagram in
figure 5. Courtesy of *GSI Grapevine.*

After first shooting a few preliminary profiles, and then filling in where necessary with additional ones,[5] the party chief
employed all the available data to construct a subsurface contour map of the selected reflecting layers under the entire lease
area. Such a map showed the shapes of the various reflecting
layers—whether they all followed the normal regional slope;
were flat lying; sloped in opposite directions, indicating an
arching structure (an "anticline") of possible interest as an oil
reservoir; or were broken into tilted sections by vertical dis-
placements (faults) along which adjacent blocks had moved up
or down. It was the party chief's responsibility to ask for additional seismic data if needed to obtain an adequate contour map
for the client. In the end, the contour map of the subsurface
reflecting layers represented the best interpretation of the seismic data that the party chief and the client's geologist, working
together, could produce. This final contour map, however,
could only indicate structures. Whether oil or gas was present
in any of the structures could be determined only by drilling. It
was the responsibility of the client's geologists to decide
whether or not, and where, to recommend drilling.

[5]"Filling-in" profiles were likely to attract the attention of observant
"scouts" from the client's competitors, because they usually meant that
some abnormal structure was showing up, and the party chief wanted
more reflection data.

GSI FIELD PARTY 310'S FIRST ACTUAL FIELD WORK

Cecil's first field assignment as party chief was carried out for Twin States Oil Company (a subsidiary of the Sun Oil Company), with R. J. Mosier as the company's local geologist. The assignment was to determine if any subsurface structural conditions favorable for the accumulation of oil and gas were present beneath a small area under lease to the company, a few miles south of Maud, Oklahoma.

He and Ida had rented an apartment in Maud soon after their arrival in Seminole, and during their first week they began to move in their few household necessities. The last hours of Sunday were devoted to unloading the remainder of their personal belongings and putting their apartment in order.

The next day, Monday, November 3, 1930, was a busy one for the Greens. Ida occupied herself with the varied activities that would become all too familiar in the dozens of stop-and-go moves that lay ahead: have the electricity turned on, find out where to buy the best groceries and other necessities, and locate the library and movie theaters. The library was important, because books borrowed from it would help pass the hours when Cecil was at work. The movies were the principal form of public entertainment in the 1930s; having become enthusiastic moviegoers, Cecil and Ida seldom missed a show.

Cecil spent his first day in Maud preparing for the arrival of his field party. In an early telephone call to Dallas he learned that the party would not be assembled until late in the afternoon, so he and two of his crew members visited the area where they would be working, to size up the lay of the land. The remainder of the crew, with three new Ford trucks, had arrived when they returned, whereupon Cecil immediately began an inventory of equipment. Discovering that some items were missing, he placed another call to Dallas for the missing components. All in all, it was a typical first day preparing for fieldwork on a new contract.

Although Beers had done his best to instruct Cecil in the most important aspects of doodlebugging, Cecil found that his knowledge of electrical engineering and his management skills would be quickly tested; having never been either a crew member or a party chief before, he was unprepared for technical and personnel problems that Beers had not mentioned. One such problem made his first twenty-four hours, as well as the two following days, unforgettable.

The very first profile gave a strong indication of things to come. With everything apparently ready, the dynamite in the first shot hole was detonated and the ground shook, but the recording equipment malfunctioned, and no clear record appeared on the sensitized paper when it was developed. Successive shots proved no more successful; try as hard as Cecil and his crew could, hour after hour, the recording instrumentation would not function properly.

There was nothing left to do but take the instruments back to the garage in Maud, where Cecil and crew members worked until midnight—fruitlessly, as tests the next day proved. Even Eugene McDermott, GSI's vice president, dropped in to help with the work. After two more unsuccessful days and nights, Cecil had no recourse but to telephone headquarters in Dallas and ask for help—an embarrassment, to say the least, for the party chief, even though McDermott himself, an experienced doodlebugger, had also been unable to discover the cause of the malfunctioning. Little wonder that Cecil recorded in his diary for these troublesome days an increase in "the pains of indigestion" (ulcers) that he had developed during his last few months at Federal Telegraph.

The technician from Dallas quickly located the trouble—the recording system had been wired up backward in the Dallas shop!

Once the recording equipment was working properly, the ten-man crew spent long hours over the weekend to make up for lost time. They experienced a few other worrisome incidents: an inadvertently missed shot point, a truck stuck in a field, sudden transfer of personnel. But in the end, Cecil, with several assists from Eugene McDermott, was able to interpret the data and submit his report to Twin States Oil Company on time. The crew was in Maud just twenty days.

This first assignment was followed by operations from bases at Seminole (a month), Prague (forty-five days), Stillwater (a month), Perry (two months), and Guthrie (two months) (see map figure 7).

A remarkable incident illustrating the versatility in management skills expected of the party chief occurred when Party 310 was conducting seismic work in a wild, hilly, and heavily wooded area not far from Prague. Prohibition was the law of the land, but a few Oklahomans chose to manufacture their own stimulants with home-made stills hidden in the more rugged parts of the countryside around Prague. So when Cecil visited his crew in such areas he had to be alert lest he be mistaken for a revenuer ("revenooer") and shot at, because of his garb— broad-rimmed hat, high leather boots and riding breeches (see Ida's account preceding figure 5).

On one occasion, when he finally came upon his crew near a creek, he found the equipment unattended and all the crew members sound asleep in the shade of the trucks. Aroused, the chagrined doodlebuggers offered the following explanation to Cecil:

"Well, we were digging these holes here when a man came over the brow of the hill yonder—said his name was Snow, we were on his land, and just what were we doing boring holes in the ground?

"We're working for GSI," he was told.—"Oh, yeah, but who gave you permission, and what are you doing here now?"—

"Well, we're digging holes."—"Yes, I can see that, but what for?"—"Oh, we're going to put a little charge of dynamite at the bottom of each hole and explode it."

"Dynamite, hell!" he shouted, "I got three kegs of whiskey buried over there. If you do your digging up or down the 'crick,' that's all right by me, and dig as many holes as you like, but not right around here."

So the crew members said they would move their equipment farther up the creek. This show of cooperation brought an invitation from the landowner to accompany him over the brow of

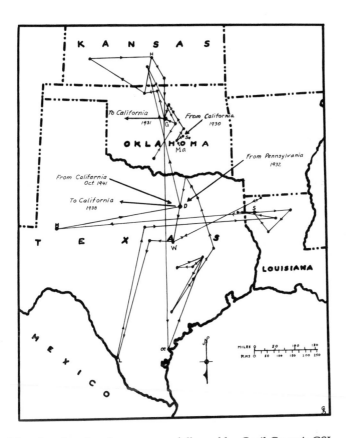

7   Map showing the zigzag course followed by Cecil Green's GSI
Field Party 310, with small dots indicating where bases of operations
were established: CC—Corpus Christi; D—Dallas; G—Guthrie; H—
Hutchinson; L—Laredo; M—Midland; Ma—Maud; S—Shreveport;
Se—Seminole; and W—Waco. Cecil and Ida, coming from California,
arrived in Oklahoma in late October 1930, joining Roland and Helen
Beers in Seminole. On November 2 they established their first base of
operations in the small town of Maud (population 265) a few miles
southwest of Seminole. There they began the odyssey which, during
the next decade, would take them on the zigzag route shown on the
map. Similar patterns were followed later in California, New York and
Pennsylvania before the Greens finally settled in Dallas and became
part owners of GSI. The move to their permanent home in Dallas in
early 1942 was the fifty-fourth that they had made since becoming
doodlebuggers back in Maud twelve years earlier.

the hill and accept his hospitality by sampling some of his corn-squeezings fresh out of the spigot from his active still.

After the observer had related this astonishing story to Cecil, one of the crew members then asked him,

"Mr. Green, we just had to go, don't you think? So we accepted his invitation, and by the time we got back we were all too sleepy to keep on working."

As Cecil told me, after relating the episode,

"If only one or two of the crew had been asleep, I would have fired them on the spot, but I couldn't fire the whole crew, otherwise our client would have caused trouble. Furthermore, the crew were so embarrassed that they worked extra hours to make up for lost time. Under the circumstances, I just couldn't fire any of the crew."

And so the incident of the sleepy crew became a part of the lore of Party 310.

Upon the completion of the work near Guthrie at the end of June 1931, Party 310's client, D. Max Morgan, terminated his contract with GSI. This action brought to an end, at least for the time being, any further work for Party 310, and hence for Cecil. Termination of the contract was the direct result of two quite different but nevertheless closely related economic conditions.

By July 1931, the Great Depression was affecting all aspects of the nation's economy and would continue to do so until the middle of the decade. And by mid-August the newly discovered East Texas Field was flooding the market with such an excess of oil that the price fell to as low as ten cents a barrel—just about as cheap as water! To restore equilibrium to the market, Governor Ross Sterling, on August 16, ordered the field closed and sent the National Guard to enforce his order.

Under such depressing conditions of reduced demand and excess supply, there was little incentive for petroleum companies to continue geophysical exploration—a decision that caused GSI's business to drop precipitously. As a result, GSI's vice president McDermott suggested to Cecil and other party chiefs, as well as to crew members, that they return to college for further training or find temporary employment elsewhere until GSI rode out the economic storm and could reemploy them.

Cecil's response to McDermott's suggestion was typical—he acted immediately. Choosing to seek temporary employment rather than to pursue further educational goals, he called Charles Litton in Palo Alto.

THE SECOND AND LAST PERIOD OF EMPLOYMENT AT THE FEDERAL TELEGRAPH CO. (JULY 1931–JUNE 1932)

The eight months that Cecil and Ida had spent as doodlebuggers with GSI had been happy and adventuresome ones. Cecil had become an experienced exploration geophysicist, had greatly enjoyed managing a field crew despite initial frustrations, and had found the kind of activity he most enjoyed— traveling from place to place and working closely with people in the open countryside. Ida, too, had enjoyed the challenge of frequent moves, the excitement of new experiences, and the close relationships with the other women who strove to provide comfortable temporary homes for their children and husbands.

Litton (at Federal Telegraph) replied curtly to Cecil's call asking if he could have his former job back:

Of course! Didn't I tell you that you were a darned fool when you decided to leave last October?

Cecil's inquiry took more than a little courage. Not only did he remember Litton's caustic remarks when informed that Cecil was leaving in October 1930; he also had not yet recovered fully from the ulcers he had developed as a result of trying to keep up with workaholic Charlie during his last three months at Federal Telegraph. Furthermore, going back to Palo Alto meant that Ida, who was enjoying the ever-changing odyssey of the exploration geophysicist, would once again have to establish a household in Palo Alto, and she might face another confrontation with Litton over Cecil's working hours. Nevertheless, her trust in Cecil's judgment and her strong loyalty to him led her to shrug her shoulders and start packing. She more than equaled

Cecil's courage in the face of what might lie ahead for both of them in California.

The Greens left Guthrie on July 4, 1931, en route to Palo Alto. After a short visit with the Beerses, they headed for California through Kansas, Colorado, Wyoming, Idaho, and Nevada. Though they elected to stay at autocamps instead of stopping wherever they could pitch a tent, they did follow their customary practice of visiting the main tourist attractions along the way—Pike's Peak, Garden of the Gods, Cody, Yellowstone Park ("Old Faithful" and the bears), and "Craters of the Moon." Dusty and muddy roads, flat tires, and a leaking gas tank were merely temporary annoyances that did not dim these seasoned tourists' enjoyment of travel. Reminiscing today, after more than sixty years of worldwide travel since 1927, Cecil and Ida look back on trips like this one as far more exciting than their recent cruises on the *Queen Elizabeth II*. Upon arriving in California on July 13, after ten days and 2,650 miles, they briefly visited the elder Greens in Tormey, then moved into temporary quarters at 375 Waverley Street, in Palo Alto.

When Cecil reported to Federal Telegraph for the second time, he learned from Litton that his first assignment was to help move Federal Telegraph's Palo Alto laboratory from California to New Jersey; he would then direct it after it was consolidated with the company's main plant in Newark. The parent company—International Telephone and Telegraph—had decided to make the move because it was too expensive to maintain its principal research laboratory on the West Coast and its head office on the East Coast.

Cecil also learned that Litton wanted no part of going back East, where he had worked earlier for Bell Laboratories, and that he wanted to stay in California and eventually start his own business. This Litton actually did in 1931 by developing a new company in Redwood City north of Palo Alto. Years afterward, when he sold his business interests for $1,500,000, the buyers named their new acquisitions Litton Industries to honor the near-genius and unreformed workaholic, Charles V. Litton, whom Cecil so greatly admired and from whom he learned so much.

In the meantime, while listening to Litton's plans to separate from Federal Telegraph and IT&T, Cecil started planning the move of the laboratory to Newark—and his and Ida's own move as well. By early September he had everything ready for both moves. Then, for the fourth time in five years, they made the long transcontinental journey—this time from West to East. There were differences, however, from all former trips: they traveled separate routes and by different modes—Ida by sea, through the Panama Canal; Cecil by land and train.

Ida, with the Oakland automobile that had long since succeeded Tillie II, sailed from San Francisco on the *S.S. Virginia* on September 12, 1931, and arrived sixteen days later. (Her brief comments regarding the voyage are recorded in appendix D.)

Cecil left San Francisco the day after Ida's departure, taking the Southern Pacific's *Overland Limited* to Chicago. From there he stopped off at Schenectady to enjoy a brief visit with the elder Flansburghs at Ballston Lake before continuing by train to New York. Arriving at Grand Central Station, very early in the morning of September 16, 1931, he taxied to Pennsylvania Station, boarded a train to Newark, and then took a taxi to the Newark office of the Federal Telegraph Company, where he reported for work at 8:00 a.m. Cecil was never one to be late for any kind of appointment, strongly believing that such inattention to promptness could well be ascribed to lack of interest in his work and taken as an indication of general laxness.

During the next twelve days, Cecil found temporary lodgings and plunged immediately into consolidating Federal Telegraph's Research and Development Laboratories. Soon after Ida's arrival, the Greens rented an apartment at 375 Clifton Avenue in Newark and started a style of life that was fast-moving and demanding of both time and energy. Cecil had to work long hours to overcome one difficulty after another in the laboratory before he could write the following triumphant entry in his diary for the 1st of December: "Built the first water cooled rectifier tube in Lab."

After a pleasant Christmas interlude with the Flansburghs at Ballston Lake, Cecil and Ida returned to an active social life that consisted of entertaining or being entertained by their new acquaintances several times a week. A typical evening included dinner, then card games (usually bridge), and finally dancing or listening to the radio—a far different life-style from what they had enjoyed as doodlebuggers. But Cecil soon began putting in more and more hours at the laboratories, even though Litton was no longer around. Indeed, by mid-Frebruary he was spending so much time at the plant, frequently returning after dinner to work until midnight, that Ida took over the diary and recorded mainly her own daily activities—teas, shopping tours, occasional movies alone, and many hours of reading. Only on weekends were she and Cecil able to enjoy walks and drives into the countryside, attend a movie together, or share a dinner party and dancing.

In spite of his work schedule, however, Cecil found time to visit the local auto showrooms and inspect the latest models, for he still retained his love of automobiles. Furthermore, the Oakland was ready to be traded for a new car. By the end of April Cecil and Ida decided on a new Pontiac 6 standard sedan: cost $870.00 minus $350.00 for the two-year old Oakland standard coupe that had cost $1,284.00.

Then following their customary practice of showing off a new car to Ida's parents, they drove to Ballston Lake, took the elder Flansburghs for a long ride, and returned home with their usual largesse of "2 dozen eggs, 2 broilers, 5 cans of fruit, wine, also mint."

Conditions at the laboratories worsened in May and early June, as the Depression persisted, and on June 10 Cecil received notice of a cut in his salary. The notice, combined with the overwork and frustrations at the laboratories during the preceding months, catalyzed the Greens' desire to return to the West or Southwest.

Then came a bolt from the blue. Only five days after his notice of a reduction in salary, Cecil received a night letter from Eugene McDermott asking him to return to GSI on a one-year contract. The same evening, the telephone rang in the Greens' apartment in Newark; it was McDermott.

McDermott explained that the oil glut that had been produced by the East Texas Field had abated considerably, GSI had developed some new techniques that produced better seismographic records, and the company wanted Cecil to return as a party chief at an attractive salary. Listening in on the telephone conversation, Ida whispered to Cecil, "Ask him for $50.00 more a month and say yes!" Cecil found that McDermott was willing to increase his original offer by the $50.00 per month. So, after considering it overnight, Cecil accepted the offer and simultaneously informed Federal Telegraph of his decision to leave. Federal Telegraph promptly offered a substantial raise in salary, but Cecil and Ida decided that it was not large enough to persuade them to change their plans. So on July 6, 1932, Cecil spent his last day at the Federal Telegraph's Laboratory in Newark, and once again Ida began packing for a trip back to the Southwest. Ida's preparations were cut short, however, when Cecil was suddenly informed by J. Erik Jonsson, then in charge of GSI's instrument laboratory in Newark, that his first doodlebugging assignment would be along the New York–Pennsylvania line for three months. This was Cecil's first direct contact with Jonsson, who soon thereafter moved his shop to GSI headquarters in Dallas to become an indispensable member of the resident staff there. During the next few years Jonsson and Cecil worked closely together—he at headquarters in Dallas, Cecil in the field—and soon became good friends. This relationship would result in their becoming co-owners in two highly successful business enterprises, and their friendship continues today.

THREE MONTHS OF GSI WORK ALONG THE NEW YORK–PENNSYLVANIA LINE

Resumption of work with GSI, along both sides of the boundary between New York and Pennsylvania, brought Cecil and Ida back to essentially the same kind of gypsy life-style that they had previously enjoyed in Oklahoma. While Cecil labored to get his field party functioning effectively, Ida resumed her role of house-hunter and social hostess to both GSI personnel and client representatives. When these activities did not require her time, she frequently accompanied Cecil when he visited his crew, even if he occasionally hauled a few hundred pounds of extra dynamite in the back of his car.

These three months of exploration were conducted first for an individual client, then for a small independent gas company. The surveys involved fieldwork near Corning, Belfast, and Campbell in New York, then Galeton and Wellsboro in northern Pennsylvania, all of which served as bases of operations for the crews and as places where the Greens and other party members found temporary housing. By the time the work was finished, and Cecil and his party had received instructions to proceed to Dallas for a new assignment, he and Ida had gained an intimate knowledge of another part of the United States.

BACK HOME WITH GSI IN THE SOUTHWEST (OCTOBER 1932)

Leaving the Northeast for the last time—they would never again return for employment of any kind—the Greens departed Wellsboro, Pennsylvania, on October 1, 1932. It was one of their typical night trips, as recorded by Cecil:

We decided to visit folks [the Flansburghs at Ballston Lake] on farm—leaving Wellsboro 9:30 p.m., arriving farm at 4:00 a.m. . . . went to bed 4:30 a.m., arose 10:10 a.m.; repacked car, changed oil, repaired tire, helped pick up a few hickory nuts; gathered some apples for the trip to Dallas.

Then at 4:30 the next morning, they left the Flansburgh farm, driving as far as the Shenandoah Caverns the first day and, of course, visiting a cave before retiring. From there they followed a route that took them through the Virginias and into Tennessee beyond Chattanooga.

In one day they drove 635 miles (a record for them) through Birmingham and Tuscaloosa, Alabama, and Meridian, Mississippi, to Shreveport, Louisiana, where they found a $3.00 room for the night at the New Jefferson hotel. (This hotel would become a frequent haven for the Greens later on.) The breakfast to start this record-breaking day so impressed Cecil that he described it in his diary (October 5, 1932) as follows:

Had breakfast in Birmingham, Ala. Menu—baked apple in cream, ham, eggs, grits, toast, and coffee, all for 25¢.

Upon their arrival in Dallas the next noon, they were informed that the California job—their new assignment—had been called off and that they should proceed north to Oklahoma City to meet J. Clarence Karcher, GSI's president, at the Skirvin Hotel at 6:00 p.m. At dinner, with Karcher, and J. W. and Isabel Thomas, who had followed them from Dallas, Cecil learned that he and Thomas were to go to Kingman, Kansas, to take over a field party whose chief, Earl Johnson, was in the hospital. According to Ida's notes, the field party consisted of "7 married couples on crew, with 8 children, and 2 pregnant; 3 with no children. Some crew!" Karcher also mentioned that all the oil men were excited over the well that Cecil had sited previously near Perry, Oklahoma, because it showed successful production in an unexpected place. Cecil has saved the account in the local newspaper, which reported: ". . .initial production 220 bbl oil/ hour and a large gas flow. Well located 14 miles southwest of Perry, Okla."

In Kingman, Ida and Mrs. Thomas were unable to locate suitable living quarters, but Cecil came to the rescue by finding ". . .an immense house for $10.00 per week, plus gas and light." Furthermore, there was an electric refrigerator, rather than the usual icebox, and an electric stove. But, as would be the case again and again in the future, such luxury did not last very long; the party had to move to Chandler, Oklahoma, after ten days. The move from Kansas to Oklahoma brought the Greens back to familiar ground, but economic conditions were very different.

By late 1932 the United States was deep into the Great Depression. Banks were closing across the nation, the stock market continued on the downward course that began with the crash of 1929, and adverse economic conditions affected Oklahoma and Texas where Cecil's crew was working. Though Franklin Delano Roosevelt had decisively defeated Herbert Hoover in the November election, it would be some time before Roosevelt's efforts to improve the economy began to take effect.

Two days after Thanksgiving, Ida spent the afternoon wandering around downtown in Duncan, Oklahoma, and was moved to write in her diary:

The streets were filled with farmers shopping for the necessities of life, dressed in shabby overalls, while wives were in faded worn garments. Pale gaunt faces with many children suffering with scurvy. Seeing these people, we realized they swayed the election [Roosevelt vs. Hoover]; and who can blame them for trying someone different with hope for betterment?

This was the period when dust storms swept across Oklahoma, destroying crops and the farms on which they grew, forcing many families to load their sparse household goods on wagons or trucks and head for California. The Greens were more fortunate than many of the farmers whose land they explored, however, because GSI was recovering from the 1931 oil glut, Cecil's job as a party chief seemed secure, and his salary was sufficient to enable them to add to their savings on a regular schedule. Most important, however, was the fact that they were together.

Although a strong wind was blowing out of the north as 1932 ended, to be followed by snow and heavy dust storms in January and February of the new year, 1933 was to be tranquil for GSI and the Greens. Five days into January, Cecil and Ida were comfortably settled in the Franklin Hotel in Medford, Oklahoma (population 1,023). For $7.00 per week they enjoyed a spacious living room, a small dressing room, and a private bath, with arrangements to take their meals in the hotel dining room. Ida's social activities got underway quickly with a party for the GSI wives and children on January 13. Six days later she was the honored guest at a party celebrating her thirtieth birthday. The month of January was also a busy one for Cecil, because the members of his crew worked especially hard in competition with the other GSI parties in an attempt to establish a record for the month. On the 28th they shot twenty-three profiles, and on the 31st, even more—"the largest number of profiles ever made in one day by a GSI crew." Obviously, Cecil had his party running smoothly and efficiently.

On February 6, the Greens celebrated their seventh wedding anniversary, which Ida specially remembers because she had to keep the windshield, which had wipers operated from inside the car, cleared of a mixture of snow and dust as Cecil drove the thirty miles back to Medford from the Youngblood Hotel in Enid.

Cecil's salary was now more than he and Ida needed, so that they were beginning to follow the stock market with the intention of buying some blue-chip stocks in the near future. By July they had purchased a few tens of shares of AT&T, Commercial Solvents, Curtiss Wright, General Electric, General Motors, Radio Corp., United Aircraft, and U.S. Steel. Their federal income tax statement for 1932 showed:

. . . total income for last year [1932] $5,400 of which $2,900 is taxed at 4% making a tax of $116.11—the largest ever.

During their seven months in Medford, the tempo of Ida's life by turns speeded and slowed. She recalls preparing food for many lunches, teas, dinners, and picnics—and always trying to be ready with appropriate supplies when Cecil unexpectedly brought home crew members, or company officials from Dallas, such as J. Clarence Karcher or Eugene McDermott. As an example of Ida's culinary skills, following is the menu she prepared for a dinner to which Cecil had previously invited his new computer, Fred Romberg:

Fruit cup, butter splits, olives; yellow tomato salad, lamb, asparagus tips, beets, browned potatoes; orange biscuits, peach gingerbread; coffee.

Ida continued serving such meals year after year while Cecil was actively engaged in GSI's operations.

Ida shopped frequently, purchasing articles that she felt suited her role as the wife of the party chief—an attitude about personal appearance and dress that was developed in her high school days and is fully as important to her today (figure 8). She has always taken pride in her appearance and deportment, whether at home or in public. Her list of purchases on a late September day included:

8   Ida M. Green and others. *Top left:* Ida (ca. 1932). *Top right:* Ida and her brother Fred (right) with GSIer Ed. J. Stulken, on a hiking trip in the Sierras in 1938. *Bottom:* Cecil and Ida at the entrance to the Texas Instruments plant in Richardson, Texas (ca. 1955). Top photos by Cecil; bottom photo courtesy of Gary & Clark, Dallas.

Dobbs black felt hat	7.50	
Nurcer black purse	2.95	
Brown Marinette Boucle dress	19.75	$39.30
Matrix shoes brown suede and lizard	7.85	
Brown hose	1.25	

Cecil was putting in long days in the field or in the office, but he always tried to take evening walks of a few miles with Ida before adding stamps to his collection or developing the film in his new camera and making enlargements. And of course there were the movies to attend, such as:

"Farewell to Arms" (Helen Hayes and Gary Cooper)

"Tonight Is Ours" (Frederick March and Claudette Colbert)

"Sharp Money" (Edward Robinson)

"She Done Him Wrong" (Mae West)

"Men Must Fight"

Ida described the last-named movie as follows:

It was a story showing how U.S. is strong for peace and disarmament, etc., and advances 10 years to 1940 and shows us involved in war *unprepared.*" (My emphasis: recall that this was written in 1933!)

Even with all of the social activities, however—bridge parties, luncheons and dinners, movies, long walks, and occasional trips to the field with Cecil, and a special weekend at the Chicago World's Fair—there were many days that Ida had to spend alone with only her thoughts to keep her company. On such days she was likely to read or reflect on events of her past life, even at times writing down the results of her contemplations:

Youth is early morning with dazzling sunbeams penetrating nature to the extent that she bursts into blatant song, like a musical comedy. *Age* is eventide when soft shadows cast their rays and nature becomes subdued murmurs as in a symphony.

My definitions of life as I would like it to be.

Amid the dull and lonesome days, certain unusual events stood out for her:

Purchased my portable electric sewing machine in Greenville, Texas. Am I pleased or am I pleased!! Paid $50 cash. Cecil thought it a fine idea as well as a good price.

Whereupon she happily began a sewing hobby that resulted in dresses, coats and other clothes for many years thereafter. Ida's delight was decreased a few days later, however, when the stock market hit a new low, and she reacted exactly as many an investor has in similar situations, by writing:

Stock market shows sharp decline. Rather sorry now we didn't sell and realize a small profit and then be able to buy now.

Her spirits soon rose, however, when she went horseback riding, in Hutchinson, Kansas:

Went horseback riding with Cecil and nine of the boys of the crew. Horse was a pacer and I seem to have missed somehow, as found myself on the ground very much frightened.

In spite of the fall, however, Ida served dinner that night to Cecil, McDermott and two other crew members who did not leave until 11:30 p.m. Ida didn't give up easily.

As 1933 came to an end, Cecil and his crew were working overtime to avoid having to return to Hutchinson after Christmas to complete their contract. Meanwhile, Ida was packing their bags for a two-week vacation trip, by train, to Vancouver. It was to be her first night ever in a Pullman car, and she looked forward eagerly to the trip as a new experience in travel.

The year 1933 had been eventful for the country. Dust storms continued to plague the plains of Oklahoma and Texas; the Depression was being felt deeply; and Roosevelt, newly inaugurated, announced the Bank holiday which closed all the banks in the country to prevent panic withdrawing of funds; a major earthquake produced considerable damage in the Long Beach, California, area; the *Akron* crashed near Atlantic City, with seventy-one passengers killed and only three survivors; the World's Fair in Chicago drew large crowds; the National Recovery Act (NRA), devised to help the economy, was meeting qualified success; the stock market continued to be unstable;

and people across the country were flocking to the Hollywood productions to see and hear their favorite movie stars.

The year ended for Cecil and Ida in Vancouver, where they celebrated the beginning of 1934 with Cecil's parents and relatives.

1934 AND 1935: A HAPPY TIME AND AN IMPORTANT PERIOD OF LEARNING

Within three weeks of their return to Hutchinson, Kansas, after their year-end vacation, the Greens moved some sixty miles south to Wellington, Kansas, near the Kansas–Oklahoma border. Here they were introduced to a recently developed new drink called Dr. Pepper—"a prune drink like Coca Cola." A week later, the party members helped Ida celebrate her thirty-first birthday as they prepared to depart Wellington for Corpus Christie, Texas, 800 miles to the south.

January 23, 1934, found the Greens comfortably settled in a two-room-and-bath apartment in Corpus Christi at Main and Chaparral Streets, from which they could see and smell the waters of the Gulf of Mexico. All of the party—crew members and their families—were soon walking, swimming, and picnicking along the mainland shore and on nearby Mustang Island, where they especially enjoyed collecting shells, a new experience for most of them. Such activities in the balmy weather of a Gulf Coast winter seemed to stimulate the poetic urge in Ida, and on their eighth wedding anniversary (February 6, 1934) she mused:

Eight years have come and gone, and still within our hearts there is a song. Oh life is passing all too soon, and short will be our eight-year honeymoon.

The romantic feeling surged again a few days later when the two were on an outing to Mustang Island, and Ida wrote:

Cecil and I wandered around the sand dunes and beach. Had a nice talk. I feel that happiness is so elusive to most, that once you have it, hang onto it for dear life, because happiness between a man and a wife can be a beautiful thing.

The devotion of Ida and Cecil to one another that is expressed in these words has endured through the decades and is as strong today as it was more than fifty years ago when they wandered among the dunes of Mustang Island. Indeed, it has been written recently that Cecil and Ida have so blended their attitudes and principles that whether they are together or apart each mirrors the other in a remarkable identity of spirit and of action.

The weekday dreaming and the weekend frivolities came to an end abruptly, however, when Cecil received orders to move his party northward and inland to the small town of Madisonville, Texas. This was the beginning of a series of moves that lasted two years, in which the Greens crisscrossed East Texas from Corpus Christi and Laredo in the south to Denison and Greenville near the Oklahoma line in the north, and from Madisonville in the east to Dublin and Hamilton west of Waco. (See figure 7.) A dozen small towns served as bases of operations whose varying duration was determined by the time required for exploration of the adjacent areas. By the end of this two-year period of "stop-here-then-go-there," the Greens were veteran doodlebuggers in every way.

Cecil and Ida had run the full gamut of living quarters—from a small, dirty shack to a twelve-room house they once sublet for a few months. Ida had prepared their meals and had offered guests refreshments ranging from snacks of cakes and hot chocolate to full-course dinners cooked on everything from a portable two-burner Coleman stove to the latest gas and electric ranges. She did their laundry, maintained their wardrobes, and managed somehow to create their successive homes with the household effects that she could pack in a single trunk, carried either on a company truck or in their own car. In the best of circumstances, she and Cecil took most of their meals in hotel dining rooms or restaurants. But there were more than a few times when amenities were simply lacking, and when their living quarters were so primitive that after eating their food from plates set on an orange crate, they could brush the crumbs through cracks in the floor to clucking chickens below.

In short, Ida had long since learned to do with whatever was

available. So had Cecil, for this two-year period was one of frequently changing circumstances involving every aspect of his
party's operations—personnel, equipment, client requirements,
office and field responsibilities, operating budgets, and final reports. A wide variety of unexpected situations also required instant and decisive action—for example, when a crew member
turned up too inebriated to work safely. He had to be fired or
sent back to Dallas for reassignment.

There were also situations of both discomfort and pleasure
that live on in the Greens' memories. One of these involved a
contract with Seaboard Oil, to fulfill which Cecil was instructed
to take his party to a site 60 miles from Laredo. The women
would stay in town, while the crew members would make their
camp on a ranch. So Cecil set about buying equipment and supplies for the twelve crew members, who helped him bring their
goods to the ranch, put up tents, build tables, and fix supper.
Cecil wrote of the first night:

Didn't sleep much on account of mosquitoes—Heard coyotes
howling for the first time—Saw lots of tarantulas.

But the next evening Seaboard Oil's chief geologist, Ray Stehr—
with whom Cecil had spent many days in the field—took Ida
and Cecil across the Rio Grande River to the Alma Latina where
they had a quail dinner with Carta Blanca beer. This procedure
was repeated in Nuevo Laredo the next two nights, with Cecil
and Ida hosting, with a different selection of drinks, and with
dinner followed by dancing. Being party chief did have its rewards! But those joyful memories were needed later, when rain
brought shooting to a premature end, and the camp was broken
up. Wrote Cecil on June 12–13, 1935:

Geo. [George P. Cloos, his computer] and I left Laredo at 2:00
a.m. in coupe managing to reach Sutton Ranch at daybreak
where we finally drowned the ignition in a water hole and had
to walk last 10 miles through wind and water to camp arriving
9:30 a.m. Found rear end of shooting truck torn up as result of
trying [to] pull R22 out of hole. Left camp right after early lunch
with remaining members of crew including Mexican cook and
boy, and Recorder and water trucks . . . Wanted Mexicans to
stay, but they refused.

It was in this "crucible of experience" that Cecil proved to himself, once and for all, that he found greatest pleasure in working with people rather than with the inanimate paraphernalia of technology in the laboratory. When Cecil joined GSI as a party chief, he had brought to his job the technical knowledge of a superbly trained electrical engineer. He was able to learn quickly how the party's instruments operated, the kind of data they produced, and how the data could be best interpreted for the oil company clients. What he did not bring to his job as chief of a seismic exploration party was extensive experience in working with men and women having different degrees of technical training, highly individual personalities, and variable manual skills. Actually, his management experience when he joined GSI was rather limited—he had only directed small groups in the stable confines of production laboratories at General Electric, Raytheon, Wireless Specialty Apparatus, and Federal Telegraph.

The problems that beset him as an exploration geophysicist presented a whole new set of challenges—finding adequate garage space and services for trucks and other equipment, handling explosives both in storage and in detonating procedures, extracting stuck and overturned vehicles from dirt roads made into quagmires by floods, mollifying farmers and ranchers who did not understand why crew members first drilled shallow holes and then tried to blow them up with dynamite, and trying to outwit the scouts of the client's competitors, who were always lurking in the background. And always foremost in the mind of the party chief was the challenge of promoting the most effective cooperation among personnel in the handling of their respective assignments and equipment.

One special aspect of managing people caused Cecil occasional distress—the conduct of his crew members and their wives with respect to consumption of liquor. Drinking in the field could not be tolerated and was cause for instant dismissal: every member of the crew had always to be alert around moving equipment and in handling explosives, lest an accident occur. On the other hand, hospitality required that the party chief offer customary refreshments not only to visiting superiors and other dignitaries but also to crew members and their wives when they gathered for dinners, picnics, and other social activities. Once Cecil unexpectedly visited his party and found all the crew members asleep instead of working, because they had imbibed too freely of a local moonshiner's product. As discussed more fully earlier on, he could only chastise the whole crew—firing them all would have stopped work instantly and would have caused Cecil trouble with the client.

During this two-year period when Cecil and Ida were moving throughout Texas, GSI flourished. Additional field parties were formed by redistributing experienced personnel. Cecil was asked to train some new recruits as replacements for the experienced members of his own crew who were assigned to newly organized parties. He was also called upon frequently to advise nearby parties on problems with personnel or equipment.

In addition to receiving these extra assignments, Cecil was being visited more and more by Karcher and McDermott, from headquarters in Dallas, and was being called to Dallas regularly for conferences. Clearly his performance as party chief, and as general troubleshooter, was being watched carefully and noted approvingly by both officers.

PROMOTION FROM PARTY CHIEF TO REGIONAL SUPERVISOR: A CRITICAL STEP IN CECIL'S CAREER (1936)

As 1936 began, with more than four years of service as a party chief in the United States and Canada to his credit, Cecil could rightly be called a veteran exploration geophysicist. He was skilled in field use of reflection seismology and experienced in its management. Furthermore, he had gained a limited but practical knowledge of geology from the professionals employed by GSI's oil company clients. He had demonstrated to his GSI superiors, Karcher and McDermott, that he was an unusually perceptive engineering troubleshooter and an outstanding manager of personnel—democratic and generous with his crew members and their spouses but firm when required even when it came to firing an employee. And he had proved himself an excellent teacher by turning the raw recruits sent from Dallas into skilled professionals. In short, Cecil had far exceeded what was normally expected of the average party chief. His performance was the result of hard work, long hours, and concern for his crew members and his employer, GSI.

Charles V. Litton also knew about Cecil's outstanding skills from having observed him as a production manager at Federal Telegraph in Palo Alto and Newark. It was only natural, therefore, that he immediately thought of Cecil when in early January 1936 the Farnsworth Television Company of Philadelphia engaged him to find them a chief engineer. When Litton asked if Cecil would consider the position at $600 per month, Cecil answered that he might, since his salary at GSI was only $550. Before giving a definite reply, however, he wanted to review his prospects at GSI with Karcher.

Late the same day, Roland F. Beers, now Cecil's supervisor, arrived to go over the records on a prospect near Jonesboro, Louisiana, Cecil's base of operations. At lunch the next day, Ida told Roland about the offer Cecil had received from Farnsworth via Litton. Roland listened attentively but said nothing yet about his own plans.

The following Sunday—the day in the week when he usually wrote letters and telephoned about personal matters—Cecil wrote to Litton that he was still interested. He also asked Karcher to call on him as soon as possible to discuss his future at GSI. Karcher arrived in Jonesboro the next day and when Cecil returned to his office from the afternoon in the field, he found Karcher waiting there. Karcher said he had been told by Roland Beers that Cecil had something on his mind. When he heard from Cecil the details of the Farnsworth offer, he immediately offered to match the $600 salary. Cecil decided to stay with GSI for the time being, even though he received a telegram from the Farnsworth Television Company requesting that he visit Philadelphia at their expense. Having made his decision to continue working at GSI, Cecil so informed the Farnsworth Company by telegram and Charles Litton by night letter. But he had yet to hear from Beers.

When Beers arrived in the late afternoon, Cecil and Ida took him to dinner at the local hotel, where he confidentially informed them of his plan to leave GSI in five months to start his own seismograph company. Roland invited Cecil to consider joining him in the venture and to let him know his decision in a month or two. Inasmuch as Karcher had not yet informed Cecil of his future prospects in GSI, other than the salary increase, Cecil only took Beers's invitation under advisement; he first wanted to hear more from Karcher.

On February 11, as Cecil recorded in his diary, Karcher dropped by his new office in Monroe, Louisiana,

. . . to say hello but didn't mention any promotion beyond a casual mention, on getting into his car, that he wished I would train Koenig, so as to enable me to spend ⅓ time with crew & rest elsewhere.

Then on February 25, 1936, came the following cryptic telegram for Cecil:

CALL ME BETWEEN ONE AND TWO FROM PRIVATE TELEPHONE   J C KARCHER

When Cecil telephoned Karcher, he was advised to start some supervision work by calling first on party chief William (Bill) E. McDermott (Eugene's brother) at Arcadia, Louisiana, about 60 miles east of Shreveport. Two days later came a firm commitment from Karcher, as recorded by Cecil in his diary:

Karcher called on me in late afternoon to advise me that he was very glad to give me this supervisor's job with early increase in salary and a permanent headquarters in Shreveport, etc. Advised me that I would have charge of . . . four [crews] . . . counting my own.

By mid-March there were eight field parties for Cecil to supervise, and Karcher suggested that he move to Shreveport from Natchitoches. Shreveport was more convenient to Dallas and to the scattered field parties that Cecil would need to visit, and as a much larger city than Natchitoches, would provide more amenities for Ida.

Already the Greens had unpacked and repacked four times as Cecil's party moved its base of operations from Jonesboro to

Marshall, Texas, then back to Louisiana, to Monroe and finally to Natchitoches. Now Ida had to put one more apartment (at 56 Rutherford Street in Shreveport) into a livable condition while Cecil took his first big step up the corporate ladder as supervisor, with a salary increase promised.

Just as Cecil began his new duties as supervisor there was another change to record: on March 3, 1936, Cecil became a naturalized citizen of the United States. Erik Jonsson and Roland Beers were his two sponsors for citizenship, and the Jonssons treated the Greens to dinner at the Dallas Athletic Club after the important event.

For the next five years and nine months until early December 1941, Cecil would be almost constantly on the go as GSI's regional supervisor. It was a drastic change in both work and lifestyle, for now Ida would often sit alone in Shreveport—or occasionally in Midland, Bakersfield, or Los Angeles—longing for the return of the time when she could share much of Cecil's daily life and work, and still wondering if they would ever be able to have a permanent home of their own.

THE LIFE OF A GSI SUPERVISOR AND HIS WIFE

By the end of March 1936, Cecil was supervising the eight field parties listed below, and he very quickly became acquainted with what would be the essential ingredients of his daily schedule for the next five and more years as supervisor—trouble, tension, travel, and training.

Base of Operations	Party Chief	Client
Minden, La.	C. V. A. Pittman	Tidewater
Winnfield, La.	R. G. Elms	Shell
Mansfield, La.	Wm. McDermott	Gulf
Atlanta, Tex.	J. W. Thomas	Stan Louis
"	E. Galey	Pure
"	M. J. Walzcak	Seaboard
Longview, Tex.	J. Gary	Sun
Carthage, Tex.	F. L. Bishop	Magnolia

Soon Cecil began to arrive late for dinner, having earlier telephoned Ida that he might be delayed. Too often the delays compounded, and there came the apologetic call that he would have to spend the night at the site he was visiting. Soon he bowed to the greater efficiency of a weekly schedule of visiting four or five field parties in succession before returning to Shreveport for the weekend. Occasionally, Cecil's responsibilities kept him away for several weeks at a time. Absences finally were so frequent during one period that Ida's women friends began to wonder if in fact she had a husband. Eating alone, taking her habitual evening walk by herself, and either foregoing altogether the movies she loved or sitting alone in the darkened theater, did not appeal to Ida at all. Sewing, reading, shopping, and weekly tea and bridge parties helped to fill the hours, but even then there were many lonely ones.

In fairness to Cecil, it should be recorded that when he did have free time, especially on weekends, he was always very attentive—eager to drive into the countryside, go on a picnic, or attend any movie in town. Furthermore he occasionally took Ida along on short visits to a crew, and twice she accompanied him on trips of a week or more. It was hardly a treat, however, for her to sit in the car or in the shade of a tree with the temperature at 100°, the sun blazing down, and the dust-laden wind abrading her face. No, being the wife of a supervisor of seismic crews was far different from being the wife of a party chief. More than once in reminiscing, Cecil has been heard to remark:

In retrospect, I often wonder why she didn't pack up her belongings and go back to Ballston Lake or Schenectady. She certainly would have had good reasons to leave me!

Ida is not the kind of person who weakens under discomfort, dejection, or loneliness, however; determination is part of her character. Otherwise, she never would have ridden those thousands of miles, four times across the continent in an open touring car, or endured the concern of a young bride asked to camp out on those long early trips. In recalling those years, Cecil is wont to say with sincere admiration, "She had what it took, and it took a lot." Fortunately for the Greens' marriage, Ida's strong character, deep sense of loyalty, and abiding love for Cecil prevailed. And in those years was formed her lasting and oft repeated desire: "To help make the world a better place to live in."

It would take another book to describe all the troubles that confronted Cecil on his weekly visits to the crews, for they were legion—a vehicle overturned, stuck in the mud, or wrecked; an intoxicated employee sleeping on the job or a crew member careless with explosives; a conflict between employees that could only be resolved by transferring or discharging one of them; a lazy employee simply unwilling to do his part; an irate farmer or rancher yelling, "Why the hell you-all shooting off the dynamite on my place?" and so on and on. Typical is the entry in Cecil's diary for May 5, 1936:

. . . at 4:30 [p.m.] word came in over telephone from Linden [Tex.] that a crew member had knocked down a negro child on highway. Called up Erik Jonsson at night to advise him of the accident and he in turn told me that shooter's helper on Gary's crew lost a cap which a native found and blew off three fingers.

or another on August 13, 1936:

A newspaper clipping describes a prolonged heat wave in Texas, Arkansas, and Louisiana. Shreveport had highest reading since Aug. 18, 1909—108.8° at 5:20 p.m. on Mon., Aug. 10, 1936. Temperatures elsewhere in region:

Mt. Pleasant, Tex.	118	Dallas, Tex.	109.6
Texarkana, Tex.	117	Shreveport, La.	108.8
Longview, Tex.	113	Eldorado, Ark.	112
Ft. Worth, Tex.	109.8		

This was long before air-conditioning in sleeping quarters, and the crews that Cecil was supervising were in the field every workday sweating it out. No wonder tempers were short and tension rose.

Fortunately for Cecil, his daily rounds of trials and tribulations were temporarily relieved by brief vacation periods which, however, always depended on cooperative arrangements with the different party chiefs. On such occasions, he and Ida headed for Ballston Lake to spend the time with Ida's parents, or to the San Francisco area, where the elder Greens were living.

The year 1936 ended with the country's partial recovery from the Depression, Franklin D. Roosevelt's election to a second term as president, and GSI's continued prosperity. During December, Cecil spent day and night visiting one party after another. On December 10, en route to Texarkana, he remembers hearing over his car radio a proclamation from London that King Edward VIII had abdicated in favor of his younger brother. The next day, on the way to Mineola, Texas, Cecil recorded in his diary that he heard the King's farewell message and his "God save the new King [George VI]." The news was a special reminder of his English birth, though he had recently become a naturalized citizen of the United States.

As the year ended, he once again recorded his and Ida's desire to settle down permanently on the West Coast (diary entry for January 1, 1937):

As the New Year starts, we naturally wonder what we shall be writing on this same page in 1938. During past year I have experienced promotion from party chief to supervisor, but so much time away from each other makes Ida and me once again yearn for the West Coast and to be settled down.

On a happier note, it can be recorded that in early December Cecil traded in his old tan Buick, with some 67,000 miles on the odometer, for a new dark green Series 60 straightback four-door Buick sedan. And a week before Christmas, Ida became the proud owner of her first car—a dark blue Oldsmobile 6 sedan bought for $920.

CECIL'S RESPONSIBILITIES WIDEN:
A TRIP TO ECUADOR

GSI had first expanded its seismic exploration to foreign countries in 1931—1934, and from time to time Cecil had assisted this foreign work by releasing some of his experienced crew members for work abroad. Now, as supervisor, his knowledge of the oil potential in the rest of the world was about to be broadened. In mid-1937, McDermott asked Cecil to fly to Ecuador to evaluate the field work of the GSI crew there under party chief E. F. McMullin.

Unfortunately for Ida, this was not a trip on which she could go along. The journey would be directly to the exploration site, where there would be nothing for her to do all day or evening while Cecil was working, and no comfortable or interesting place for her to stay. She had just enjoyed three successive weekend excursions with Cecil, after settling an apartment in Midland, but now she would be left alone in a strange town for at least two weeks.

Cecil's assignment required his first trip on an airplane. Starting in Dallas, he flew to Atlanta in a Lockheed two-engine Electra, changing there to a Douglas DC-2 for Miami. While spending the next day (May 31) in Miami waiting for his final typhoid shot, he saw Amelia Earhart in the cafe at breakfast and noted in his diary:

She too starts on a trip tomorrow at 5 a.m.—only around [the] world.

From Miami, Cecil took a four-motor Sikorsky flying boat to Barranquilla, Colombia, via Cienfuegos, Cuba, and Kingston, Jamaica; from Barranquilla he traveled by two-motor Sikorsky plane to Colón, Panama Canal Zone, and then southeastward along the Panama Canal and the Colombian coast to Buenaventura and Tumaco, Colombia and finally to Guayaquil, Ecuador, where McMullin awaited him.

McMullin was working for the Anglo-Ecuadorian Co. on the Santa Elena peninsula, where his party was investigating the Ancón oil field. The field was of special interest because of oil seeps that resulted in open tar pits on the surface. These pits had become known to the *Conquistadores* three centuries earlier, and had provided the tar for caulking their galleons.

Leaving Guayaquil, McMullin took Cecil on a special rail car—a Ford V-8 beach wagon mounted on railroad wheels—75 miles to Santa Elena; then they drove in a 1927 seven-passenger Buick touring car by dirt road to the Ancón oil field. Cecil noted that the weather at the end of the peninsula where the oil field is located was cold enough at night to require four blankets, thanks to the Humboldt Current coming from the Antarctic Ocean. Unfortunately for Cecil, he had ignored the advice of several colleagues to take warm clothing, thinking that Santa Elena, at three degrees south of the equator, would be hot.

The problem that had led to Cecil's visit was the Crew's inability to obtain the data needed to advise their client in which direction to extend the field. After considerable experimenting, Cecil found that he could not help, and so had to recommend that GSI terminate the contract. Then after four days of frustrating work, Cecil flew back to Midland via Colón; San José, Costa Rica; Managua, Nicaragua; Tegucigalpa, Honduras; Guatemala City; Mexico City; and Brownsville, Texas.

MIDLAND TO BAKERSFIELD TO LOS ANGELES
There followed a month of intensive catch-up work for Cecil before he and Ida could start a much-needed month's vacation with Cecil's parents in Tormey. They had spent only two weeks in California when business intervened again; a telegram from McDermott summoned Cecil to Bakersfield, California, to supervise the work of two party chiefs: Al McCluney, based in Bakersfield, and Robert C. Dunlap, Jr., working out of Ventura on the coast.

The increase in Cecil's salary to $700 a month at the end of July was solid evidence that he was impressing GSI's top administrators in Dallas with his handling of personnel, his overall managerial skills, and his constantly improving judgment of prospects for oil and gas; it also implied that he would probably be asked to assume additional responsibilities. The latter were not long in coming.

Two significant notes appear in Cecil's diary for August 1937. While in Ventura checking Dunlap's party, Cecil mentions that he spent an entire day "writing a long letter to Erik Jonsson and Eugene McDermott to help them form some opinion [concerning the] importance of this area [the Ventura area] over West Texas." A few days later he was writing "letters to Stanford and

University of California asking for men to apply for geophysical work." Clearly, Cecil was working more closely than ever with the GSI executives in Dallas.

Unfortunately for Cecil, however, California was no different from Texas when it came to the trouble-travel-tension-training schedule. When a careless driller dropped 300 feet of drill pipe down a deep hole, the party chief and his crew had to spend more than a week recovering the pipe. Following this episode, Cecil spent a week trying to contact company geologists in the Los Angeles area, but they seemed never to be in their offices. Back in Bakersfield, he learned about another problem on the same crew: the party chief had found two crew members fast asleep while the others were trying to load a deep hole with explosives. Cecil decided then and there that the two sleeping crew members had to go—one to be transferred to another crew, the other to be fired. He had, however, to write a letter to Jonsson in Dallas explaining his action. That paper work done, more recruiting efforts had to be initiated.

By mid-September 1937 Cecil was beginning to wonder just what plans Karcher and McDermott had for his future with GSI. A query to McDermott brought a telegram: "Return to Midland to resume work in West Texas." But Cecil deferred his and Ida's departure from California for one last round of visits to the crews he had been supervising, a farewell visit with his parents and other relatives in Oakland and San Diego, and a final word with certain petroleum geologists he had become acquainted with, particularly those with Standard of California—George Knox, George Cunningham, and William Kew.

Hurrying back to Texas from Bakersfield, the Greens covered the 1,580 miles in three and a half days, arriving at their Midland apartment at midday on October 29. Even before unpacking the car Cecil called McDermott and received his next assignment: an immediate trip to Oil City, Pennsylvania, to evaluate the records Bill McDermott was getting. So off to Dallas and Oil City went Cecil by train late the same evening, leaving Ida alone in Midland, unhappy that they had had to return to Texas from California. Two weeks later Cecil returned to Midland only long enough to greet Ida before rushing off by auto the next day (November 14) to meet Eugene McDermott in the small town of Spur, near Lubbock. In Spur, the two geophysicists spent three days carefully going over data that party chief M. P. Jones had obtained in four months of shooting on the Swenson Ranch for the Gulf Oil Company. After McDermott left, Cecil stayed on for two more long days to help Jones with interpretation.

Finally, on November 19, Cecil was able to drive back to Midland to spend the weekend before Thanksgiving with Ida— their first together in the Midland apartment in three weeks. After Thanksgiving in Midland, Cecil again had to do considerable traveling during December, but did get home for Christmas.

At the end of 1937, he wrote one of his brief year-end summaries:

Well, another New Year and still in the old wandering around game. The past year has been quite eventful in point of travel— changing location—so that the New Year will have to go some to surpass this feature. Naturally we still have that ambition for our own diggings but still not in sight.

Owning a home anywhere was clearly inadvisable for the Greens at the time, in the light of the many moves required of Cecil, but they nevertheless were planning for the future in other ways. As their income increased during these years, they followed their longtime practice of putting a substantial portion of Cecil's salary into savings bank deposits or securities— mainly blue-chip stocks. They bought life insurance and annuities for protection, and had their first wills prepared as early as 1937.

Yet prudent and foresighted as they were in these earlier years of their marriage, Cecil and Ida never hesitated to share their time and limited means with others. Wherever they lived, even if only for a few weeks, they quickly became known for their friendliness, hospitality, and generosity.

Having assumed responsibilities greater than those usually

assigned to a regional supervisor, Cecil had become the one field geophysicist in GSI most likely to be called upon when a troublesome operational or personnel problem arose anywhere in the field. As Cecil's business travel increased, Ida was left alone more, attempting to fill her days and weeks with activities that helped to reduce the loneliness that sometimes overcame her. She did accompany Cecil whenever possible, on short trips of a few weeks, but these were never as frequent as she and Cecil desired.

After spending twelve days in Colorado at the beginning of 1938, to evaluate a party chief's data, Cecil returned to Midland, completed his income tax return ($220 for 1937), and then with Ida drove to Dallas for a five-day business session with GSI officials. The Greens continued on to New Orleans for the 1938 Annual Convention of the Society of Exploration Geophysicists (SEG), at which Cecil delivered one of his earliest papers, "Determination of Velocities by Reflection Profiles." On the way back to Midland they stopped at Dallas so that Cecil could confer with Karcher and Jonsson on personnel problems connected with the breaking up of crews because of recently terminated contracts. One week after their return to Midland from their eighteen-day, 2,100-mile auto trip, McDermott suggested that the Greens move from Midland to Dallas, so that Cecil would be nearer the parties operating in the Shreveport, Louisiana, area. Accordingly, on Easter Sunday (April 17, 1938), they loaded their two cars with their small household items, and the next day drove to Dallas and began to search for housing. Finding the ads for rentals in the *Dallas Morning News* to be of no help, they started a blind search up and down the streets in North Dallas, finally renting a place at 5452 Longview Avenue for $55 a month. The next morning they were off to visit crews near Texarkana and El Dorado, Arkansas. After a week, they returned to Dallas and spent their first evening back visiting the Jonssons. The Greens have a vivid memory of this evening, because a thief opened a front window of Cecil's Buick, which was parked in front of the Jonssons' home, and took their two suitcases, their coats, and the typewriter.

Cecil's next assignment demanded two months of very intensive work, with Dallas as his temporary base, and permitted more nights at home with Ida. Then suddenly he was instructed to return to Bakersfield for an indefinite period to resolve some personnel problems and organize a fourth crew for work near Stockton.

This time Cecil chose Pittsburg (California) as his base because of its proximity to his work site in Martinez, on the river near the Carquinez bridge. Cecil would lead the new field crew requested by George Cunningham of Standard Oil Company of California. The crew, which had been working for Karcher in Texas, comprised two drills; recorder, water, and shooting trucks; and six crew members. Cecil took immediate charge, noting that it was the first time he had used sixty-four seismometers. After three weeks as acting party chief, Cecil had the crew working together effectively, but he sensed some tension. So when he called an evening meeting to introduce his successor, Bill McDermott, Cecil forcefully reminded the crew of the absolute necessity for all members to cooperate fully, then treated them to drinks at the corner fountain. On such occasions, which fortunately were rare, Cecil pulled no punches; he made it quite clear that a noncooperative member could expect to be sent back to Dallas for transfer or termination. Noncooperation could not be tolerated on a seismic party using explosives.

As Cecil's California assignment ended, he and Ida parted company for a time—Ida going by train for a month-long visit with her parents in New York, Cecil making a brief visit to Tormey to celebrate his thirty-eighth birthday with his parents. He then started on a five-week, 6,500-mile auto trip on which he visited GSI parties in Nebraska, North Dakota, Montana, and Alberta.

Ida rejoined him in Medicine Hat, Alberta. Together again, they drove to Vancouver to visit Cecil's relatives there, and then continued south to Pittsburg, where Cecil was soon back to night work analyzing data recorded by party chiefs Bill McDermott, in Pittsburg, and R. D. Terry, in Bakersfield. This turn of events did not please Ida, who was impatient to get back to

Dallas and resettle their apartment. The situation worsened for her when Cecil was asked to take over E. D. Gaby's crew for two weeks while he went on vacation. Worse still, soon thereafter Cecil was informed by the Dallas office that he would have a longer-term assignment in California. He and Ida should return to Dallas to vacate their apartment and dispose of their furniture before moving back to Bakersfield.

Forced to abandon their comfortable apartment in Dallas, Ida was deeply disappointed and began to wonder just how much more of the crazy geophysical odyssey she could tolerate. At their usual company parties in Dallas, she could not resist expressing her growing discontent with having to move around so much. Laconic under ordinary conditions, Ida was driven to add a few choice words to her usual comments when the opportunity presented itself at the parties, but in the end, despite her discouragement, she set her teeth and once again prepared to move.

During their brief interval in Dallas, Cecil was briefed about a number of new projects under development by GSI, so that he could be a more effective salesman for the company when he returned to the West Coast. Among these projects were the new chemistry laboratory for soil analysis under Jim Toomey's direction—a special interest of Eugene McDermott; new developments on amplifiers, cameras, and camera drive motors; and the plan to build a recorder on a Ford truck chassis. Strategies to absorb personnel from returning crews were also discussed.

With their furniture sold, their books and other bulky items prepared for shipment, and their lighter items packed in their two cars, the Greens found departing harder than ever before. They called on the Jonssons, Karchers, and Fishers, and finished the day by taking their longtime friends, Ray and Sally Stehr, to dinner at the Golden Pheasant where they had dined many times before. The next day they turned off the utilities in the apartment, lunched with Eugene McDermott, and were saddened when Mrs. Grace Wallace in the downstairs apartment cried at their leaving; not an unusual reaction for the many friends that the Greens have left behind in their ever-lengthening odyssey.

On arriving in Bakersfield, Cecil plunged into his work. For example, he quickly learned about the lack of leadership in one of the parties. He immediately decided to replace the neglectful party chief and wired Eugene McDermott in Dallas to that effect. Cecil had to direct the party himself until he could find another chief—just one of the many personnel woes of the supervisor.

But there were compensations. Cecil and Ida especially remember the last two months of 1938, because they were able to spend Thanksgiving and Christmas in their own apartment at 1719 Palm Street in Bakersfield with the elder Greens their house guests for both festive occasions. Dancing at a GSI party in Bakersfield's El Tejon Hotel they saw the old year out and the new year in. Later in the day they watched the Tournament of Roses from the corner of Colorado and Hill Streets in Pasadena, then saw the University of California defeat Duke University, 7–3, in the Rose Bowl.

Cecil completed his diary for 1938 with the following comment:

Somehow or other we feel pretty thoughtful this New Years Day—for there seems to be at least a meager chance that we shall go on living in California. Also our fortune and health has been excellent during 1938. Here's hoping that 1939 will be as good.

As it turned out, 1939 brought world-shaking events in Europe and a more unsettled life for the Greens in California. Cecil continued to attend to problems of the parties he was supervising locally, and he was also called upon to visit other parties as far away as Colorado and Pennsylvania. On one long trip of seven weeks, involving visits to those two states, among others, Ida accompanied him most of the time. But it was to be their last long period of travel together anywhere for many months.

Back from his fifty-day trip away from Bakersfield, Cecil discovered that GSI's seismic business was on the decline. The invasion of Poland on September 1, 1939, and the simultaneous declaration of war on Germany by England and France, had serious repercussions for GSI. There were far too few men qualified for foreign work to replace those who were returning to the States for one reason or another. Meanwhile, to generate new business, Cecil began visiting petroleum companies in Los Angeles—Superior, Continental, Ohio, Richfield, Texaco, Shell, Standard of California, and Union—seeking contracts for the new geochemical program of soil analysis that McDermott, Toomey, and others had developed in Dallas. Thus yet another responsibility was added to Cecil's work load.

Since his added responsibilities took Cecil to Los Angeles more and more frequently, McDermott advised him to move his office from Bakersfield to the Subway Terminal Building in downtown Los Angeles. Before leaving Bakersfield, however, Cecil initiated one more activity—a Class Study Program in which he invited company geologists and geophysicists to meet for evening lectures on subjects of mutual interest. He opened the first meeting on Wednesday, September 20, 1939, in the El Tejon Hotel, after a dinner for the group of forty-five men. At this organizational meeting, Cecil, acting as chairman of the geophysical section, emphasized the importance of cooperation and collaboration between geologists and geophysicists. Three weeks later he gave the first scheduled lecture, to ten program participants, in Frank Morgan's Richfield office in Los Angeles on the subject of "Soil Surveying" ("which seemed to be well received," wrote Cecil). This successful Class Study Program was the first of many that he would organize and lead in later years, all with the main purpose of bringing together experts in diverse fields to exchange ideas.

October 2, 1939, is a date especially remembered by the Greens because on that Monday they rented an apartment furnished with "back East" antique furniture purchased by the owner, an Episcopal bishop. Here, in the Westwood Village District of Los Angeles, they would live for two turbulent years, often uncertain about their future with GSI. That uncertainty only ended when they returned to Dallas and took the greatest gamble of their lives—purchase of a one-fourth interest in GSI.

On the first day in his new office in Los Angeles, Cecil interviewed a number of applicants for the position of "stenographer" and hired a Mrs. Alvina Dearth, his first secretarial assistant. In the late 1980s it is difficult to imagine how Cecil accomplished all his letter and report writing before 1939 without the assistance of such a person, but he did. The first task he assigned Mrs. Dearth was to type his opinion of a report (by Burt Beverly) on Saudi Arabia that George M. Cunningham, exploration manager of the Standard Oil Company of California (SOCAL), had asked him to evaluate. Cecil and George had become good friends as a result of their business relations— SOCAL was one of GSI's most important clients, and George was their geologist—and they had developed respect for one another's judgment and integrity. Shortly after receiving Cecil's evaluation, George invited Cecil to accompany him on a trip to the Near East, if GSI approved.

Cunningham was planning a short trip to Egypt and Saudi Arabia to check on SOCAL's exploration and drilling activities in the two countries, and he wanted Cecil's assessment of the sites. Since GSI had two crews working for CASOC (California Arabian Standard Oil Co.) in the Persian Gulf area, Cunningham knew that Cecil could visit them, as well as advise on SOCAL's operations. GSI officers in Dallas approved the trip, and Cecil accepted George's invitation with enthusiasm. Cecil bought himself a tuxedo and other clothes, took his second typhoid shot, completed his dental work, and otherwise prepared for what was planned as absence of two or three months.

The dreaded day came for Ida on October 20, 1939, only eighteen days after she and Cecil had moved from Bakersfield to an apartment at 1622 Greenfield Avenue in the Westwood Village District of Los Angeles. Not having lived in Los Angeles before, and without friends for companionship, Ida foresaw a long period of loneliness and knew that she would have to seek activities that would keep her busy and make the time pass more quickly. So it was with foreboding and tears in her eyes that she bade Cecil goodbye that Friday evening, wondering if she would ever see him again: war had already started in Europe, and Cecil had to travel by ship across the Atlantic Ocean, where German submarines patrolled.

AROUND THE WORLD IN EIGHT MONTHS (1939–1940)

Traveling across the continent on the Southern Pacific's *Argonaut*, Cecil made a three-day stopover in Dallas to confer with GSI officials and purchase $1,000 in traveler's checks before continuing on to New York. On arriving in New York in the early afternoon of October 26, he registered at the Governor Clinton Hotel, met Cunningham and his wife, Helen, and spent the rest of the day with them.

The next day Cecil and George attended to arrangements that had to be made before they could board their ship: obtaining the necessary travel papers, getting their third typhoid shots, and seeing to the shipment of their instruments to Saudi Arabia. All preparations completed, Cecil left the Cunninghams for the time being and spent the remainder of the day with other friends in Newark. Arriving back at his hotel by subway at 3:00 a.m., he wrote good-bye letters until 5:00, snatched two hours of sleep, then arose, and within an hour was telephoning Ida's parents in Ballston Lake. As a doodlebugger, Cecil had long since become accustomed to such hectic nights.

After his telephone call he and the Cunninghams taxied to Pier 7 in Hoboken. By 11 a.m., Cecil and George were aboard the American Export Lines' SS *Exeter*. They waved to Helen Cunningham from the deck as the ship finally headed out to sea at 1:00 p.m. on Saturday, October 28, 1939.

As the *Exeter* cleared the harbor and reached the open sea, the ship's purser surprised Cecil and George by remarking that the war made such a trip downright dangerous:

Don't you fellows know that there is a war going on in Europe and that there are German U-boats prowling around in the Atlantic Ocean?

Cecil responded by pointing out that there was a large U.S. flag painted on each side of the ship and that each flag was brightly illuminated. The purser's rebuttal: a submarine commander might shoot before noticing the flag.

Fortunately for Cecil and George, the trip across the Atlantic went without incident. When land was sighted off Cape St. Vincent on the coast of Portugal, and the *Exeter* was approaching the Strait of Gibraltar, Cecil by previous arrangement was called so that he could go on deck and enjoy the view of the British-held Rock. It was a sight he did not want to miss, even though he was now a naturalized American citizen. Wrote Cecil in his diary:

A clear moon shining as we went on deck to see a British destroyer arrive alongside us. After boarding us for a short conference, we were instructed to head for the harbor where we anchored opposite to the Rock of Gilbraltar . . . I took pictures all day in the clear sun. British sailors came aboard to remove mail—one of them taking a letter to mail [to] Ida. The American crew threatened to strike, but were talked out of it by the captain. Pulled up anchor at 3:30 and saw a beautiful sunset back of the Rock.

The next stop was in Marseilles, where Cecil watched copper slabs being unloaded from the *Exeter* for dispatch to a French munitions plant, where the copper would be used for shells. Suddenly realizing that his ship had been carrying war material for France, now at war with Germany, he wondered if the U-boat commanders in the Atlantic might have known about that copper. No wonder the purser had called Cecil and George foolhardy!

The following day's run took them to Naples, then on to Alexandria, to be met by R. C. Wedemeyer, the representative of the Standard Oil Company of California (SOCAL) in Egypt. He drove them over the 150-mile British military road to Cairo, where they registered at the Continental-Savoy Hotel.

In Cairo, for the next month Cecil and George reviewed the refraction records that a Texas Company crew had made for CALTEX (Standard Oil Company of California/Texaco). When time was available, they visited the Sphinx, the pyramids, and other monuments of ancient Egypt.

Early in the morning of December 21, Cecil and George departed for Baghdad. Cecil recalls that his Leica camera was taken from him by a British security officer as he boarded the plane. En route, they stopped once at Lydda in Palestine for breakfast and once at H-3 on the Iraq pipeline to refuel. The next day, after seeing some of the attractions of fabled Baghdad, they boarded a British Airways four-motor Hannibal for the flight down the valley of the Tigris-Euphrates River to Basra, at the head of the Persian Gulf. The next day they continued southeastward to Bahrain Island, where they were met by GSIer Neil Mann and petroleum engineer Phil McConnell. A courtesy call on the several British advisors to the sheik of Bahrain was followed by a cocktail party at which Cecil and George, as guests of honor, met a number of the resident petroleum personnel.

One of the objectives of the trip was to determine how best to place the explosives being used by a GSI marine seismic crew operating off Bahrain Island under the direction of Mann as party chief. To do this, Cecil and George accompanied Mann on a launch from the island. From Mann's house-office barge, Cecil and George experimented by drilling shot holes into the bottom of the Gulf.

One night, while the crew was engaged in his work, a 25-mph north wind (a "shemal") arose, causing some of the GSI craft to drag anchor and drift with the wind. Cecil was marooned with Mann's crew for several days before the waters quieted enough that the fleet could be reunited. Some of the craft, including the dynamite jolly boat, and the house-office barge, had drifted southeastward as far as Qatar, where it took some vigorous negotiating with the local authorities before Mann was able to regain possession.

On the last day of 1939, Ellis W. Shuler, Jr., chief of a GSI seismic party in Saudi Arabia, took Cecil and Cunningham on a round trip by car from Dhahran to inspect his camp about 150 miles north. Their route took them by Abu Hadriya, where CASOC (California Arabian Standard Oil Company) was drilling a well on a structure which had been located previously by James Gary's GSI seismic crew (see figure 9).

A few weeks afterward, in late January 1940, Cecil and George again visited Shuler's camp, this time to observe experimental recording in an area between Jebail and Hinat, south of Abu Hadriya. It was on this trip with Cunningham that Cecil first met Max Steineke, whom he and many of his geology friends regarded as one of the world's foremost oil finders. Steineke had been appointed chief geologist of CASOC in 1936 and continued in that position until 1950, two years before his death. He had been in Saudi Arabia as early as 1934 and with Tom Koch had reported certain surface structures that later proved to be sites of some of the world's largest oil fields.

G. C. Gester,[6] who was vice president for production of SOCAL, the parent corporation from 1920 to 1943, in 1951 wrote the following about Steineke's geological success as an oil finder in Saudi Arabia:

The greater part of this reserve [Saudi Arabia's vast oil reserves] can be credited to Max Steineke for his vision, optimism, persistence, and friendly leadership in the field of reconnaissance and detailed geology. The method he developed in the area probably resulted in the discovery of greater reserves than the work of any other single geologist.

The well at Abu Hadriya, which was being drilled on one of GSI's postulated structures when Cecil met him there, had already reached 9,600 feet, far below any known producing strata along the Persian Gulf, and no oil had been found. Although skeptical of GSI's promising geophysical work, Steineke had reluctantly directed that drilling continue only until all the casing or drill pipe on the site was used.

Having achieved their several objectives in Egypt and Saudi Arabia, Cecil and George now considered returning home.

[6]Gester, G. C., "Max Steineke, Sydney Powers Medalist," *AAPG Bull.*, vol. 35, no. 7, pp. 1696–1697, (1951). Also see Owen, E. W., "The Trek of the Oil Finders: A History of Exploration for Petroleum," Mem. 6— Semicentennial Commemorative Vol., pp. 1–1331, *AAPG*, (1975).

9   Cecil and companions in Saudi Arabia. Cecil (*center*), George M.
Cunningham (*left*) and Ed Skinner in Arab dress for visit to Amir Ben
Jalui at Hofuf, Saudi Arabia, in 1940. Photo courtesy of GSI.

Since they were halfway around the world, they decided they would gain nothing by simply retracing their route back to California. It would be more useful to continue eastward to India and Indonesia, where both SOCAL and GSI had operations. From Indonesia they could cross the Pacific to San Francisco.

Accordingly, in mid-February 1940, the two men flew from Bahrain Island to Karachi, in northwestern India (now Pakistan). There Cecil met officials of the Indian Oil Concessions Ltd. (IOC) and looked over the seismic reflection data obtained by GSI party chief Robert P. Thompson. A few days later he and Cunningham visited Thompson's camp some 500 miles up the Indus River valley at Tanabulha Khan in Bahawalpur. Here Cecil watched the crew conduct ground roll and velocity experiments in sand, shooting with separate groups of seismometers—eight in each group. Fred J. Agnich, who later became a president of GSI, was a member of Thompson's crew.

While still far up the Indus River with Cecil, Cunningham received two radiograms: one from SOCAL's home office in San Francisco, the other from Saudi Arabia. Which should he open first?

He decided to open the one from his home office first, since it was probably from G. C. Gester. "Before going on to Indonesia," Gester wrote, "you should return to Saudi Arabia to make a post-mortem as to why we ever became involved in that $1.5 million dry hole at Abu Hadriya."

"Somebody's sure going to catch hell," said Cunningham, after reading the radiogram, "and now I'll have to go back to Saudi Arabia."

Cecil suggested, "Why don't you read the other radiogram, George?" This one, from Steineke, read:

Abu Hadriya well just came in for 15,000 barrels per day at 10,115 feet.[7]

[7]Edgar W. Owen, in his monumental *Trek of the Oil Finders: A History of Exploration for Petroleum*, published in Tulsa in 1975 by The American Association of Petroleum Geologists, states on page 1330: "A well was spudded at Abu Hadriya in 1938 and was completed as an oil producer at a depth of 10,115 feet in March 1940" (Lebicher et al., 1960).

"Well, what do we do now?" asked Cecil.

"Let's have a beer," replied George. The good news meant, among other things, that he could continue on with Cecil to Indonesia.

Back in Karachi after his visit to Thompson's camp, Cecil conferred with the IOC officials, and as a result decided that the seismic work in which Thompson's crew was engaged should be terminated at the end of the month. Cecil advised the Dallas headquarters of his decision and asked Erik Jonsson for instructions regarding the reassignment of Thompson and his crew. Then after two weeks of field trips with Cunningham, investigating anticlinal structures around Karachi, Cecil and George flew to Medan, in Sumatra, arriving there on April 2, 1940 (see figure 10).

During the next few days in Medan, Cecil was busy reassigning the members of Thompson's India crew, as directed from Dallas, and communicating with three GSI crews operating in Sumatra for CALTEX. Two of the crews were working in the flooded jungle area of the eastern coastal plain of the island under A. E. Storm and Sam Stoneham, while members of the third crew were in hospitals up in the mountains near the western coast, recovering from malaria contracted in the jungle.

Before leaving Medan, Cecil and George visited the two operating crews several times. Cecil remembers these extraordinary visits clearly. Each party consisted of the dozen or more GSI crew members accompanied by some 700 native workers, who cut trails through the jungle, built bridges over flooded areas, built structures for sleeping, eating, and office work, and moved the entire party, equipment, explosives, and supplies from one shooting area to the next. The helpers had to carry everything because of the flooded jungle terrain—hence the large number of native employees (see figure 11).

On one trip in to visit an established crew, Cecil decided he would lead the train of bearers who were carrying food and other supplies from the staging point on the nearest river. He stepped out at a good clip down the narrow trail cut through the jungle, the bearers streaming along behind him, with the oil

10   George M. Cunningham and Cecil H. Green (*right*) at the Equator
near Medan, Sumatra, during a visit to GSI crews in Sumatra in 1940.
Photo courtesy of GSI.

11   GSI operations in Sumatra in 1940, at the time of Cecil's visit. *Top:*
Crew camp in the flooded jungle of Central Sumatra. *Bottom:* The observer (seated), an instrument operator with the telephone, and a few
of the 700 crew members of one of the GSI field parties. Photos courtesy of GSI: C. H. Green.

company's geologist somewhere in the middle of the column and Cunningham bringing up the rear. As he emerged from the jungle and entered a patch of savanna covered with high grass, Cecil was surprised that the bearers suddenly accelerated their pace and walked faster than he felt like walking until they reached the other side of the savanna and were about to enter more jungle. Here they seemed suddenly to have trouble with their packs, shoelaces, or whatever, and during the brief halt, Cecil was able to catch up and resume his lead position. The carriers again fell in behind him in orderly fashion and kept their positions, with Cunningham and the geologist somewhere behind, until they reached the next stretch of savanna. Then, as before, the carriers quickened their pace, streamed past Cecil, and again seemingly developed a variety of troubles just as they reached the next area of jungle.

Sensing something peculiar about the behavior of the bearers, Cecil asked Oscar Van Beveren, GSI's client geologist, who had always stayed in the middle of the whole train, what was actually going on. Answered the knowledgeable geologist:

Well, I hate to tell you, Cecil, but there are tigers in this jungle, and the tiger finds it easiest to pick off the first man or the last one in the column.

Between visits to the two field crews and to the convalescing members of the third crew, Cecil was kept busy with GSI business and with the socializing that was so necessary to maintain good public relations wherever he traveled.

Finally, after spending the month of April in Medan with Cunningham, Cecil embarked alone on the last leg of his long trip—the voyage across the Pacific. Cunningham, advised to stop over in Saudi Arabia and Egypt before returning home, left Medan on a KLM plane on May 3, 1940. Cecil, with another GSIer, Toby Lovinggood, departed the next day on the Dutch passenger-freighter *Van Heutz* bound for Singapore and Hong Kong. A brief stop at Georgetown, Island of Penang, permitted a little sightseeing before the ship sailed for Singapore, where she dropped anchor off Clifford Pier on the afternoon of May 7.

During a two-day visit ashore, Cecil and Toby took a room at the Adelphi Hotel and visited some of the night spots. During the day they had ample time to visit museums and to enjoy the sights, including the Tiger Balm Gardens.

On May 9, 1940, as the *Van Heutz* lifted anchor and steered around a mine field, Cecil and Toby watched planes from a British aircraft carrier maneuver around the mother ship. The next day, as the *Van Heutz* entered the South China Sea bound for Hong Kong, the news came over the radio that Germany had invaded the Netherlands and Belgium: Cecil was now traveling on a belligerent ship. A further reminder of the war in Europe was a 10,000-ton French cruiser out of Saigon that his ship passed on the second day. The *Van Heutz* arrived off Hong Kong harbor after dark on the third day from Singapore. Searchlights from the land played on the ship as a British naval craft opened the boom and let her into the harbor, where she tied up to a buoy offshore with the myriad city lights in full view.

Cecil and Toby had a week ashore before it was time to board the *President Taft* bound for Honolulu, where Cecil had earlier invited Ida to meet him. En route, in Shanghai and Kobe, Cecil enjoyed the experiences of the first-time tourist—riding cable cars up to mountaintops, and rickshas around the streets, buying silk gowns for Ida and stamps for himself, and noting the vast variety of people, dress, customs, architecture, and the like. Pleasant as these experiences were, however, Cecil recorded his deeper feelings when he wrote in his diary, "Can hardly wait for Honolulu!"

At long last, on the afternoon of May 27, 1940, the *President Taft* sailed out of the inner harbor at Kobe, past the German battleship *Scharnhorst,* and into the Pacific, headed eastward. Four days later, to Cecil's keen disappointment, the purser advised him of a radio message from the *President Pierce* that "Mrs. C. H. Green did not travel on that ship to Honolulu." Cecil wrote in his diary, "I sure hope Ida will be waiting for me." Happily, Ida was on the dock when the *President Taft* finally arrived at Honolulu on the 4th of June. It was a joyful reunion after a separation of seven and a half months, the longest period of time that Cecil and Ida had spent apart from each other since their marriage more than fourteen years earlier. Never again, in the forty-six years that remained before Ida's death in 1986, would they be separated for so long.

Fortunately for both of them, their ship to San Francisco did not sail for ten days, so they had enough time to enjoy the many attractions of the Hawaiian Islands while staying at several different hotels. Indeed, on one occasion Cecil was so relaxed after an afternoon on Waikiki Beach that he forgot his wallet when they went to dine at the Green Lantern and had to ask the manager to trust them until the next day. Most important were the hours of conversation, during which Ida shared the feeling of dread with which she had watched Cecil leave, and the deep loneliness of the ensuing months. The Greens had moved from Bakersfield to Los Angeles only two weeks before Cecil's departure, so Ida had been on her own in making new friendships and becoming familiar with the sprawling metropolis. Although she had owned her own automobile for almost two years, Ida seldom drove long distances alone. But she did not yield to self-pity and melancholy. To occupy her time and satisfy her need for activity, she registered for a course in interior design at UCLA and became involved in social work in her immediate community. Through these and other activities she gradually made a few friends. Nevertheless, especially when she was informed that Cecil's trip would be considerably extended by the assignments in India and Indonesia, loneliness increased, and she began to wonder if she had only the life of a grass widow to look forward to. She had already felt forsaken when Cecil visited Ecuador; now it would be months instead of weeks before he returned from halfway around the world, and the war in Europe was spreading. In short, to put it in her own words, she was "getting fed up with being alone!" She had just about decided to call it "quits" with Cecil, when she got his April cable from Medan in Sumatra, inviting her to join him in Honolulu. Only in recent years, when he has had more time for reflection on their long life together, has Cecil fully realized how near he came to losing Ida, and what strength and courage she showed in her loneliest hours.

On June 14 they boarded the Matson ship *Lurline* for the five-day voyage to San Francisco. As their reunion voyage was coming to its happy end, Cecil arranged for them to be called as their ship approached the California coast, and he and Ida were on deck to watch as they passed under the Golden Gate Bridge at 5:45 a.m. on June 20, 1940. Three hours later they disembarked to be greeted on the dock by the elder Greens.

A PERIOD OF UNCERTAINTY FOR GSI (1940–1941)

Upon his return, Cecil learned from SOCAL's office in San Francisco that the two GSI crews in the Persian Gulf area and the one in New Guinea, were being withdrawn because of the turn of events in the war in Europe. When he visited Phil Gaby's party near Stockton, he was informed furthermore that this was the only GSI crew still working in California. At GSI headquarters in Dallas, Cecil learned that one of the crews in Sumatra that he had visited only two months earlier had also been laid off because of the war scare abroad. No more than eight GSI crews were now active worldwide.

Cecil was feeling rather blue about GSI's future prospects when he returned to Los Angeles at the end of June. His pessimism was hardly reduced when Ida voiced her resentment that he had not received a raise (and some GSI stock at a lower price) from McDermott for having made his long foreign trip. Cecil's grim mood deepened after he had spent the month of July trying with little success to sell new contracts and worrying meanwhile about how to find employment for the key members of crews that had been laid off.

One petroleum company, already under contract with a GSI competitor, expressed interest in a GSI contract, but only at a cheaper price; another told Cecil they "might be able to steer work in GSI's direction, but it would have to be at much less than the usual monthly price of $6,500." Cecil met similar discouragement from both geophysicists and oil company geologists elsewhere in the United States and Canada. Dismayed, he even called on five aircraft companies in Southern California to inquire about a job for himself in case GSI collapsed. But everywhere he heard the same answer—no job! Little wonder, then, that Cecil wrote on July 10, 1940:

Oil business certainly looks bad, and I feel a little worried about the future of seismograph. Even at best there appears to be a real "danger" of moving back to Dallas.

During the remainder of 1940, as the war raged in Europe, and it became increasingly evident that the United States would ultimately be drawn into the conflict, GSI's future became more and more uncertain. Cecil was faced with a variety of problems: shrinking business, malfunctioning instruments, poorly performing employees, former clients' reluctance to renew contracts, and the necessity of seeking new kinds of service contracts. Looking for work to do, Cecil even prepared a bid on behalf of GSI on a drilling job on Highway 101 south of San Luis Obispo—20,000 feet of 18-inch drain holes at a dollar a foot. But when it was sent around to a dozen contractors, the bid proved too high. It was the very next day (August 16, 1940) that Cecil noted in his diary:

W. Willkie gave acceptance speech for Republican candidacy . . . News over radio indicates that Hitler's 2,500 plane attack on Britain yesterday was a failure so that we may consider Hitler has received his *first* setback.

As 1940 was ending, GSI's future seemed more uncertain than ever. The Greens feared even more that they might have to move back to Dallas. With that fear in mind, Ida had urged Cecil to investigate the possibility of getting work at the Scripps Institution of Oceanography in La Jolla, or perhaps at the Naval Electronics Laboratory in San Diego—after all, he was trained as an electrical engineer. Cecil's "West Coast Fever" had never completely subsided, and, like Ida, he dreamed of the day when they could settle down somewhere along the Pacific Coast. The prospect was more than ever attractive to Cecil because several of his uncles and aunts (his mother's siblings) had already moved to San Diego. Furthermore he and Ida had purchased a two-year-old house for his parents on 4091 45th street in San Diego just four days after Christmas. Cecil remembers that as they drove back to their Los Angeles apartment, he and Ida listened to President Roosevelt's ominous preparedness speech on the car radio.

3
THE CORPORATE YEARS (1941–1975)

1941: THE YEAR OF DECISION
The difficulties that developed for GSI in the second half of 1940 continued unabated into 1941. As Cecil became increasingly involved with officials in Dallas in regard to GSI's future, he was simultaneously confronted with a growing number of operational problems. His usual tasks as supervisor not only continued but increased—visiting the different crews to help them solve their personnel problems and to remedy malfunctioning instruments, testing new equipment and procedures, and meeting increasing pressure from client companies for quick oral and written reports on latest exploration work.

Superimposed on these responsibilities were two problems engendered by the war overseas—the efforts to get new contracts as old ones were terminated and to find new assignments for key personnel returning from both domestic and foreign crews whose contracts had ended, and the need to adjust to new government regulations and restrictions imposed in anticipation of the United States' entry into the war. These problems engaged more and more of Cecil's time and energy as the number of crews decreased, experienced personnel left, and geophysical exploration worsened in general. Indeed, by mid-1941 Cecil began to wonder if the seismograph would continue to be an important exploration tool. The cause for his pessimism went deeper than the many problems arising in the field parties and the difficulty of securing new contracts. Cecil was becoming increasingly concerned over the management policy being followed by Everette L. DeGolyer and J. C. Karcher, the two top officials of GSI in the Dallas office.

Karcher and DeGolyer had organized Geophysical Research Corporation (GRC) in 1925 as a subsidiary of Amerada Petroleum Company. Karcher as vice president of GRC had the responsibility of exploring for Amerada only. In a few years, however, he had GRC parties exploring for other major oil companies—a practice that soon aroused concern among the client companies about the affiliation between GRC and Amerada. Because of this concern Karcher and McDermott resigned from Amerada and GRC in 1930 and started GSI as a service company completely dedicated to exploration service, hence with no production interests. Soon, however, Karcher was using idle GSI crews to explore for his own account. When this activity was successful in finding oil, he quietly organized Coronado Oil Company to manage production, making GSI a wholly owned subsidiary. This action again aroused concern, leading Karcher to decide to sell all of Coronado's assets, including GSI.

GSI's success in finding some new oil fields for Coronado stirred Cecil's concern about GSI's future. In short, he asked himself, how can the current GSI maintain its integrity and credibility as an independent geophysical exploration service company if it works on its own account while simultaneously working for Coronado's competitors.

When Erik Jonsson made a visit in late April, Cecil learned that he, too, was worried about the policy being followed by DeGolyer and Karcher, and how that policy would affect GSI's future.

Then on June 12, during a brief visit to Bakersfield, Mc-Dermott told Cecil that Karcher was selling all of Coronado's assets—which of course included GSI—to the Stanolind Oil and Gas Company of Tulsa, Oklahoma. McDermott added that there might be a chance for GSI employees to buy GSI from Stanolind.

Since Stanolind had its own exploration division, Cecil immediately foresaw that Stanolind, after buying all of Coronado's assets, would decide to sell its GSI subsidiary. If such a sale were made, what then would happen to GSI and to all its employees, himself included?

So he was not greatly surprised when, on June 14, Jonsson and H. Bates Peacock—another GSI supervisor—telephoned him in Los Angeles, asking if he would join them and Mc-Dermott as an equal partner in the purchase of GSI from Coronado before the latter closed its deal with Stanolind. Cecil agreed to join his three GSI associates, after gaining Ida's reluctant approval, but only if GSI sold its unsuccessful soil analysis laboratory.

The four potential partners met in Dallas, and during a two-day conference—August 29–30—agreed to buy the assets of

GSI and then sell the soil analysis laboratory as soon as possible. In agreeing to this plan, Cecil committed himself and Ida to provide $75,000, their fourth of the purchase price of $300,000. But by no means did the Greens have that amount readily available.

By a judicious program of saving, Cecil and Ida had bought life insurance and retirement annuities and had accumulated a small portfolio of a dozen high-quality stocks. In addition, they had a substantial checking account, and Cecil had an annual salary of $9,600. Nevertheless, even when all their assets were combined, the total fell far short of their share of the purchase price.

Cecil solved their financial problem temporarily by securing from the Republic National Bank of Dallas a loan of $35,000 (at 4.5 percent interest), by putting up $15,000 in cash, and by committing his and Ida's stocks and life insurance as collateral. In fact, it was the willingness of the bank's president, "dear old Mr. Fred Florence," as Cecil called him, to accept the collateral that made the loan possible.

Ida, however, had misgivings. Her agreement, which made possible Cecil's joining his three longtime friends in the purchase of GSI, did not come easily. Like Cecil, she had developed a great liking for California and was hoping that they could settle down permanently somewhere in the San Diego area. Buying their share of GSI would surely mean that they would have to move back to Dallas sooner or later. Nevertheless, as she had done so many times before, she put aside her fondest dreams and consented to the gamble, by far the greatest risk they had ever taken.

But Ida did not put aside her misgivings, for the world news was ominous; it seemed certain that the United States would soon be drawn into the European conflict.[1]

[1] England's HMS *Hood* was sunk off Iceland by the German battleship *Bismarck* in late May. Germany invaded the Soviet Union on June 22, and by July 4, war with Germany seemed imminent. One of the features of the national holiday was a debate between Senator Pepper (Florida) and Senator Wheeler (Montana) as to what the United States should do about the war in Europe.

As Cecil and Ida continued to discuss his agreement in the light of world events, he himself began doubting that he had acted wisely in joining his potential partners. Finally, he decided to take action, as he recorded in his diary for the weekend of October 10–11:

I wrote a letter to Mac[McDermott] and Jonsson in which I expressed [my] intention of resigning from GSI rather than joining the pending venture to buy GSI.

The letter brought an immediate telephone call from Jonsson informing Cecil that his resignation would not be accepted until he came to Dallas two weeks later for yet another conference. The Greens arrived back in Dallas on October 23, and the next day Cecil reported to the GSI headquarters, lunched with McDermott, and informed him that he had decided to stay in the proposed deal after all, final details to be worked out in early November.

So it was that the *Dallas Times-Herald* soon afterward informed the public of purchase of GSI by Green, Jonsson, McDermott, and Peacock, and of Karcher's sale of Coronado's capital stock, as reported in the news items reproduced in figure 12.

The purchase price of the GSI capital stock (not mentioned in the news item) was still $300,000, with the four partners sharing equally in the ownership. In his diary entry for the day the purchase was announced, November 10, 1941, Cecil noted that he and his associates "went to bank [Republic National Bank of Dallas] to sign up for $300,000 loan to purchase GSI from Coronado . . ." This commitment required that Cecil and Ida raise their $75,000 share on rather short notice. Fortunately, the arrangements they had made in August were still in place. Cecil also noted in the same day's entry, "Made Vice President today." His position in GSI had changed from employee (as a regional supervisor) to co-owner and corporate executive (vice president). This was his first step up GSI's corporate ladder, and it led ultimately to the position of board chairman.

Then came the climactic day—Friday, December 5, 1941—for consummation of the purchase of GSI's assets by the four part-

FOUR TEXANS BUY GEOPHYSICAL FIRM CAPITAL STOCK

Eugene McDermott, president of Geophysical Service, Inc., Tuesday announced the purchase of all the company's capital stock by himself and three associates, H. B. Peacock of Houston, Cecil H. Green of Dallas and J. E. Jonsson of Dallas.

Geophysical Service, Inc., has been engaged in geophysical exploration work in the United States and in countries all over the world for many years. Its home office and research laboratories are located in Dallas, and branch offices are located in Houston and Tulsa.

Management of the company will be continued under the leadership of Eugene McDermott as president, with Mr. Peacock and Mr. Green serving as vice presidents. J. E. Jonsson is secretary and treasurer.

KARCHER SELLS OIL COMPANY

CAPITAL STOCK OF CORONODO CORPORATION BOUGHT BY STANOLIND

Dr. J. C. Karcher, president of Coronado Corporation, and his associates, Wednesday announced the sale of all the capital stock of Coronado Corporation to the Stanolind Oil and Gas Company of Tulsa, Okla., at a price in excess of $5,000,-000.

The Coronado Corporation was founded in 1930 and owns properties and operates in Texas, Louisiana and Alabama, its principal properties being in Southwest Texas.

The Coronado Corporation produces approximately 2,000 barrels of oil per day from the La Rosa and Welder Fields, and also operates a 60,000,000 cubic foot recycling plant in the La Rosa Field. It owns producing royalties in Luby, Chapman and Placedo Fields, all located in Southwest Texas.

The transaction was consummated in the banking rooms of the Republic National Bank of Dallas.

12   Announcements of the purchase of Geophysical Service Inc. (*Top left*) and sale of Coronado (*top right*) by the *Dallas Times-Herald*, in November 1941. *Bottom:* The four co-owners (*from left to right*), H. Bates Peacock, Eugene McDermott, Cecil H. Green, and J. Erik Jonsson, formed a partnership to purchase GSI on December 5, 1941. Top photos courtesy of the *Dallas Times-Herald;* bottom photo courtesy of GSI and Texas Instruments.

ners, which involved the payment by Cecil of $75,000 for his 1,250 shares. Fortunately for this biography, Cecil recorded the events of that memorable day in his diary as follows:

Float personal bank loan for GSI. Day in Dallas following break-fast down town with Bates Peacock. Spent session with three partners as well as Tom Knight (contract attorney); then went to Republic Bank. I received check from GSI in amount of $37,500 which I endorsed to Republic Bk. for deposit, then signed note for $22,500, then wrote check for $15,000 to bank, finally writing check payable to board for $75,000 for my 1,250 GSI [shares] which they hold as security. I wrote check of $156.24 as share of interest on previous $300,000 company note.

In the evening of that same day the four partners and their wives gathered at the Dallas Petroleum Club to celebrate the day's events. Now they were the owners of Geophysical Service Incorporated and would be responsible for its future performance.

It was probably at this particular party—though no one now is certain—that Ida made her oft-quoted remark, "I'd like to be a philanthropist!" when Erik asked her what she would like most to do if they all got rich.

Two days later, while relaxing in their new apartment at 4223 University Boulevard near the SMU campus, Cecil decided to telephone GSI party chief K. E. Burg, then in Alberta, to discuss having him bring his crew home before Christmas. He was astounded when Burg told him he had just heard by radio that Japan had bombed Pearl Harbor, and had declared war on the United States. Like so many who heard the fateful news that Sunday afternoon—December 7, 1941—Cecil has never forgotten how he learned of the attack. The next day after Pearl Harbor, the United States declared war on Japan, and two days later, on December 10, Germany and Italy declared war on the United States. Although there is no hint of panic in Cecil's daily entries in his diary, surely he and his three partners must have had acute misgivings about what lay ahead for their newly purchased company. The chilling fact was that before the new owners of GSI could transact their first day of business, the United

States was at war across two oceans in widely separated theaters.

Without hesitation the four partners immediately agreed to several courses of action. Cecil, in charge of field operations, would make every effort to increase the number of field parties and expand their activities in geophysical exploration. Peacock would continue fieldwork as a regional supervisor. At the same time, McDermott and Jonsson would seek government contracts to make military equipment in the Dallas laboratories for the Armed Forces, and would assist Cecil with his personnel problems when possible. As Jonsson has expressed it in conversation, "We rolled up our collective sleeves and went to work with a will."

CECIL AS VICE PRESIDENT OF OPERATIONS: THE WAR YEARS (1941–1945)

Hard times lay ahead for the new GSI and Cecil took on the responsibilities of co-owner and vice president of operations. Even before the United States' involvement in World War II, oil companies had temporarily cut back their exploration programs, and the number of GSI crews that GSI had in the field had fallen from twenty-eight to six by December 1941. This drastic reduction in operations resulted in substantial losses of income in late 1941, and the question arose: Can the new owners reverse the downward trend?

No sooner were the Greens settled in their University Boulevard apartment than Cecil was involved with Jonsson in selling the soil analysis laboratory—one of the provisions for entering the GSI deal. Although the use of soil chemistry to locate oil deposits had seemed at first promising, little business had resulted from the work of the laboratory. Its sale (for $11,000) brought in some much-needed cash.

When the decision was made to expand seismic exploration by seeking new contracts and organizing additional field parties, Cecil found himself confronted by two closely related problems: how to revive former clients' interest in new contracts, and how to keep key personnel and recruit new crew members to replace those being inducted into the Armed Forces. Indeed, GSI found itself in competition with Selective Service, especially for extremely valuable crew members. When war was declared, Selective Service had gone into action, declaring that exploration for oil was not essential to the war effort. So companies like GSI were advised to get their technical people into some other more important work or Selective Service would draft them. This threat made more difficult Cecil's efforts to keep GSI's field parties intact and operational, and drove McDermott and Jonsson to Washington to seek war work for GSI's technical people in the Dallas laboratories.

Had this original decision of Selective Service regarding exploration for new oil been enforced, the result would have been disastrous for GSI. Crew members would have been inducted without regard for their experience, proficiency, or importance to a party's field operations, and it would have been impossible to keep existing crews operational, let alone recruit for additional parties.

Fortunately, a wiser view prevailed. Secretary of the Interior Harold Ickes created the Petroleum Administration for War (PAW), with Everette L. DeGolyer, then rated the world's outstanding petroleum geologist, as director, assisted by Ralph Davies, a vice president of Standard Oil Company of California. The PAW finally convinced Selective Service that the war could last longer than expected, and that it would be ill-advised to damage existing oil fields by opening valves wider; wider valves would reduce pressure below the level for maximum recovery. Following PAW's recommendations, the government rescinded the earlier edict and declared that exploration for new oil could be considered essential to the United States' military effort.

This action was favorable for GSI, so far as exploration for oil was concerned, but Cecil still had to plead as hard as ever with local draft boards for exemption of key personnel, and he lost as many cases as he won. To quote Cecil:

"We had to fight to hold each key field man, and in my case, I recall working with over 90 draft boards plus some state headquarters."

Even GSI's president, McDermott himself, had to go to Washington frequently to try to get key GSI crew members reclassified or deferred.

In one case, in which it was extremely important that a certain key employee be exempted lest the party of which he was a member have to end operations altogether, Cecil pressed the appeal through required Appeals Board channels until it landed on the desk of General E. L. Compere in Little Rock, Arkansas. After pleading his case with the officer, Cecil finished by saying,

Sir, you may well think I am unpatriotic, but this man is so critical to the continuation of our operations that I feel he can contribute more to the war effort as our employee than as a member of the Armed Forces. That is why I feel duty bound to press his case to the limit.

General Compere replied,

Mr. Green, we are going to fight to draft this man; but if, down in your heart, you feel this man is critically important to your exploration operation, then it is your duty to fight harder than we do to keep him.

As the war continued, and United States Armed Forces were deployed around the world, GSI continued to lose one experienced employee after another until Cecil was reduced to hiring almost any man he could find "so long as he had two arms, two legs, and a head, even if there wasn't much intelligence in the head or flexibility in the limbs." There were constant problems with drunkenness, laziness, avoidance of tough jobs, absenteeism, and many of the other troubles that arise when such people are pressed into service. After four years of trying to keep the numerous crews fully enough staffed to do a day's work, Cecil concluded,

. . . there can never be such a situation as *zero unemployment*. There are just too many people who are totally unemployable, and I ought to know, because I hired some of them during the war years. I could describe enough cases of such men to make a whole book in itself.

Fortunately for Cecil, Dudley Coursey was brought in to Dallas from the field in 1943 and made personnel manager to keep field and shop crews staffed. Even with Coursey's invaluable help, however, Cecil still had to be constantly concerned about personnel.

Another major problem which confronted Cecil during the war years, as he tried to keep existing parties operational and to organize new crews, was the shortage of essential materials and equipment because of wartime rationing—photographic recording paper, supplies, chemicals, dynamite, tires, and even hydraulic jacks for trucks. Crews had to beg and borrow parts and supplies to keep moving, even resorting to salvaging components from discarded equipment. Nevertheless, in spite of all these problems, they did keep going somehow, largely due to the ingenuity and experience of the few old-timers on each crew.

Immediately after Pearl Harbor, GSI had only six parties and one hundred employees. But by the end of 1945, as the war ended, Cecil and his associates had increased the number of field parties from six to fifteen, and GSI was ready for expansion worldwide. This accomplishment was only possible because Cecil was away from Dallas and from Ida so much of the time, traveling by whatever means was available—automobile, train, ship, or airplane. Ida could not often accompany him, so she was particularly annoyed when he telephoned that he would have to supervise a crew for a week or two while a party chief took a much-needed vacation, or stay with a party for a few unanticipated days to help train new recruits.

Cecil definitely matured as a result of his wartime activities, and demonstrated management and administrative skills that were not overlooked by McDermott and Jonsson in Dallas. In his different activities, Cecil's outstanding ability to work with people at every level manifested itself repeatedly. Little wonder that he so often remarks that "getting ahead depends not on *working* people but on working *with* them!" This attitude goes back to his years as a fledgling electrical engineer when he discovered that he would much rather work with people than with inanimate equipment in a research laboratory. It explains why he found seismic exploration such a happy combination of technology and people, and why his experience and confidence in relationships with colleagues and workers expanded under the pressure of wartime operations.

In his new role as corporate executive, Cecil was exposed for the first time to the world of high finance. As a party chief and later a regional supervisor, his financial experience had been limited to making out weekly or monthly expense accounts and those of the personnel he supervised, calculating the immediate expenses of field operations and equipment, and setting up contract budgets. Ownership in GSI and the varied duties of a company officer greatly extended and enhanced his knowledge of finance. It was the beginning of an understanding that would be central to his later philanthropic activities.

As vice president of operations, Cecil had to increase his social activities in order to advance the interests of GSI. Special lunches, dinners, and other entertainment for prospective employees and clients required increased work and involvement on Ida's part, too. When Cecil and Ida bought their first permanent home at 3621 Caruth Boulevard, a new style of life began for them. Here their odyssey of more than fifteen years and many thousands of miles came to an end, and at long last Ida had a house of her own to decorate and furnish, while Cecil could indulge in occasional horticulture.

The prospect for a more relaxed life-style in their first home soon faded, however, for Cecil's new responsibilities called for more travel than ever. Although she accompanied him on many of his shorter visits to the numerous crews scattered over the western United States and across the border in Alberta, Ida was often left alone. Not one to be idle, she joined a number of women's organizations, participated in money-raising campaigns to support social and war-related causes, enrolled in classes at Southern Methodist University, and read omnivorously as had been her habit since girlhood. Furthermore there were always luncheons, cocktail parties, and dinners for Cecil's GSI business associates and potential clients, both outside and inside the home, and the many more informal gatherings with close friends. The Greens have been social individuals from the first days of their marriage, and their evident pleasure in socializing continued throughout their years together. Moreover, now that Cecil was a co-owner and vice president of GSI—rather than merely a salaried employee—he and Ida had an entirely different attitude toward the company and an increased incentive to help it expand and flourish.

GSI'S WORK FOR THE ARMED FORCES DURING THE WAR YEARS (1941–1945)

In 1942, while Cecil and his associates concentrated their efforts on expanding GSI's usual geophysical service, Jonsson and McDermott went to Washington to seek contracts with the military establishment. The four partners reasoned that since the GSI instrument laboratory in Dallas had been making the seismographs and other equipment for their geophysical field parties for eleven years, the personnel of the laboratory should have the expertise to make military equipment.

An early result of Jonsson's search produced some small engineering tasks for the U.S. Signal Corps. Soon thereafter, McDermott learned that the Navy's Bureau of Aeronautics (Bu Aer) had taken an interest in a reconnaissance geophysical exploration device recently developed by Victor Vacquier at Gulf Research and Development Corporation. The idea was that a slow, low-flying airplane tow a Magnetic Aerial Detector (MAD) at the end of a 200-foot electric cable in order to detect and outline sedimentary basins in new foreign concessions. The Bureau of Aeronautics had decided that the MAD's ability to detect minute changes in the strength of the earth's magnetic field could be very useful in locating masses of magnetic material such as submerged enemy submarines, which were then operating off the Atlantic coast and posing an ever-present menace to ships in the United States convoys bound for Europe.

GSI was asked to make a number of the MADs, and Jonsson was assigned to negotiate the contract and direct manufacturing. Other contracts for military equipment, also arranged by Jonsson, followed, and by the end of the war income from these contracts far exceeded that from geophysical exploration.

In negotiating the MAD contract, Jonsson met First Lieutenant Patrick E. Haggerty, then in charge of electronics procurement for the Bureau of Aeronautics. The two became good friends. Jonsson recognized the potential in the young lieutenant, and at the war's end he induced Haggerty to join GSI in charge of the electronics division.

TEXAS INSTRUMENTS IS BORN OF GSI (1951)

In reviewing the activities of GSI during World War II, it becomes clear that what seemed to be a disaster for the company on Pearl Harbor Day actually turned out to help its recovery and long-term growth. The four partners who had made their big gamble in buying GSI at its lowest ebb in 1941 had, as a hardworking cooperative team, brought their company safely through the vicissitudes of the war. Cecil and Peacock had regained a leading position for GSI among its principal competitors in geophysical exploration; Jonsson, McDermott, and their dedicated technical staff in Dallas had developed a manufacturer of electronic components and devices in the company's instrument laboratories and shops. And it can be said that in its impressive performance every aspect of GSI was tested to the limit—personnel, field operations, research and development, manufacturing, and finance.

Thus a greatly strengthened GSI emerged from the war ready and eager for bigger challenges. The dominating question became: Where do we go from here? It was the question that preoccupied the "Executive Suite" in the immediate postwar years as many of the government contracts were terminated. The four GSI owners, together with Haggerty (who had joined GSI in November after V-J Day), recognized that the corporate structure of GSI needed to be changed. There was unanimous agreement that in the restructuring, GSI should not only continue its worldwide geophysical service, but it should also enlarge and diversify that service. The geophysical service should, of course, go forward under the leadership of Cecil as vice president of operations.

But a much more important question than expanding the geophysical service remained for the GSI partners to resolve: What was to be done about GSI's greatly expanded instrument laboratory and the expertise that its employees had developed during their war work?

Fortunately for industrial historians, GSI's then secretary and treasurer, J. Erik Jonsson, who played the dominant role in directing the company's manufacture of military equipment, recorded the details of how this question was answered. In 1956, before the New York Society of Security Analysts, Inc., he discussed how GSI's initial instrument laboratory ultimately became Texas Instruments Incorporated (TI), long since a familiar name on Wall Street. Jonsson's story of Texas Instruments Inc. is told in a twenty-nine-page pamphlet produced and distributed by TI in 1956; it is from this highly informative document that the excerpts below have been taken.[2]

Perhaps the best way to tell the TI story is first to delve briefly into its history, reviewing those events which may put in proper perspective the company's present aspirations, motivations, composition, and posture. Founded in March of 1930 as Geophysical Service (GSI), the company pioneered the use of the reflection seismograph in exploration for sub-surface structures favorable to the accumulation of oil and gas. Its clients were the major oil companies. Successful from the first, GSI grew rapidly and soon moved into exploration for its own account. Here, too, fortune smiled upon it as the company quickly made a number of substantial discoveries.

By 1939, it seemed wise to separate GSI's oil business from exploration activities. With a change of name to Coronado Corporation, the oil division took over the old corporate structure, and a new GSI was formed as its wholly-owned subsidiary. In late 1941, Coronado's owners decided to sell out to a major pro-

---

[2]Jonsson, Erik, "The Story of Texas Instruments Inc." Presented by J. E. Jonsson, president, before the New York Society of Security Analysts, Inc., Thursday, December 13, 1956 (and printed in a small pamphlet). A second well-told history of TI, complete to 1986, is included in Dolores Proubasta's informative profile of Jonsson that appeared in the June 1986 issue of *Geophysics: The Leading Edge* (Vol. 5, No. 6, pp. 14–23) published by the Society of Exploration Geophysicists.

ducer. GSI, the wholly-owned subsidiary, was not included in this transaction, but was sold on Saturday, December 6, 1941, to a group of employees who had been with it since its beginning. Thus, before a full day's business could be transacted by GSI's new owners, the United States was at war. This might have been a mortal blow to many another endeavor, but petroleum was so vital to the country's defense that there was little question that exploration efforts must be maintained.

However, because of wartime uncertainties of the preceding two years, oil companies had temporarily cut back their programs, and GSI field parties tapered from 26 to 6. Profits disappeared, and during late 1941, GSI incurred substantial losses. The question was: Could the new group reverse this trend? Financially the new venture was weak. Company borrowing power was small; reserve capital of the new owners was not substantial. To a casual observer it might well have seemed that a foundation had been built, not for a growth company, but for an industrial fatality. . . .

To those of us who knew the facts, the outlook seemed rough, but we had good people, the job was worth doing, and we were in Texas, where the expression, "You can't get there from here," has still not been heard. So, in spite of the clouded outlook, expansion of geophysical operations was begun, and during 1942 the company undertook some small engineering and development tasks for the Signal Corps as well. These were soon supplemented by various projects, including 'submarine hunting' aerial magnetometers for the Navy's Bureau of Aeronautics.

By VJ Day, GSI had built slightly more than a million dollars' worth of military equipment and had restored its exploration crew level to about 16. Cancellation of war contracts was not too serious a problem since the need to redesign and refurbish all geophysical instruments and equipment would pick up the slack of manufacturing.

So much was learned from exposure to wartime production and engineering in fields closely allied to our own that it was decided to return to military manufacturing as a permanent part of our industrial career. Many of the techniques of submarine hunting, of artillery sound ranging, of radar applications have counterparts in exploration. Thus, a manufacturing business with a high engineering content might logically be added to our business with a resultant cross-fertilization of scientific knowledge. In addition, a common pool of manufacturing and test equipment, inventories, and manpower could be used for much of the potential production.

To aid this venture, GSI sought the services of a group of Navy men after their discharge from military service. [These included the First Lieutenant Patrick E. Haggerty, procurement officer on the MAD equipment project of Navy's Bureau of Aeronautics; Robert W. Olson, civilian desk officer of the Engineering Department; and Walter Joyce, another civilian employee of Bu Aer who also worked on the MAD project.] In the spring of 1946, these men established the Laboratory and Manufacturing Division of GSI, signalling a period of rapid and substantial growth. For five years [1946–1951] the story was work, build, expand, hire, train, borrow, earn, get ready to grow again. . . .

By 1951, history repeated itself. Manufacturing and exploration activities had increased so much that it was necessary to set them up in separate corporate structures. The manufacturing unit became General Instruments Inc. and, again, a new GSI was organized as a wholly-owned subsidiary. Because of a conflict in name with another company, the name General Instruments was changed to Texas Instruments Incorporated in late 1951 . . . [3]

TI's management and production team had been augmented and developed until now it began to take on today's look. Most of the group were engineers, thirty to forty years of age. They were hard-working, ambitious, eager to go, and ready for expansion on a major scale.

Once the name Texas Instruments Inc. was selected for the manufacturing unit of GSI in 1951, TI became the parent company

[3] When, in 1951, it was learned that the name General Instruments Inc. had been preempted, a new name had to be found for the Manufacturing Division that had been established in 1946. So Jonsson appointed a small committee to select a new name for the rapidly developing and expanding division, instructing Haggerty, the committee chairman, to be sure to have Texas included in the name. Cecil recalls that after they had rejected a number of names, and were getting a bit weary, Haggerty finally suggested, "Since we are a Texas company, why don't we just call ourselves Texas Instruments?" Thus was coined the now-familiar name, Texas Instruments Incorporated—TI for short.

pany in the corporate structure, and a new GSI was organized as a wholly owned subsidiary.

Jonsson, as president, and Haggerty, as executive vice president, immediately took charge of the newly created TI. By a series of ingenious and insightful actions over a two-year period, Jonsson brought about the merger of Texas Instruments and the International Rubber Company. Using the merger and some very confident statements on his part as arguments, he succeeded in getting Texas Instruments Incorporated listed on the New York Stock Exchange. Thus, on October 1, 1953, when the NYSE opened for its usual day's business, Texas Instruments Inc. appeared for the first time, as *TXN*, and J. Erik Jonsson, standing on the floor of the Exchange, bought the first 100 shares of the new offering at $5¼ a common share! Those who knew about Jonsson's efforts to raise the finances and make the other arrangements to get TI listed on the NYSE could only marvel at his genius in personal persuasion and entrepreneurship.[4]

But there was another equally outstanding player in TI's dramatic emergence as a "good little company" promising enough to get listed on the NYSE. He was the young naval officer, Pat-

[4]When J. Erik Jonsson retired on December 31, 1966, as chairman of the board of directors and as an employee of Texas Instruments Incorporated, he was designated Honorary Chairman. TI's 1966 Annual Report included the following tribute to Jonsson on page 20:

Mr. Jonsson's association with Texas Instruments began with its predecessor company, Geophysical Service Inc., shortly after the founding of that company in 1930.

In the succeeding 36 years, he has been a principal influence in developing the particular character of Texas Instruments. Employees of the company, its shareowners, and its customers are deeply indebted to him for his leadership in establishing the basis for the growth and development of Texas Instruments from a single line business to a highly diversified, technologically based corporation; for his sound judgment, managerial skills, and financial direction; for the influence of his own high principles and integrity upon the policies and principles to which the company is committed and by which it operates; and for the recruitment, inspiration, and encouragement of the accomplishments of others.

rick E. Haggerty, who had joined GSI in 1945 at Jonsson's urging, and whom Jonsson had selected as his executive vice president in 1951, when the new name was adopted. Jonsson was so impressed by Haggerty's extraordinary management style that in 1958 he decided to move on to board chairman and give Haggerty free rein as president of TI.[5]

The first trade of TI common stock at $5¼, which gave Jonsson great satisfaction, attracted little attention during the following few months—it was bought at $5½ a share on January 7, 1954. However, as TI began to produce high-quality germanium transistors that made it possible to manufacture the first small transistor radios at a reasonable price, and particularly after the company announced its newly developed silicon transistor in May 1954, the price of the common stock began to soar, reaching $200 by 1960, and peaking in that year at $256 a share—a fantastic forty-nine-fold increase in only seven years! And this was only the first surge in TI's financial progress; more growth came later. As the common stock has split several times, and has increased still more in value after each split, the wealth of the original shareholders has grown steadily.

Thus the answer to the question—Whence came the Greens' wealth?—is quite evident. It came from their original quarter ownership of GSI which gave birth to Texas Instruments Inc., now one of the giants of the electronics age.[6]

[5]The extraordinary success of Haggerty's management style is described by David Allison—with intercalated quoted remarks by Haggerty himself—in the commentary titled "Can a System Perform Like a Hero?" that appeared in 1969 in *Innovation*, 8, pp. 16–31.

[6]The phenomenal growth of Texas Instruments Inc. has been recorded extensively, not only in its inhouse publications but also in many newspapers, journals, magazines, and the like. A brief comment on TI, headlined "Newcomer's Growth" appeared in *Time* magazine on pages 79–83 of the April 8, 1957 issue. *Business Week*, on page 49 of its May 16, 1970 issue, discussed TI's manufacture of a special sandwich wire of aluminum coated with copper under the title "A Wire That Slices the Cost of Copper." And the *Wall Street Journal* for Monday, January 30, 1984, cites TI's recent troubles and recovery in an article headlined "Profit of Texas Instruments Increases 82%." These are only a few examples of the publicity that has attended TI's phenomenal growth.

CECIL BECOMES PRESIDENT, THEN BOARD CHAIRMAN, OF GSI (1951)

As explained earlier, it was decided in 1951 to separate the manufacturing and exploration activities of GSI into separate corporate structures—General Instruments Inc. (changed to Texas Instruments Inc. in 1951) and Geophysical Service Inc., respectively. At the same time there were important changes in both official assignments and ownership.

McDermott became chairman of the board of both TI and GSI; Jonsson was appointed president, and Haggerty executive vice president, of TI; Green moved up to president of GSI, at the same time being appointed a vice president of TI as well as a member of its Board of Directors; and Fred Agnich succeeded him as vice president of GSI.

Then, when TI stock went public in 1953, three of the four original GSI partners—Green, McDermott, and Jonsson— bought the holdings of their fourth partner, Peacock. These holdings of Peacock they sold to Haggerty, Agnich, and other members of the old GSI family, at the same time retaining their respective holdings in the original offering.

Not only did Cecil's advancement to president of GSI bring added responsibilities for the future of the new subsidiary, but as a vice president, board member, and major stockholder of TI he had also to devote time and energy to the development of the parent company.

All of this followed the familiar pattern of what had happened so many earlier times when Cecil changed jobs—more hours and new work experience in the office; more miles of travel to places both near and far; and more social affairs, many of which involved Ida directly. Indeed, such social activities—mostly luncheons and dinners—came to require more and more of her time and energy as the Greens' circle of business friends widened. Reluctantly, as she had done before under similar circumstances, Ida gave up some of her own special interests, such as course work at Southern Methodist University, in order to devote time to GSI and TI business. Prosperity, which came in full measure, exacted a further price. The hours alone with Cecil grew fewer and fewer, and therefore became more and more precious. For example, the Greens' carefree walks diminished in both number and distance—a short walk around a block had to suffice. All too often Ida had to walk alone.

It is probably true that Cecil and Ida never worked harder than during the five years he was president of GSI. Cecil's primary responsibility was to see that GSI grew and contributed its proportional part to the revenue from TI's annual overall program. And the record shows that GSI did, indeed, grow under Cecil's able leadership. New crews were organized, trained, and sent abroad in increasing numbers; new and more accurate field equipment was designed and put into operation; and much more sophisticated methods of data analysis produced commensurately better interpretations for clients. Exploration expanded from the dry land into the shallow waters offshore when special research vessels were built to work at sea. For example, the 159-foot shallow draft vessel, the *Sonic*, was an important addition to GSI's Marine exploration activities in 1953. Likewise, GSI crews took their specially built equipment into the snow-covered Arctic, the sand-covered deserts of the Near East, and the tropical rain forests of the East Indies, where they had been exploring for the past twenty years.

The result was that when Cecil advanced from president to chairman of the board in 1955, on GSI's twenty-fifth Anniversary, he turned over to Fred J. Agnich, who succeeded him as president, the responsibility for more than a thousand employees. These individuals either supported or were crew members of more than sixty field parties that were using reflection, refraction, or gravity methods of geophysical exploration throughout the world.[7]

In no way, however, did moving up to board chairman dimin-

[7]The interested reader can find a much more complete discussion of GSI's first twenty-five years in the company's little magazine, *The Grapevine* (11/8, Sept. 1955). The *Dallas Morning News* of Sunday, May 17, 1970 also carried a brief history of GSI under its business and finance section (11B).

ish Cecil's devotion to the affairs of GSI, or to those of TI. For the next four years, 1955–1959, until he retired to become honorary chairman of GSI's Board of Directors, he kept as busy as ever, even though no longer bound to the daily office schedule that he had followed previously. Now he and Ida had time to engage in other activities—including for Cecil an extension of his interest in the training of young earth scientists and the development of a much broader involvement in the overall education process.

For Ida, it was now possible to engage in that philanthropy in which she had expressed keen interest when Jonsson had asked her, so many years before, what she would like most to do if the part-ownership of GSI made her and Cecil rich. Indeed, for the past 35 years, both Cecil and Ida have been spending most of their time, energy, and means in a program of innovative philanthropy that has benefited education and medical research throughout the English-speaking world and has won them the gratitude of a multitude of admirers.

4

CECIL AS EDUCATOR, COMMUNICATOR, ORGANIZER, AND CATALYZER; IDA AS DEVOTED WIFE AND STRONGLY SUPPORTIVE PARTNER

INTRODUCTION

Cecil's potential as a teacher was recognized early when he was selected to instruct in GE's School of Advanced Engineering in Schenectady as a part of his first job after completing college. A few years later, at Raytheon and then at Wireless Specialty Apparatus Company, he discovered that he preferred directing people in the production line to working alone in the research laboratory. His next job, as production manager with Charles V. Litton at Federal Telegraph in Palo Alto, imposed on him the awesome task of trying to keep up with a near-genius technician and confirmed workaholic, but it also gave him the exceedingly valuable experience of learning new techniques by watching Litton in action—information that he could then pass on to others. That experience with Litton also left Cecil with the indelible impression that there is a definite physical limit for everyone, even the most dedicated worker.

That Cecil could learn quickly, teach and direct effectively, and manage successfully was soon demonstrated after he joined GSI of Dallas in 1930 as a party chief. He had the basic training of an electrical engineer and some laboratory and manufacturing experience in designing and overseeing the production of condensers and neon signs, but he knew nothing about geophysical exploration. As party chief he had to manage a complex outdoor operation involving a high degree of cooperation among eight or ten individuals, each doing his special job with his particular piece of equipment. It also meant that Cecil had to interpret the data obtained by his crew and finally report conclusions to GSI's client, an oil company. All these aspects of geophysical exploration that Cecil had to become familiar with in a hurry he learned from Roland F. Beers, who had induced him to leave Federal Telegraph and join GSI. Beers was an excellent teacher, Cecil an apt student, and after a few days of observing field operations, followed by instruction in Beers's office, Cecil was on his own as a party chief.

During Cecil's five years as chief of GSI's Party 310, he successfully mastered the art of geophysical exploration and developed the strong leadership qualities that led to his appointment as regional supervisor in 1936. As both party chief and supervisor, Cecil himself continued to learn from experience and also taught others by his own successful innovations.

Highly sensitive to the personal attitudes and opinions of both his GSI associates and of the geologists who worked for GSI's industrial clients, Cecil soon recognized an important lack of understanding between geologists and geophysicists. Many geologists regarded exploration geophysicists like Cecil with suspicion, because they based their interpretations on data obtained from mysterious instruments ("black boxes"), the principles and operations of which they did not understand. Many geophysicists, for their part, disdained most geologists for their ignorance of applied mathematics and physics. Unlike the geophysicists, the geologists depended largely on what they could actually see on the surface, such as oil seeps and slightly inclined strata, in order to locate possible oil reservoirs.

Though he was trained in engineering, Cecil's instincts and curiosity as a student led him to seek as much knowledge as possible about geology and applied seismology, first from Roland Beers and then from friendly clients' geologists like Ira Cram of Pure Oil, Raymond Stehr of Seaboard, and George Cunningham of Standard of California. Cecil decided he could do something about the gap in understanding between geologists and geophysicists.

Soon after he moved his base of operations as supervisor to California, in 1939, Cecil interested a number of petroleum geophysicists and geologists in the Bakersfield–Los Angeles area in forming a study group to exchange ideas and information of mutual interest. The first meeting, which Cecil opened and chaired, took place in Bakersfield on Wednesday evening, September 20, 1939, with forty-five men attending. A week later, W. S. Olson of the Texas Co. replaced Cecil as chairman of this Bakersfield group, since Cecil was moving to Los Angeles and would soon be leaving on a trip to the Middle East with Cunningham.

On his return from this trip, Cecil organized a similar Los

Angeles geophysical study group, the first meeting of which was held in the Subway Terminal Building on November 12, 1940. During the next five months there were nine meetings of this group, at the last of which Walter English replaced Curtis Johnson as chairman for the next season. Though Cecil's active participation in the California study groups ended because of his move to Dallas later in the year, his initiative in starting the geology-geophysics study groups in the two California localities is an example of his early commitment to educational efforts that proved beneficial to geophysical exploration specifically and to the petroleum industry generally.

CECIL AND THE *GSI GRAPEVINE*

One of Cecil's most successful efforts in promoting communication began in November 1944 with the birth of the *GSI Grapevine*, confidently announced as a *News-Gossip Sheet*, Vol. 1, No. 1.

The story has it that numerous GSI employees, scattered around the world, had asked for some sort of communication medium to be enjoyed by the entire GSI family. A "news and gossip sheet" by and for employees seemed to be an apt description of the envisioned medium. But who had the interest and time to arrange for such a publication, who would be responsible thereafter for gathering the news to keep the proposed sheet alive and growing, and who would actually deliver it?

The responsibility finally came to rest on Cecil's desk, because he had been one of the prime supporters of the sheet from the very beginning. However, already fully occupied with his vice presidential duties, and possessed of the appropriate authority, Cecil asked contract administrator W. C. (Bill) Edwards, Jr., a fellow GSIer, to bring forth the first issue and to serve thereafter as editor, with the responsibility of delivering monthly issues of the sheet.

Fortunately for GSI, Bill knew exactly where to turn for the items that could be included in the first issue. Ima Poindexter and Maxine Coursey, each with a GSI husband, served as eyes and ears for the administrative hierarchy. They would provide the desired "news and gossip" items. And because Mrs. Coursey and Mrs. Poindexter were considered the reliable "grapevine" for information of all sorts on GSIers everywhere, why not call the envisioned publication the *GSI Grapevine!* The first issue appeared in November 1944 during the difficult years of World War II, when GSIers were scattered worldwide either as Armed Forces personnel or as exploration geophysicists.

Successive volumes of the *GSI Grapevine*, with decade anniversary issues in 1954, 1964, 1974, and 1984, record the impressive growth of GSI from a struggling company to one of the world's leading companies now conducting geophysical exploration from the air, on the ground, and at sea in both hemispheres. Since this exploration now uses state-of-the-art methodology and instrumentation, some of which has been developed in GSI's own research laboratories in Dallas, future historians interested in recording the development of geophysical exploration for petroleum will find a rich source of relevant material in the forty-three volumes thus far published. Much of the earliest history of GSI's offspring, Texas Instruments Inc. (TI), is also to be found in the earlier volumes.

In short, thanks to Cecil's instinct for sharing information, the *GSI Grapevine* was there at the very beginning of the electronics revolution in which its own TI would become one of the world's leaders.

ORGANIZATION AND DIRECTION OF THE GSI STUDENT COOPERATIVE PLAN

By the time he became part owner and vice president of GSI, Cecil was already acutely aware of the importance of well-trained and imaginative younger personnel in maintaining leadership in geophysical exploration. Furthermore he realized that he must employ more such scientists and engineers if GSI were to grow and still keep its leading position as an independent geophysical service company. The limited success of his recruiting efforts, particularly during the years immediately following Japan's surrender, convinced him that he must somehow intensify the search for qualified employees—and perhaps take initiatives to actually develop the superior employees he sought. One crucial event in this process occurred in 1950, when a visit by Cecil to his undergraduate Department of Electrical Engineering at MIT failed to arouse departmental interest in his recruiting problem. So he turned to the Institute's Department of Geology and Geophysics, where he and I discussed GSI's need for summer employees. Out of that discussion came the MIT-GSI Student Cooperative Plan, described more fully in appendix E, which Cecil initiated in the summer of 1951 and directed for fifteen years before turning it over to his GSI associate, geophysicist Kenneth E. Burg. Burg directed the Plan for another two years before it was terminated.

In that seventeen-year period, this imaginative and innovative training program, soon dubbed the *GSI Co-op Plan* trained more than 360 carefully selected students from a total of 87 North American colleges and universities.[1] Today, twenty-four years after inception of the Plan, a substantial number of former trainees occupy leading positions in academia and in the petroleum industry.

As GSI president, Cecil also arranged that his company support MIT's Geophysical Analysis Group (GAG) research program.[2] He closely followed GAG's progress, quickly realized how the results being obtained could be used in geophysical exploration, and at the same time deliberately gained acquaintance with certain of the student participants *and* their wives, by taking them out to dinner when he visited Cambridge. Since he regarded the wife of an exploration geophysicist as a critical influence in determining how happy her husband would be as a new GSIer, he wanted to be certain that the wives, as well as the husbands, had an opportunity to ask him questions about employment with GSI. Then when the students completed their graduate work he hired several of them. These new employees

[1] See R. R. Shrock's *A Cooperative Plan in Geophysical Education*, 143 pp., published by Geophysical Service Inc., Dallas, Texas (1966).
[2] See R. R. Shrock's *Geology at MIT 1865–1965*, Vol. 2, pp. 657–678, MIT Press, Cambridge, Mass. (1982).

confirmed Cecil's astute judgment by their contributions to making GSI an acknowledged leader in processing and interpreting the rapidly increasing flow of seismic data from field operations.

THE GRADUATE RESEARCH CENTER OF THE SOUTHWEST (GRCSW), SOUTHWEST CENTER FOR ADVANCED STUDIES (SCAS), AND THE UNIVERSITY OF TEXAS AT DALLAS

After the entrepreneurs Green, Jonsson, McDermott, and Peacock bought the assets of Geophysical Service Inc. in 1941, and were faced with guiding its operations through the difficult years of World War II, they soon realized that to be successful entrepreneurs they would have to recruit many highly competent engineers, scientists, and business personnel.

This need became especially evident and urgent when the rapidly growing General Instruments Division so far outstripped the geophysical exploration program, in number of personnel and size of budget, that it had to be reorganized as the company parent, Texas Instruments, with GSI as a wholly owned subsidiary.

Trained personnel of the kinds needed by GSI and TI were not only in short supply in the Dallas–Fort Worth area, they were not even being trained in any significant numbers by the region's educational institutions. Furthermore the few persons who did complete advanced training were quickly snapped up by the high technology companies that had suddenly developed in the Dallas–Fort Worth area during and following the war years.

Up to that time the economy of Texas had been based on cattle, cotton, and oil; hence there had been little encouragement for the development of higher education in engineering and the basic physical sciences. This situation was quite in contrast to that of other centers in the United States—such as Los Angeles and the Bay region, Chicago and Detroit, and the New York–New Jersey and New England sector—where industrial technology and higher education thrived in a symbiotic relationship. So GSI and TI, as well as their competitors in the rapidly developing Dallas–Fort Worth area, had to look to the universities of the North and Northeast and the West Coast for the kind of highly trained personnel they would need if their organizations were to grow into industrial leaders.

Cecil, Jonsson, and McDermott, all much concerned about their respective personnel problems, joined a group of similarly concerned industrial and educational leaders in an historic evening meeting at Southern Methodist University on May 27, 1960. The purpose of the meeting was to debate a series of proposals developed by Lloyd V. Berkner, renowned scientific research administrator from the East Coast. Berkner had first learned of the Southwest's changing economic and social conditions from Jonsson, during a 1958 flight from Dallas to New York, and was so impressed that he immediately made a thorough study of those conditions. The proposals resulting from his study were the subject for discussion at this meeting.[3]

The following excerpts were taken from Berkner's report of that meeting:[4]

Present with Mr. Jonsson at this discussion were:

Claude C. Albritton	C. J. McCarthy
Harvey Gaylord	Robert McCulloch
Cecil H. Green	Eugene McDermott
Hemphill Hosford	Eugene McElvaney
J. Erik Jonsson	H. Gardner Symonds
Stanley Marcus	Willis M. Tate
Lewis W. McNaughton	C. A. Tatum, Jr.

The essence of the proposals was that a regional institution dedicated solely to high scholarship should be founded at Dallas,

---

[3]Berkner, L. V., *Graduate Education in the Southwest*, SMU Press, May 1961. See also "Whither Graduate Education?" *Physics Today*, July 1963, and "The Technical Revolution of Today," *Franklin Institute*, June 1963.
[4]Berkner, L. V., "Report of the President" in the first Annual Report: Graduate Research Center of the Southwest, June 30, 1963, Dallas, p. 4.

with the objective and ideal of scholarly research and post-doctoral education, to aid in the development of a regional tradition of academic attainment. This institution was to work with universities and their faculties in building graduate opportunities as rapidly as possible. It was to work with industry as a source and storehouse of fundamental thought out of which innovation would be encouraged.

The institution would devote itself entirely to fundamental research and advanced teaching and would not be in competition with the universities in their pre-doctoral work. It would be organized as a university with the highest standards for faculty selection; while qualified to grant degrees to ensure its academic stature, it would not do so in practice, in order to encourage university development within the region. . . .

The May evening meeting ended with the clear determination to undertake the founding of such an institution. In the following December, I [Berkner] joined with the group to bring that institution to life.

Action quickly followed this decision. The first step was to obtain the funds required. These funds were guaranteed by the same three entrepreneurial couples who had taken their chances with GSI in 1941 and founded Texas Instruments Inc. in 1953: Cecil and Ida Green, Erik and Margaret Jonsson, and Eugene and Margaret McDermott. And so the Greens, together with the Jonssons and McDermotts, became co-founders of what would later develop into an important component of the University of Texas system. Berkner was appointed president in December 1960, to oversee the creation of the institution, and soon thereafter, in the first GRCSW Annual Report he wrote:

The Graduate Research Center of the Southwest came into being on paper February 14, 1961, when it received its Charter from the State of Texas as a non-profit educational institution with degree-granting authority—an authority it does not plan to use.

J. Erik Jonsson became chairman of the first Board of Directors, and also chaired the Executive Committee, which included Cecil H. Green, Eugene McDermott, C. A. Tatum, Jr., A. Earl Cullum, Jr., L. V. Berkner, and Ralph N. Stohl, secretary.

The committee selected a site for the new center about eighteen miles north of downtown Dallas and purchased a tract of 1,200 acres with a central core of some 340 acres reserved for the future campus. Berkner recommended purchase of such a large tract in the belief that it might be needed if a nuclear reactor were ever constructed.

Development of the future campus began immediately. Construction started on the first building, the Laboratory of Earth and Planetary Sciences (later named the Founders Building), as recorded on a plaque in that structure:

> The Founders Building
> made possible by
> the vision and generosity
> of
> Cecil and Ida Green
> Erik and Margaret Jonsson
> Eugene and Margaret McDermott
> who provided the stimulus
> and the initial guarantees
> to ensure the establishment
> of the
> Graduate Research Center
> of the Southwest
> and
> by the volunteer leaders and donors
> who gave liberally
> of their time, energy and substance to
> The Dallas Founding Fund
> October 29, 1964

As the center developed from the grand plan of campus, buildings, faculty, staff, and supporting facilities, there evolved simultaneously the typical hierarchy of groups at different levels of authority and responsibility. Different committees were created to assume specific responsibilities, such as academic and administrative policy (see figure 13). In the foreword of the first Annual Report [of GRCSW] of June 30, 1963, Erik Jonsson, chairman of the Board of Directors of both TI and GRCSW, wrote as follows about the collaborative thinking that had led to the establishment of the Center:

Six years ago [1957], some of us associated with technical industry became acutely conscious of the need to expand the scientific and technological competence of the Southwest. We
realized that this problem could be solved only by an aggressive
program for enlarging and upgrading advanced education
within the region. Some of the universities were already working toward solutions. At that time, we had the seed of an idea
for an additional approach which would augment the efforts of
the universities and colleges. This germinated in February 1961,
when the Graduate Research Center of the Southwest received
its charter.

In June 1963, two years after GRCSW was chartered, Jonsson
could report that

The faculty and staff of the Center now comprised about one
hundred. The research goals of this Community of Scholars at
present are oriented toward the science of the Earth, the physics
of the atmosphere and space, the nature of gravitation, and
other intriguing subjects. The research results are being reported in leading journals and at meetings of the world's scientific societies. . . .

We have been particularly fortunate in obtaining eminent scientists to head these efforts; and they, together with their talented
staffs, are attracting other superior men. Upon this group, the
success of the entire venture depends. We feel that we have
reached the critical mass; the proper scientific climate has been
established and the growth from here on will be both steady
and sound.

In the same 1963 Annual Report, President Berkner described
his idea of what GRCSW should become:

The Graduate Research Center of the Southwest is founded on
the ideal that man's fundamental resource for the future is the
trained, inquiring and creative mind. Through the establishment of a "community of scholars," dedicated to the exploration
of the most advanced complexities of nature, the progress of the
region can be accelerated by greater utilization of the resource
of the mind—the only major resource remaining for development to man's benefit.

Cecil recalls (in one of my taped recordings) that Berkner was
not only a politician's scientist but also a good salesman, who

13   The University of Texas at Dallas. *Opposite page, top:* The Founders
Building, so named to honor the Greens, Jonssons, and McDermotts
who originally guaranteed the funding for the 1961 founding of
GRCSW-SCAS, precursor of UTD. *Opposite, bottom:* The three
husband-founders of GRCSW-SCAS (*from left to right*)—Cecil H.
Green, J. Erik Jonsson, and Eugene McDermott—at the dedication of
the Founders Building, October 29, 1964. *Above:* The Cecil H. Green
Center, funded by The University of Texas and named in Cecil's
honor, dedicated in October 1976. All photos courtesy of the University of Texas at Dallas.

could raise the three founders' spirits when they became discouraged with their costly project because of its slow progress.

There were many times when Erik, Mac and I used to get discouraged and not get anywhere. Maybe we should give the whole project to one of the local colleges, we suggested! By that time Lloyd would have scheduled another meeting, and by the time he got through the session, we had forgotten all about our concern and were off on another enthusiastic start. But we soon realized that the first idea we had would not work. That idea was that SCAS (Southwestern Center for Advanced Studies, the name that replaced GRCSW in 1965) would develop a faculty of highest competence and facilities of the most modern kind, whereupon the surrounding universities in Texas and elsewhere would want to send their best graduate students to GRCSW to complete their doctoral thesis work and continue on in post-doctoral research—in short to "polish them off" and send them out as SCAS's "finished product."

But how many came from surrounding schools? None! Very simply put, no students came, so our original plan didn't work at all the way we had expected.

We soon concluded that the three of us couldn't continue to provide the financial support needed to build the kind of super graduate school we had envisioned. So what should we do?

We finally concluded that SCAS, as GRCSW was soon named, could best grow as a component of one of the leading state universities. Finally, the University of Texas System agreed to accept SCAS, "lock, stock and barrel" and to provide funding for developing it into a full-fledged university.

Thus in 1969 appeared a new university in Dallas, created and funded in large part by the Greens, Jonssons, and McDermotts and aided and supported by the academic, industrial, and civic leaders of the Dallas–Fort Worth metroplex.

THE UNIVERSITY OF TEXAS AT DALLAS (UTD)

A NOTEWORTHY COLLECTION OF 30,000 BOOKS
To show their continued support of the GRCSW-SCAS enterprise, even after it became UTD and a component of the University of Texas System, the Greens, in 1973, made a donation of $100,000—approximately 40 percent of the total cost of an

impressive collection of 30,000 books on Central America. The collection is now housed in UTD's McDermott Library building.

THE CECIL H. AND IDA M. GREEN HONORS CHAIR IN THE NATURAL SCIENCES

To give further support to the new UTD, the Greens, in 1974, endowed the Cecil H. and Ida M. Green Honors Chair in the Natural Sciences in the hope that an outstanding scientist would be appointed to the chair as soon as possible. Francis S. Johnson, a distinguished physicist, was selected to occupy the chair in 1974, and has served continuously to the present (1988) except for one short leave of absence.

During his leave, the income from the chair's endowment was used to fund the Cecil H. and Ida M. Green Distinguished Lectures Series, which—according to UTD pamplets—was established "as a forum for sharing ideas representing each of UTD's six schools." Among the fifteen luminaries who served as Green Distinguished Lecturers during 1980–1986 were Philip Handler and Frank Press (National Academy of Sciences presidents), Michael Croft (National Youth Theatre of Great Britain), John Mervin (*Forbes* magazine), and Fred H. Hechinger (*New York Times*), Sir Michael Tippet (composer and conductor), Vernon E. Jordan (civil rights leader), Harry Messel (head of the University of Sydney's School of Physics), John Diebold (computer expert), and Margaret L. A. MacVicar (MIT Green Professor of Physical Sciences).

By 1976, the University of Texas at Dallas had been expanded to include degree programs at both graduate and undergraduate levels, and the campus had grown with the construction of six major buildings, which were dedicated on October 18 of that year (see figure 13). Three buildings were named for the founders: the Cecil H. Green Center, housing the School of Human Development, the School of Management and Administration, the School of Social Sciences, and the School of General Studies; the Erik Jonsson Center, serving principally as a classroom and office building for the School of Arts and Humanities, but also including space for the School of Management and Administration and the School of Natural Sciences and Mathematics; and the Eugene McDermott Library, housing the university's library collections and facilities, with space available for 750,000 volumes. This building also houses the bookstore, health service, student counseling service, and the office of admissions and the registrar.

That Cecil continues to be involved in UTD activities is shown by the fact that he was the guest speaker at the dedication of the Berkner Building, as well as that held for the new student union building, on April 23, 1982, the first official alumni event held for graduates of UTD. He also participated in the first commencement ceremonies. Ida, for her part, left UTD a generous bequest.

ORGANIZATION OF THE ASSOCIATION FOR GRADUATE EDUCATION AND RESEARCH OF NORTH TEXAS (TAGER)

The failure of the GRCSW-SCAS program to attract advanced graduate students from other Texas and neighboring universities did not shake Cecil's faith in the benefits of intercommunication and collaboration between academia and industry. Thus in 1965 he led a group of prominent Dallas–Fort Worth educational leaders—Gifford K. Johnson (GRCSW-SCAS), Willis M. Tate (SMU), James M. Moudy (TCU), John D. Moseley (Austin College), A. Earl Cullum, Jr., Thomas L. Martin, Jr., R. W. Olson, and Beeman Fisher—in organizing The Association for Graduate Education and Research in North Texas (TAGER).

Whereas GRCSW was chartered as an independent, privately funded academic institution to produce doctoral and postdoctoral scientists and engineers from its own students, TAGER was organized to stimulate and accelerate the development and growth of higher education in the Dallas–Fort Worth region by establishing a consortium of all existing colleges and universities to pool their individual faculty resources and to induce professors to think in terms of their discipline rather than of campus boundaries.

It was envisioned that the consortium would also provide education to off-campus students throughout the industrial sector of North Texas by establishing classrooms in industrial plants where young scientists and engineers could work toward advanced degrees on company time, the degrees to be granted by authorized member universities of the consortium. In this way, a particular industry could greatly increase the number of its employees with advanced training without having to conduct a full-fledged graduate school of its own. Moreover it would encourage younger employees of the local high-technology industries to upgrade their training without seeking leaves of absence.

An early question to be answered was how to bring students and teachers together. First came the suggestion that professors travel from campus to campus, or plant to plant, but that proposal was soon deemed impractical; next, busing students was considered, but that, too, was judged unfeasible, as well as a waste of time. At this juncture, one of Cecil's MIT fellow alumni, A. Earl Cullum, Jr., who was also a member of the TAGER organizing group, suggested that closed-circuit TV could be used for intercommunication. He knew about such a TV educational system in operation at the University of Florida. Why not go over and have a look?

So Cecil, Cullum, and several others flew to Gainesville where Dean Thomas L. Martin, Jr., showed them his system, which extended from Gainesville to Orlando, with connections to various industrial outlets including Cape Canaveral. Cecil recalls,

We were so enthused about what we saw that we came back not only with the idea, but, also with Thomas L. Martin himself to serve as a new Dean of Engineering at SMU where he took on the important task of building a strong graduate School of Engineering.

Convinced that a closed-circuit TV system was the solution to TAGER's problem, Cecil, together with Ida, pledged the funds for the control center, to be established on the campus of GRCSW. The TV system itself, with its relay stations between

Dallas and Fort Worth, and also between Dallas and Sherman, soon became known as the Green Network. The Greens' funding of the TV network was crucial to the success of the whole TAGER program, and their gift is fully acknowledged in the plaque that hangs in the control center of the Green Network on the UTD campus (see figure 14).

The history, purpose, and nature of TAGER can be summarized best by quoting several paragraphs from a recent pamphlet titled "Talkback Television at the SMU Institute of Technology":

The organization of TAGER was the culmination of planning that began in 1965, when seven educational institutions—Austin College . . . Bishop College . . . Southern Methodist University . . . Southwest Center for Advanced Studies [SCAS] . . . University of Dallas . . . Texas Christian University . . . and Texas Wesleyan College—made the decision to pool their resources, both physical and academic, in a mutual association to promote graduate education and research in the North Texas area. With the substantial contribution in both influence and funds from Dallas industrialist Cecil Green, and his wife Ida, a firm foundation for TAGER was created and the construction of the TAGER Microwave Network was assured. TAGER was then able to implement its new and innovative ideas on graduate education which included the incorporation of industries and educational institutions into a larger, more complex network.

The aim of the educators and industrialists who signed the TAGER charter was to enhance graduate education in the North Texas area. The initial emphasis was on increasing the number of doctoral degrees granted in the sciences and engineering, but disciplines in the humanities and social sciences are now being added. . . .

The TAGER TV system is a closed circuit system with "talkback." In other words, all communication channels are one-way video and two-way audio. Thus, every student has the capability of asking questions just as though he were in the classroom. All students hear all questions and all answers. Therefore, despite geographical separation, modern communication techniques can be used to successfully reproduce the fundamental student-teacher relationships.

14   The Green Network of the Association for Higher Education of
North Texas—The Association for Graduate Education and Research
of North Texas (AHE/TAGER). *Top:* TAGER's 250-foot broadcasting
tower (the Green Tower), located in the northwest corner of the UTD
campus in Richardson, Texas, transmits signals over northern Texas,
as shown in figure 15. *Bottom, left:* Entrance to the TAGER Network
Building, adjacent to the Green Tower. *Bottom, right:* The dedicatory
plaque on the TAGER building. Photos courtesy of AHE/TAGER.

The accompanying graphic (figure 15) shows the current geographic layout of the TAGER system.

The talk-back feature was a triumph for the electrical and electronics engineers in the TAGER development committee. While students in the same room with the lecturer merely raised a hand to ask a question, distant students, watching the lecture on television, spoke into the receiver on a telephone which was connected to a phone beside the lecturer. The questions as well as the answers were heard by all the students listening to the lecture, wherever they were in North Texas.

When Cecil took me to a TAGER lecture several years ago, I was amused to note that many of the students in the room where the lecture was being given riveted their attention on the small monitoring screen in one front corner of the room rather than looking at the live speaker directly in front of them. Cecil remarked that he had noticed the same thing, and assumed that it was because the students had become accustomed to a TV screen at a very early age and preferred it to the live action in front of them.

Cecil was deeply involved in every aspect of the development of the TAGER system. He offered innovative ideas, pledged generous donations at crucial times, and served as chairman of the Board of Trustees for the first seven years (1965–1972), promoting a successful collaboration between academic and industrial leaders. He worked especially closely with the four principal educators who took leading roles in organizing the course work that was to be offered—President Willis M. Tate and Dean Thomas L. Martin, Jr., of SMU, Chancellor James M. Moudy of TCU, and President Gifford K. Johnson of GRCSW.

In due course the network was built, the first courses were prepared, and in 1967, two years after the TAGER consortium was chartered, its program went on the air over a closed-circuit TV network to seven academic institutions and four industrial plants. At this time Cecil could rightly say, "We achieved an environment that would support the high technology that was coming," as come it did. He could also say that the inception of

MICROWAVE RADIOS

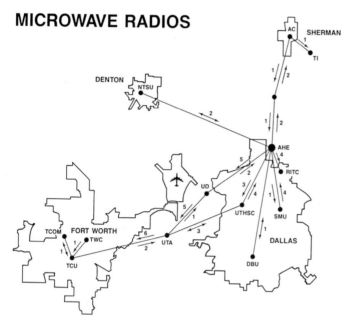

ITFS TRANSMITTERS

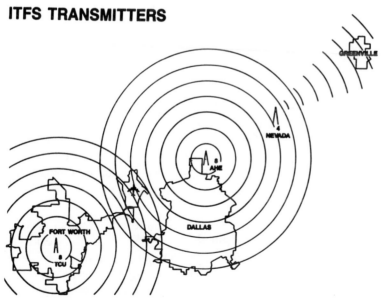

15   The AHE/TAGER System, consisting of microwave radios (*top*) and ITFS transmitters (*bottom*). Graphics courtesy of AHE/TAGER.

the TAGER program satisfied one of his long-standing desires, namely, "to encourage innovative and nontraditional methods of bringing academic and industrial young people into direct contact with one another and with their talented superiors in industry and academia."

TAGER'S first president, R. W. Olson, stated in the Fall 1967 Catalog of Graduate Courses in Science & Engineering:

Special recognition should go to Mr. Cecil H. Green for his vision and leadership and contributions, in time and resources. Recognition should also be given to Mr. A. Earl Cullum, Jr., for the engineering "know-how" which has helped make the physical network take shape in such a short time.

SMU's president L. Donald Shields wrote in the 1981 fall issue of the *SMU Digest*:

Green is credited with creating the linkage between high-technology industries and area colleges and universities which has resulted in advanced academic work for some 18,000 student/employees since 1965. The TAGER program, as this linkage is called, was one of the earliest efforts in the nation to provide classroom instruction via television.

ASSOCIATION FOR HIGHER EDUCATION OF NORTH TEXAS (AHE)

By 1980 TAGER had become a resounding success. In that year the Association for Higher Education of North Texas was incorporated to serve the interests of TAGER in the seventeen North Texas institutions of higher education. The AHE, comprising a consortium of all colleges and universities in the North Texas region, now operates two TAGER Network services—the original program started in 1967 and the more recently developed TAGER Cable TV Service.

In 1981, with the TAGER TV program in full swing, and the microwave TV system called the Green Network, Cecil was called the "father" of the TAGER TV system, and credited with its creation. On October 10, 1981, as the TAGER program approached its fifteenth year of operation, SMU's President Shields wrote:

TO CECIL H. GREEN

For your early recognition of both the need and the opportunity to use non-traditional media to enhance the education of engineers and scientists employed by the High Technology Industries of North Texas.

For your taking the lead in developing the Association for Graduate Education and Research, known as TAGER, which already has enabled thousands of individuals, who could not otherwise have done so, to earn advanced degrees by television.

And for the pioneering spirit, visionary leadership and commitment to community that have long set you apart from others, Southern Methodist University expresses its affection and deep appreciation.

TAGER has come a long way since its chartering in 1965 and its inception as a private, course-offering TV network in 1967. The growth in the number of participating industries, from the original four to the current nineteen whose employees enjoy the opportunity of enhancing and extending their education, bears witness to how fully the high-technology industries of North Central Texas have taken advantage of TAGER network's program. Its remarkable expansion is best indicated by the following two lists of current participants (as of August 1985):

Principal participants (17, *indicates original participant)

*Austin College

 Baylor College of Dentistry

*Bishop College

 Dallas Baptist University

 Dallas County Community College District

 East Texas State University

 North Texas State University

*Southern Methodist University

 Tarrant County Junior College District

*Texas Christian University

 Texas College of Osteopathic Medicine

*Texas Wesleyan College

Texas Women's University

The University of Texas at Arlington

*The University of Texas at Dallas [SCAS]

The University of Texas Health Science Center–Dallas

*University of Dallas

[*Note:* More than 50 percent of the courses offered by the above institutions have been given by SMU (50 to 65 percent, after 90 percent at the beginning); the remainder by University of Texas–Arlington (12 to 16 percent), University of Dallas (10 to 13 percent); TCU (2 to 3 percent), UTD (1 percent), and UTHSCD (1 percent).]

Associate participants (24, *indicates original participants)

Baylor University

Midwestern State University

Dallas Public Library

Fort Worth Public Library

ARCO Oil and Gas Company

Bell Northern Research/Northern Telecon

Bell Helicopter Textron, Inc.

DSC Communications Corporation

E-Systems, Inc.

Electronic Data Systems Corporation

Electrospace Systems, Inc.

*General Dynamics

Albert H. Halff Associates, Inc.

Honeywell Optoelectronics Division

International Business Machines Corporation

Johnson & Johnson Products, Inc.

*LTV Aerospace and Defense Company

MCI Communications Corporation

Mostek Corporation

Motorola, Inc.

Recognition Equipment Incorporated

Rockwell International

Tandy Corporation

*Texas Instruments Incorporated

Cecil has told me that one of the most satisfying rewards for helping to get the TAGER program organized and under way was becoming acquainted with the academic and industrial leaders of the institutions and industries that ultimately joined the consortium. He pointed out that in discussing TAGER with them he also learned of their problems and needs. As a consequence he and Ida were better able to determine the nature and extent of possible future donations. Moreover Cecil continues to follow the ever-expanding program with unabated interest.

Cecil received a standing ovation when he concluded his "Comments" to the audience assembled in Dallas, on November 13, 1987, to celebrate the twentieth anniversary of the TAGER television network. The celebration was " . . . dedicated in loving memory of Ida Green . . ." who with her husband Cecil had the foresight and the philanthropic generosity to play the major role in founding TAGER in 1967, and who left the Network one of her numerous generous 1987 bequests.

The program pamphlet distributed at the celebration, a portion of which is reproduced on a following page (figure 16), gives the names of individuals, colleges and universities, corporations, and hospitals that have participated in the TAGER partnership, which is one of the leading collaborative academic-industrial enterprises in the nation.

Further proof of the success of TAGER lies in the following statements included in the *TAGER-AHE Bulletin* for the fall of 1987:

Since becoming operative in the fall of 1967, TAGER has logged approximately 30,000 registrations in higher education, business and industry, primarily through course offerings in the var-

The TAGER Partnership

Chairs of AHE Board of Trustees

AHE appreciates the tireless efforts and unceasing support of those who have chaired its governing board over the years.

Cecil H. Green	1967-1977
Philip R. Jonsson	1977-1981
C.J. Thomsen	1981-1984
Betty Jo Hay	1984-1985
Alexander R. Bolling, Jr.	1985-1987
Rolf Krueger	1987-

Special Recognition

The following individuals have contributed 20 years of teaching, administrative, or technical service to the TAGER Network.

Kenneth Ashley	Southern Methodist University
Charles B. Baker	Southern Methodist University
Frances Brooks	General Dynamics
Jerome Butler	Southern Methodist University
Someshwar Gupta	Southern Methodist University
Kenneth W. Heizer	Southern Methodist University
Mandyam Srinath	Southern Methodist University
Don Sudbrink	AHE/TAGER Staff
W. Walter Wilcox	Southern Methodist University

Colleges and Universities

These institutions have all participated in sending courses over the TAGER network.

Austin College
Bishop College
Dallas Baptist University
North Texas State University
Southern Methodist University
Texas Christian University
Texas Wesleyan College
The University of Texas at Arlington
The University of Texas at Dallas
The University of Texas Southwestern Medical Center at Dallas
University of Dallas

Corporations

These companies have all participated in receiving TAGER educational programs for their employees.

ARCO Oil and Gas Company

Bell Helicopter Textron, Inc.

BNR/Northern Telecom, Inc.

Continental Electronics,
A Division of Varian Assoc.

DSC Communications
Corporation

E-Systems, Inc.

Electrospace Systems, Inc.

Frito-Lay, Inc.

General Dynamics Corporation

General Instrument Corporation,
TOCOM Division

Albert H. Halff Associates, Inc.

Honeywell Optoelectronics

Johnson & Johnson

LTV Aerospace and Defense
Company

MCI Telecommunications
Corporation

Mead Office Systems

Mobil Exploration & Producing
Services, Inc.

Motorola, Inc.

National Advanced Drilling
Machines, Inc.

Oscar Mayer Foods Corporation

Recognition Equipment,
Incorporated

Rockwell International

Sun Exploration and Production
Company

Tandy Corporation

Teledyne Geotech

Texas Instruments Incorporated

Thomson Components —
Mostek Corporation

Xerox Corporation

Hospitals

These hospitals participate in the Dallas Area Hospital Television Service, with telecommunications support provided by AHE/TAGER.

All Saints Episcopal Hospital

Doctor's Hospital

Ft. Worth Osteopathic Hospital

Green Oaks Psychiatric
Hospital

Harris Hospital

John Peter Smith Hospital

Kaiser Permanente Medical Care
Program

Medical City Dallas Hospital

Methodist Hospital of Dallas

Parkland Memorial Hospital

Plano General Hospital

Presbyterian Hospital

St. Joseph's Hospital

Texas College of Osteopathic
Medicine

The University of Texas
Southwestern Medical Center
at Dallas

Veterans Administration Medical
Center

16   The TAGER Partnership 1987. Courtesy of AHE/TAGER.

ious applications of computer science, engineering and management. This is the equivalent of 2,500 degrees based on an average of 12 registrations necessary to complete a degree.

Annually, TAGER now averages more than 2,500 credit and non-credit registrations in more than 180 courses, delivering as much as 10,000 hours of advanced level education and training programming.

ON THE IMPORTANCE OF ORIENTATION AND COOPERATION

As a recruiter of personnel for GSI, and later as a featured speaker and advisor on career development, Cecil soon developed his own innovative techniques in discussing professional careers in engineering and science with college students. His decision to abandon the traditional individual interview, in favor of gathering all interested students together in one group, came during one of his earliest recruiting trips to the Colorado School of Mines. He described the incident to me, as follows:

I was told, at the outset, that six students wanted to talk with me, and since the total time for interviewing could not be more than 60 minutes, that would mean 10 minutes per student.

My response was—"No, I would rather have all six students together for the 60 minutes, which would give me a much better opportunity to tell my story, and in the end would more likely result in students asking me for a job, rather than my serenading them."

Cecil has continued his concern about the quality of new and young employees of GSI, even though he retired from the company more than a decade ago. Recently he came up with the idea that if top GSI personnel could describe some of their problems and procedures to advanced students and to professors of mathematics, physics, and electrical engineering, these academics might become interested enough to suggest that some of their own students consider exploration geophysics as a career. In order to try out this innovative notion, Cecil suggested that MIT's Department of Earth and Planetary Sciences organize a one-day conference. Professors of mathematics, physics, and electrical engineering, together with selected advanced students, were invited to attend the conference and to comment on presentations by several of the top GSI staff, including the president, Dolan McDaniel. The conference produced lively discussion between the professors, students, and GSI personnel, not just after each lecture, but also during coffee breaks, lunch, and the closing cocktail. Several participants remarked, "We should have more meetings of this kind."

Considering his initial experiment a success, in that the professors were impressed by the sophistication of GSI's research and applied work, Cecil suggested that similar conferences be held at the Colorado School of Mines and at Stanford. As at MIT, the conference was hailed at both schools as an excellent way of acquainting academic people—both professors and students—with the day-to-day problems of the exploration geophysicist.

The conferences have also resulted indirectly in repeated invitations to Cecil himself to present slide shows about his own experiences as a doodlebugger in many parts of the world. These presentations have given him a unique opportunity to compare the modern methods and facilities used by GSI with those he used as a fledgling doodlebugger in the 1930s.

Thus in a subtle way he impresses the students in his audiences with the compelling need for thorough training in science and engineering, and for development of a cooperative and collaborative attitude toward their fellow exploration geophysicists—a relationship that is absolutely necessary for successful fieldwork.

INTERDISCIPLINARY CONFERENCES TO ADVANCE MEDICAL RESEARCH AND HEALTH CARE

Cecil's ability to imagine, and then to catalyze, an innovative project with a deliberate educational purpose was never more evident than when he decided to organize a special conference on biological imaging. The idea came to him while undergoing special tests in Scripps Clinic and Research Foundation (SCRF) to determine the cause of his ulcers.

Observant and inquisitive, Cecil soon learned about the institution's diagnostic research techniques and instrumentation. As he and Ida became better acquainted with the clinic's staff, and gained more knowledge about the scope of its research and health-care activities and its plans for future expansion, they decided to support the clinic's research program in diagnostic radiology as discussed farther on.

As he watched SCRF expand into new areas of diagnostic research and innovative health care, Cecil noted with special interest the increasing use of electronic instrumentation. Here were medical personnel using highly sophisticated instruments, about which they knew little, which had been designed by physicists and electrical engineers who had limited knowledge of the human body that was being exposed to the radiation from their instruments. Wouldn't it be worthwhile, thought Cecil, to invite the medical researchers to meet with physicists and electronics engineers for the purpose of discussing problems of mutual interest?

When it became evident in the late 1970s that the standard X-ray methods were going to be supplemented, possibly even replaced, by new electronic techniques in biological imaging, Cecil's interest was further stimulated. At that time it seemed that Proton Emission Tomography (PET) would become the leading technique; Nuclear Magnetic Resonance (NMR) was also promising, although not yet fully developed.

As a result of discussing these exciting new developments in medical research with his friends at Scripps Clinic and at the several medical divisions of the University of Texas System, Cecil, always a strong supporter of the cooperation and collaboration of experts with a common interest, came up with the novel idea: Why not organize a national conference on biological imaging?

The obvious participants in such a conference, Cecil reasoned, should be eminent physical scientists and engineers, together with equally skillful medical research personnel and clinicians. Forthwith he sought and secured the assistance of Andrew D. Suttle, Jr., a special friend and a nuclear scientist on the staff of the University of Texas Marine Biomedical Institute (UTMBI) in Galveston. Scripps Clinic, in turn, agreed to host the conference, to which the prominent persons listed on the following page were invited. The final result of Cecil's and Suttle's planning was the conference "Biological Imaging: Contributions from Contemporary Physics and Engineering," held at Scripps Clinic and Research Foundation on November 16–18, 1980. Positron Emission Tomography and Nuclear Magnetic Resonance were the subjects discussed.

Cecil and his associates learned some valuable lessons from this first conference, which they regarded as an experiment: some speakers were clearly better than others; hoped-for discussion after each presentation did not always take place; representatives from industry and selected graduate students could have profited from attending; and one or two postconference manuscripts were not submitted promptly. All in all, however, the first conference achieved its major goals.

During the three years following the first conference, NMR clearly outstripped PET and other biological imaging methods, and by 1983 it had become the leading technique in imaging because of its greater precision and sharper resolution, its superior discernment of variations in tissues, and the fact that it had no harmful effects such as result from excessive exposure to X-rays. In the light of these advances in NMR, Cecil and his associates decided to organize a second meeting on biological imaging in order to update all participants on the current state of the technique. Cecil offered not only to fund this second gathering, as he had done with the first conference, but also to serve as chairman and general organizer. Frank Press, recently elected president of the National Academy of Sciences, consented to be a sponsor and also offered to have the Academy host the second conference in Washington, D.C., sometime in 1983.

In planning the second meeting, Cecil and Suttle saw to it that the best possible speakers were secured. This time they also invited a known expert in each speaker's special field to act as a "commentator," whose responsibility was to conclude each presentation with questions or comments designed to provoke

SPEAKERS AND SPONSORS OF THE CONFERENCE ON BIOLOGICAL IMAGING: CONTRIBUTIONS FROM CONTEMPORARY PHYSICS AND ENGINEERING SCRIPPS CLINIC AND RESEARCH FOUNDATION, HOST NOVEMBER 16–18, 1980

Speakers

Jose R. Alonso, Ph.D.
Director of Bevalac Experimental Operations
Lawrence Berkeley Laboratory

Gordon L. Brownell, Ph.D.
Professor of Nuclear Engineering
Massachusetts Institute of Technology
Director, Physics Research Laboratory
Massachusetts General Hospital

Thomas F. Budinger, M.D., Ph.D.
Henry Miller Professor of Medical Research
Professor of Bioinstrumentation
Electrical Engineering & Computer Sciences
University of California, Berkeley

Alexander Gottschalk, M.D.
Professor and Vice Chairman
Department of Diagnostic Radiology
Yale University School of Medicine

Waldo S. Hinshaw, Ph.D.
Associate Professor,
Harvard Medical School
Associate Physicist, Department of Radiology,
Massachusetts General Hospital

David E. Kuhl, M.D.
Professor of Radiological Sciences
Chief, Division of Nuclear Medicine
UCLA School of Medicine

Paul C. Lauterbur, Ph.D.
Professor of Chemistry
Research Professor of Radiology
State University of New York at Stony Brook

Patrick L. McGeer, M.D., Ph.D.
Professor of Neurosciences
University of British Columbia

George K. Radda, Ph.D., F.R.S.
Department of Biochemistry
Oxford University

Lawrence A. Shepp, Ph.D.
Member, Technical Staff
Bell Laboratories

Michel M. Ter-Pogossian, Ph.D.
Professor of Radiation Sciences
The Edward Mallinckrodt Institute of Radiology
Washington University School of Medicine

Earl H. Wood, M.D., Ph.D.
Professor of Physiology and Medicine
Mayo Medical School

Y. Lucas Yamamoto, M.D.
Director, Neuro-isotope Laboratory
Montreal Neurological Hospital

Sponsors

William O. Baker, Ph.D.
Chairman of the Board
Bell Laboratories

Charles C. Edwards, M.D.
President
Scripps Clinic and Research Foundation

Cecil H. Green, Sc.D.
Honorary Director
Texas Instruments Incorporated

William C. Levin, M.D.
President
The University of Texas Medical Branch —
Galveston

Walter A. Rosenblith, Ph.D.
Institute Professor and Provost Emeritus
Massachusetts Institute of Technology

David S. Saxon, Ph.D.
President
University of California System

Glenn T. Seaborg, Ph.D.
University Professor of Chemistry
Director of Nuclear Chemistry Research
Lawrence Berkeley Laboratory

Frederick Seitz, Ph.D.
President Emeritus
Rockefeller University

John L. Smith, M.D.
Senior Vice President, Director of Clinical Services
Scripps Clinic and Research Foundation

Charles C. Sprague, M.D.
President
University of Texas Health Science Center — Dallas

A. D. Suttle, Jr., Ph.D.
Professor of Nuclear Physics and Radiobiochemistry
Special Assistant to the Director
Marine Biomedical Institute
University of Texas Medical Branch — Galveston

Courtesy of Scripps Clinic and Research Foundation

lively discussion and audience participation. They also made
certain that representatives of several of the leading manufac-
turers of NMR equipment were invited. IBM and GE were
among the companies represented. Lastly, ever mindful of the
need to discover the "naturals"—as Cecil likes to call the students who have shown an early and determined interest in a
special field—he collected the names of highly recommended
graduate students from around the country, then selected half
a dozen of these noted for their ability and interest and invited
them to the conference in the hope that a few of them might
consider careers in some aspect of biological engineering.

The result of all the planning was the "National Conference
on Biological Imaging II: Clinical Aspect." The meeting was held
in Washington, D. C., at the National Academy of Sciences as
the host institution, on October 16–18, 1983, with Cecil as chairman. The program pamphlet announced that the conference
would " . . . explore the most recent advances in imaging technology with particular emphasis on nuclear magnetic resonance, positron emission tomography, and the application of
computers to the reduction of data." Eighteen sponsors and
twenty-one speakers and commentators participated in the
three-day conference which, like the first meting in La Jolla, met
with widespread approval. Comments on the meeting soon appeared in professional journals.[5]

When in 1984, Cecil was asked if he had any plans to organize
a third conference on biological imaging four or five years
hence, he replied,

I expect it will happen again but I don't know whether I'll be
involved in it or not. I think I'm going to be a little bit like Alfred
P. Sloan who used to say—"You don't have a very good idea

[5] A report, edited by Cecil Green and A. D. Suttle, Jr., titled "National
Conference on Biomedical Engineering, NMR Clinical Aspects," will
presumably be published in the National Academy Press, Washington,
D. C., in due course. A special review of NMR imaging motivated by
the Washington conference recently appeared—"Nuclear Magnetic
Resonance Technology for Medical Studies," by Thomas F. Budinger
and Paul C. Lauterbur, *Science*, 226, pp. 288–298.

unless somebody else is willing to take it over on down the line"
. . . No, I am getting to the point where I am beginning to think
I ought to avoid getting involved in these things. It takes up an
awful lot of time, not only in visitations, but a tremendous
amount of time on the telephone and letter writing. I'm thinking about avoiding such assignments in the future.

Despite these intentions, however, Cecil informed me in July
1987 that he had provisionally agreed to act as a sponsor of a
third conference on imaging and related subjects, tentatively
titled "New Horizons in Medicine: Contributions from Science
and Engineering," and scheduled to be held at Scripps Clinic in
the fall of 1988. But he will act only in an advisory capacity.
Richard Lerner, M.D., director of The Research Institute of
Scripps Clinic, and Peter Wright, Ph. D., the Green Investigator
of The Research Institute, have agreed to take the responsibility
for chairing the meeting, developing the program, and securing
the speakers. Cecil's assumption that the two previous conferences had been important enough to justify a third is supported
by the following statement included in the recent letter of invitation sent to potential sponsors and speakers: "The two previous conferences in this series . . . have been extremely
prophetic in suggesting the importance of new directions in science and technology."

SPONSOR OF THE NATIONAL ACADEMY OF
SCIENCES' PROGRAM TO BROADEN ITS SCOPE OF
SERVICE TO INCLUDE BUSINESS AND INDUSTRY

Late in 1984, after informing Cecil of his desire to broaden the
activities of the National Academy of Science, in order to develop a more cooperative relationship with business and industry, Frank Press asked him if he would serve as a member of the
Academy's Industry Advisory Committee. Not only did Cecil
agree to serve, he also made a generous donation to help support the Academy's program.

Then in his mid-eighties, and with other requests for his advice and services pending, Cecil almost certainly would have
declined the request from Press except for the fact that he had

been vigorously promoting the same kind of cooperation and collaboration, among his academic and business friends, for the past fifty years.

Soon afterward a specific request came from Press: Would Cecil sponsor (that is, organize and fund) a luncheon meeting in Dallas at which Press could present his ideas to a select group of high-technology leaders? Cecil set promptly to work. Sixty-five invitations went out, mainly to top corporate officers of high-technology industries in the Dallas–Fort Worth area, but a few academic leaders were also invited. Some thirty people accepted Cecil's invitation and were his guests at a luncheon in the Dallas Petroleum Club on the 7th of November. Following the luncheon, Cecil first invited Press to comment on his plans for expanding the scope of the Academy's activities, then asked Fred Seitz, retired president of Rockefeller University, to speak for academia. Thus, once again Cecil made a substantial commitment of time and energy, as well as financial support, to the kind of collaborative effort he has so long supported.

ON LECTURING AND LECTURESHIPS

One of Cecil's outstanding attributes has been his ability to communicate his innovative ideas to others so successfully as to arouse their enthusiastic interest and then gain their full cooperation and support. He has developed his own special way of writing and speaking, and an alert listener can learn much from his rhetorical techniques. His conversation in a small informal group is relaxed, unhurried, and direct, but never trivial. Possessed of remarkable ability to recall incidents, dates, and the names of persons and places, he has long since become a master at story-telling, embellishing the narrative just enough to give maximum effect to the punch line. And his stories seldom lack a trenchant point.

Even when speaking to large audiences he still maintains an amiable informality that quickly establishes good rapport with his listeners. If the situation is such that more formality than usual is required, he rises easily to the occasion by reading the remarks that he has carefully prepared beforehand. A practiced speaker, at ease with his subject matter, he can read a speech just as effectively as giving it extemporaneously. Whatever the subject matter of his discourse, he has the charming habit of calling attention to some of his special friends in the audience, and always in a complimentary manner. And he never neglects to express his gratitude to those who have aided him in any way.

If there was reason, on some unusually pleasant occasion, to suspect that Cecil might become a bit too loquacious, he was likely to hear his longtime friend, the late Eugene McDermott, precede his introductory remarks with the muttered admonition, "Make it short, Cecil!"—a friendly give-and-take that always delighted the audience and got the speech off to a good start. It might be added that Ida, whose spoken comments were characteristically brief and to the point, has also been heard on occasion to echo McDermott under her breath when Cecil was becoming a little too talkative.

A central theme runs through all of Cecil's speeches—the necessity of "getting along" with one's fellow employees, regardless of where they stand in the organizational hierarchy. The nouns *cooperation, collaboration, tolerance, integrity,* and *honesty* appear frequently in his discourse, and also the adjectives *imaginative, innovative,* and *adaptive.* A good sense of humor, along with a natural regard for people, are emphasized as qualities that make a tough job much easier. Students can gain much helpful advice, indirectly given, if they listen carefully to Cecil.

Cecil's engaging style of storytelling, supplemented with excellent color slides taken during his world travels, has added to his popularity as a lecturer. Although he is an expert photographer, particularly skilled in the selection of subject matter, when complimented on the high quality of his slides, he modestly attempts to downplay his skill by remarking, "Oh, I always throw away the bad ones!" According to his own log, he has presented his illustrated lecture on 1939 fieldwork in the Middle East to a dozen or more groups at MIT, Colorado School of Mines, Stanford, University of British Columbia, GSI, TI, Southern Methodist University, University of Texas at Dallas, Woods Hole

Oceanographic Institution, and to the past presidents of the Society of Exploration Geophysicists at an annual meeting. It is no accident that when a university bulletin board announces a Cecil Green lecture or slide show, most seats are occupied.

It would be difficult to determine, even approximately, the number of occasions when Cecil has been asked to deliver a formal lecture, citation, or memorial; to comment on some subject under discussion; to act as a master of ceremonies or director of activities at a professional or fund-raising function; to introduce a distinguished speaker; or to meet with students to discuss a career in geophysical exploration. Fortunately, for this aspect of his lifelong educational activities, he has followed the practice of preparing written drafts of many of his more important commentaries. These drafts, together with his diaries, appointment books, and photographs have been preserved as integral parts of his professional papers (see Appendix H).

Three major North American universities have recognized Cecil's accomplishments in advancing geophysical exploration and in training beginning exploration geophysicists by appointing him to an honorary academic position. MIT and the University of British Columbia each named him an Honorary Lecturer in Geophysics, and Stanford appointed him a Consulting Professor of Geophysics. The circumstances leading to these unusual appointments are discussed in a later section (chapter 9).

An important aspect of Cecil's activities as an educator and philanthropist has been the variety of services that he has rendered academic and medical institutions, professional societies, civic organizations, and social agencies, when they requested special assistance or elected him to a specific office. Inasmuch as these special services have almost always been connected with major donations to the recipients involved, they will be included in the discussion of the Greens' philanthropy that follows.

In summary, Cecil has been an innovative educator, effective communicator, and successful organizer, catalyzer, and initiator during the sixty-five years since his graduation from MIT in 1924. And although now past eighty-eight years old, he has lost none of his enthusiasm for new projects as he goes right on planning for the years ahead. Though the activities of his long professional life of six decades have been diverse and quite satisfying, he has repeatedly declared that he takes greatest pride in what he and Ida have accomplished in advancing education, medical research, health care, and civic activities at every level, for people of all ages, throughout the English-speaking world.

IDA AS DEVOTED WIFE AND STRONGLY SUPPORTIVE PARTNER

Ida's name is linked with Cecil's in the title of this biography, which is also dedicated to her, because for more than sixty years she was Cecil's devoted wife and strongly supportive partner in every respect. I have been uniquely privileged to witness the unusual degree of devotion and collaboration between Cecil and Ida since first meeting them more than thirty-five years ago. Their happy relationship endured because they were so much alike in so many different ways and because their deep devotion to one another never wavered for long even under the most trying circumstances. Furthermore, because they had no children to divert their attention from one another, they were free to plan and act with only themselves in mind. Their forbearance of one another's foibles and their mutual trust and loyalty had much to do with that cooperation that made possible their impressive entrepreneurship and philanthropy.

No one has described better the final result of the Greens' long married life together than their admiring MIT friend Vincent A. Fulmer who in 1978 wrote,

Cecil and Ida Green are a remarkable couple in that during their now more than fifty years [sixty years in 1986] of married life their values, standards, attitudes and actions have combined in such harmony that they blend into a single personality.

Throughout the years they have worked as a team in all of their endeavors. At the same time they have maintained a deep respect for each other's independent attitudes and concerns. (Quoted from p. 2 of "An International Tribute to Cecil and Ida Green," a pamphlet prepared for the festive gathering at the National Academy of Sciences on November 9, 1978.)

Ida was an independent person, as was shown by her decision to leave high school and seek employment in Schenectady. At General Electric she demonstrated her ability as an accountant and statistician on a par with male associates. It was only after frequent dates with Cecil that she consented to marriage, and only after he accepted an attractive job at Raytheon. Giving up her position at GE, without knowing whether she could find a comparable one in Boston, and assuming the responsibility of repeatedly locating and managing living quarters for herself and her new husband, showed courage and self-confidence that matched her deep affection for Cecil.

During the first four years of their marriage Ida faced the challenges of four transcontinental auto trips as Cecil vainly sought permanent and satisfying employment. Then when Cecil joined GSI as chief of Party 310, and from 1930 to 1936 engaged in an odyssey of geophysical exploration, Ida cheerfully supervised the details of fifty-four moves, helped to maintain the morale of other party crew members' wives, and developed the ability to work cooperatively with diverse and constantly changing groups of people. She and Cecil ever after regarded their odyssey while doodlebuggers as one of the happiest and most exciting periods of their sixty years of marriage.

When Cecil, in 1936, became regional supervisor and frequently had to be away from home supervising several crews in addition to his own Party 310, Ida's life changed in several important ways. Semi-permanent residence—in Shreveport, then in Midland, Bakersfield, and finally Los Angeles—meant an end to the chore of moving and establishing a temporary home every few weeks and to strong ties with crew wives that Ida had previously enjoyed. It meant prolonged absences from Cecil and periods of intense loneliness—a new test of Ida's loyalty and self-sufficiency. Ida rose to the challange, involving herself in community social work, attending college classes, and reading omnivorously in an attempt to improve her education. Then came the move to Dallas and the purchase of part-ownership of GSI on the last working day before Pearl Harbor. At long last Cecil and Ida could have their first permanent home.

Always aware of the importance of her support of Cecil in his professional activities, as soon as Ida had furnished and decorated the rooms, she immediately started a program of social activities—luncheons, teas, dinners, parties, and the like—that involved members of the GSI family. As she and Cecil became acquainted with more and more of the employees of GSI and, after 1950, of TI, both the number and size of her parties increased. Not only did she have to look to her wardrobe; she also had to employ cleaning women and kitchen help, and, in the larger affairs, even engage a caterer. Nevertheless, in spite of the large scale of many of her parties, she never failed to prepare Cecil's breakfast of oatmeal, toast and coffee, the menu he still desires before he starts the day.

Ida and Cecil had long since become known as gracious and generous hosts elsewhere, and they soon established the same reputation in Dallas. The result, as during their wanderings, was an ever widening circle of people with whom they could develop life-long friendships. The Greens regard loyal friends as one of their most valued assets, and such friends are to be found throughout the world wherever Cecil and Ida have traveled.

Once they were permanently settled in Dallas, the Greens soon became involved in a number of activities other than those directly concerned with GSI, and later with TI. Ida's interest in the education of young children led her to support Cecil in his efforts to improve the St. Mark's School of Texas, where she was at first an active member of the Mothers' Club, then an Honorary Member. Strongly supportive of more and better education for women, especially in the sciences and in engineering, she became an active member of the Dallas branch of the American Association of University Women (AAUW), serving as vice president of the Dallas branch and as implementation chairman for the study group on "Science: A Creative Discipline." Her activities in the AAUW led her and Cecil many years later to endow the association with a graduate fellowship for specially talented Texas women. Pleased with the success of AAUW's Ida M. Green Fellowship, and hoping to induce more women to prepare for professional careers, Ida and Cecil subsequently endowed a major fellowship program at MIT for outstanding women students starting graduate work at the Institute.

A year after moving to Dallas, Ida entered SMU. Long before the days when universities made provision or offered support service for nontraditional older students, she took courses toward a bachelor's degree in Industrial Psychology from 1942 to 1953. When, however, her responsibilities as the wife of a rapidly rising officer in GSI became unduly demanding of her time and energy, she sacrificed her hopes of completing the 120 semester hours required for an earned degree.[6]

Even though unable to complete those requirements, Ida maintained her relationship with SMU, serving as a member of its Board of Trustees and of the Board's Committee on the School of Continuing Education. Additionally, she joined Cecil in deciding to donate funds for remodeling a building, for endowment of two professorships, and for advancement of science and engineering education.

Ida's interest in leading an active life prompted her to involve herself in the programs of a variety of civic and professional organizations. As Cecil's wife she shared joint membership with

[6]In a June 24, 1987 letter to Cecil, Dr. James E. Brooks, president of SMU's Institute for the Study of Earth and Man, provides the following information about Ida's academic experience at SMU,

Between the fall semester of 1942 and the fall semester of 1953, Ida successfully completed 84 semester credit hours of academic work. In addition, she audited several more courses . . .

There was a heavy concentration in Sociology and Psychology. Additionally, Ida took courses in English Composition and Social Science, Speech, Religion and Biology.

I think it was remarkable that she was able to do this work and all of the other things that you and Ida were involved in in that period of time and maintain a 1.39 out of 3.0 grade point average. On a 4 point scale, that would convert to about a 2.5 out of 4 points—a very respectable average for anyone and certainly for a mature student who is carrying a full-time responsibility in support of a very busy husband.

him in many organizations because she had the same interests. In addition, she became affiliated with a number of exclusively women's groups. Believing, with Cecil, that "one should seek membership and be active in organizations with worthy objectives," she solicited major contributions for the United Fund, helped to raise financial support for the blind and for a home for juvenile delinquents, aided Cecil in raising funds for the Dallas Symphony, and joined him in selecting valuable objects of art for the Dallas Museum of Art. She served on the boards of the League of Women Voters, the Dallas Children's Medical Center, and the Dallas Symphony Association, of which she became an Honorary Member. She combined membership with service in the Dallas College Club (chairwoman of social affairs); Dallas Symphony League; Dallas Children's Medical Center (vice president of its three hospitals; life member of the Women's Auxiliary); Women's Council of Dallas County; Women's Auxiliary of the Dallas County Hospital District (life member); the Texas League; Dallas Council of World Affairs Auxiliary; Dallas Women's Club; Dallas Cadence Club; the Delphians; Women's Auxiliary of the Dallas Geological and Geophysical Society (past president); and Scripps Clinic Cora Baker Aid Service (life member, honorary member; Hospital Auxiliary); and St. Mark's School of Texas (honorary member of Mothers Club).

As a final comment on the part that Ida played in her sixty years of married life with Cecil, she was a companion of incomparable devotion and loyalty, and a steady, supportive, and cooperative partner during every stage of Cecil's entrepreneurial and educational activities. She willingly shared the risk in purchasing part ownership in GSI; then when affluence resulted from the creation and growth of Texas Instruments, she happily became the philanthropist that she once said she would like to be if she and Cecil ever became rich. Finally, her bequests totaling more than $40 million demonstrate her declared desire to help improve the quality of life for all mankind.

5
THE GREENS AS PHILANTHROPISTS
(1950–1988)

INTRODUCTION

Cecil and Ida Green did not become philanthropists by chance; rather their sociability, compassion, and generosity, and the shared desire to improve the lives of people around them made it a strong likelihood that they would develop into philanthropists if they ever acquired the means.

Growing up during the first two decades of this century, the Greens witnessed extraordinary advances toward an easier and more rewarding life that came with the widening availability of electric power and light, the automobile, the radio, and the airplane. They met and married when the nation's energy was directed to rapid production of technology-based devices designed to improve the quality of life. Accustomed to the steadily changing conditions of life in the twentieth century, they were well aware of technology's role in shaping the future and of the importance of an education that would prepare young people for that future.

They spent the first fifteen years of their young adulthood in an unceasing odyssey through a North America that was still essentially rural. Their odyssey took them tens of thousands of miles, as they crisscrossed the continent, establishing temporary residence more than fifty times. Sensitive to others' living conditions, they noted with concern that the poorer people in the villages and small towns along the routes they traveled lacked even the simplest amenities—and, moreover, the education and mobility to obtain them. The Greens developed a genuine compassion for those who were less fortunate than themselves. When their odyssey ended in 1941 with the simultaneous purchase of part ownership in Geophysical Service Inc. and their first permanent home, in Dallas, they began to accumulate the means that would enable them to help improve the lives of others.

Early in their marriage, Cecil and Ida discovered that they shared generally serious attitudes and conservative standards, respectively; liked the same activities—going to movies, reading books, and walking; and found it easy to make and enjoy the company of friends. They soon became known as gracious and generous hosts to a rapidly growing circle of enduring friends. Neither, however, had a suppressed desire, which could now be satisfied, for the usual trappings of affluence—a seaside cottage or a mountain ski lodge, a yacht, a collection of art, or the like. Indeed, for eighteen years after they acquired part ownership in GSI, they were much too busy working for GSI and, later, TI—Cecil traveling, Ida entertaining clients and serving on the committees of charitable organizations—to indulge the usual whims of wealth. In fact, Ida was entirely sincere when asked by Erik Jonsson, after the purchase of GSI, what she wanted to do if part-ownership of the company made Cecil and her rich, she replied, "I want to be a philanthropist."

How best to help the greatest number of people? The Greens could personally make life more pleasant for a few thousand of the country and small-town dwellers they had met or seen during their doodlebugging years, but that would be only drops in a bucket. Better, they felt, to provide and expand educational opportunities for many young people who could later help to improve the lives of thousands, or even millions. Funds to pay for treatment of sickness in its diverse forms could be directly given, but, again, it seemed wiser to encourage and support the training and research of physicians and health-care specialists who, working together, could discover and administer effective methods of alleviating the illnesses of millions.

Since 1950 the philanthropy of Cecil and Ida Green has primarily supported both training and research in traditional education, including medical training, diagnostic research, health care, and civic activities. Their benefactions have been made to schools, at every level; medical institutions; and libraries and museums. Paramount for the Greens has always been the desire to assist organizations and institutions that are explicitly devoted to improving and extending human potential and to making human existence more pleasant and productive. Cecil and Ida customarily begin by making a generous donation to help fund a project that they have chosen to support. Then they follow the way the recipient applies their donation and how completely that donation achieves its purpose. They monitor in part by letter or telephone and in part by occasional visits to the recipients' premises. The visits are especially important because they give the Greens an opportunity to make their own observations, question responsible officials directly, and draw their own conclusions.

Because the Greens usually have spent considerable time, effort, and thought[1] on a project they like before pledging a substantial donation, their monitoring has almost always led to further involvement in the affairs of the recipient. If they liked what they saw being accomplished with their initial donation, they frequently pledged additional funds for the original project or for a related purpose. When this happened, both Cecil and Ida—but more often Cecil alone—were often invited to perform some voluntary service for the recipient such as membership on a board or committee to advise, raise funds, visit departments, plan future developments, or search for new top officials. This, of course, required more involvement.

The Greens' philanthropy has been diverse, timely, extraordinarily generous, widespread geographically, imaginative, innovative, and perceptive—all as a result of the substantial personal effort that they have exerted in selecting the recipients of their donations. Moreover Cecil has made it clear in his speeches that "We give because we want to, not because we are

[1] When the Awards Committee of the Dallas Historical Society chose Cecil and Ida as separate recipients of their 1982 Award for Excellence in Community Service, in the category *Philanthropy,* Cecil gratefully acknowledged receipt of the honor for both himself and Ida, including in his written response of September 22, 1982—the day the award was presented—the following comments:

I would add that intelligent giving is not easy. In fact, Ida and I have discovered that lots of time, effort and thought are necessary to make sure that our giving would have a multiplying effect by "triggering" new and healthy growth in relation to the finally selected project.

All of this adds up to the ultimate and rewarding result—that our giving has been virtually *investments in pleasure and satisfaction.* So, I can truly say that this Dallas Historical Society Award comprises a wonderful dividend—of pleasure—for which we both feel most grateful.

pressured to do so." For the Greens, their benefactions are indeed, as Cecil puts it, "investments in pleasure and satisfaction." The extent to which they have given of both their time and their means has truly earned them the title "philanthropists extraordinary."

Much can be learned about the interests and personal characteristics of Cecil and Ida by observing how they choose the recipients of their philanthropy. Their lively interest in the world and the people around them, their innate generosity, and their delight in the companionship of friends, have brought to their attention a broad range of needs from which they could select those that aroused their deepest interest.

As was to be expected, when some of their earliest major benefactions were publicized, the Greens began to receive far more requests for financial help than they could possibly respond to favorably. The careful method of evaluation that they developed requires a great deal of time and effort. The steady flow of appeals, moreover, has meant that Cecil and Ida, both considerate by nature, have carried on a voluminous correspondence that has often called for diplomacy of rather high order.

RESULTS OF THE GREENS' BENEFACTIONS

The magnitude and diversity of the Greens' philanthropies are demonstrated by the following summary of the results of their benefactions throughout the English-speaking world:

1. *50 Academic, Medical, and Civic Buildings:* of which 20 were wholly or partially funded by and named for them; to the remaining 30, otherwise named, they made large challenging donations.

2. *14 Special Rooms:* (in numerous academic, medical and civic buildings) named for them or their designates because of their donations.

3. *20 Special Instructional and Research Facilities:* in the preceding buildings and rooms, funded by them.

4. *28 Endowed Chairs in 15 Institutions:* distinguished professorships (14), visiting lectureships (5), career development appointments (2), investigatorships (2), and master teacher

appointments (2), all named for them; and 3 similar chairs, honoring others, which they funded entirely (2) or partially (1).

5. *Endowed Awards to Students:* contributions to a scholarship fund at St. Mark's School of Texas (Dallas); Gold Medal, awarded yearly to a senior in geophysics at CSM (1); undergraduate scholarships at Suffolk University (Boston) (3–4 per term); a fellowship to a woman graduate student from Texas, awarded annually through the AAUW Foundation (1); fellowships to beginning women graduate students at MIT (6–8 annually); graduate fellowships in geophysics at Stanford (2–3); graduate fellowships at TCU (5); fellowships and internships to visiting scholars at UTHSCD's SMS (variable number); postdoctoral fellowships in physics at Australia's University of Sydney (2) and in clinical medicine at Oxford University's Cecil H. Green College (1); and support of a Residency Program at SCRF.

In summary, by the end of 1987 at least 32 different professors have enjoyed the honor and privileges of a distinguished chair; more than 300 outstanding scholars have been able to advance their professional careers by limited appointments at half a dozen colleges and medical institutions; more than 150 undergraduate and graduate students have been able to start or continue their college education; and a multitude of individuals have used the spaces and facilities that the Greens have made possible.

The full nature and extent of the Greens' overall philanthropy can be evaluated best if their many individual benefactions are categorized, as in the following list (full titles for acronyms and abbreviations are given in appendix A):

LAND

1. Partial funding of purchase of 1,200 acres for GRCSW-UTD campus. (1961–69)

2. Supplementary funds for purchase of 480 acres for the Cecil and Ida Green Piñon Flat Geophysical Observatory operated by UCSD-IGPP. (1979–80)

3. Gardens surrounding the twenty-four-room mansion—Cecil Green Park at UBC, Vancouver. (1967)

4. Major donation to the Dallas Arboretum and Gardens project. (1984)

BUILDINGS (listed in chronological order of donation)

1. Raising of funds and donations for the following buildings on the campus of St. Mark's School of Texas. (1958–86)

Science and Mathematics Quadrangle

H. Ben Decherd Art Building

Cecil H. and Ida Green Library Study Center

Gymnasium and Natatorium

2. Donations to other secondary schools in Dallas–Forth Worth area:

Trinity Valley School (Fort Worth)—building funds. (1968)

Greenhill School—building funds. (1968–70)

Cistercian Preparatory School—building funds. (1971–84)

Ursuline Academy—building expansion funds. (1972–84)

3. MIT

The twenty-story Cecil H. and Ida F. Green Center for the Earth Sciences. (1959–64)

The Ida Flansburgh Green Hall for Graduate Women—dedicated in 1983.

The Cecil H. Green Center for Physics. (1987)

4. Colorado School of Mines (CSM)

Cecil H. and Ida Green Graduate and Professional Center—a three-story building dedicated in 1972.

5. Southern Methodist University (SMU)

Partially funded the Science Information Center. (1961)

Partially funded the renovation of the McFarlin Memorial Auditorium, so that the Dallas Symphony Orchestra could play there. (1961)

6. Austin College, Sherman, Texas

   A building specially designed for communication and the performing arts—the Ida Green Communication Center, dedicated in 1972.

7. Texas Christian University (TCU)

   A major donation to expand the Mary Couts Burnett Library, dedicated in 1983.

8. University of Texas at Dallas (UTD)

   Donated one-third the cost of UTD's Founders' Building, dedicated in 1964.

9. Bishop College

   Made principal donation for the new Arts Center.

10. University of Dallas

    Made one of three major donations for the Patrick E. Haggerty Science Center, dedicated in 1985.

11. KERA-TV Channel 13, Dallas–Fort Worth

    A donation for the Ralph B. Rogers Administration Building.

12. University of California San Diego (UCSD)

    The principal donation for the Cecil H. and Ida Green Faculty/Community Club Building, scheduled to be completed in 1988.

13. Stanford University

    A donation for the Ruth Wattis Mitchell Building (for the earth sciences, 1970)

The principal donation for expansion of the main library, now named the Cecil H. Green Library, dedicated in 1980.

A 1987 pledge for a new building to be named the Cecil H. and Ida M. Green Earth Sciences Research Building.

14. University of British Columbia (UBC)

Gift of Cecil Green Park—a twenty-four-room mansion and surrounding grounds—to UBC for town-and-gown activities. Officially opened in 1967.

15. Scripps Clinic and Research Foundation in La Jolla (SCRF)

Gift of house and rose garden located at 338 Via del Norte to SCRF in 1964.

Funding of a new health care center on Torrey Pines Mesa, including the Cecil H. and Ida M. Green Hospital of SCRF. (1972–75)

Pledges to Fiftieth Anniversary Fund for building. (1976)

Funds for the Northwest Garden expansion program. (1980)

16. The University of Texas Health Science Center at Dallas (UTHSCD) and its Southwestern Medical School (UTSMSD)

Funding of the new Cecil H. and Ida Green Science Building, dedicated in 1975.

Partial funding of the nine-story Cecil H. and Ida Green Biomedical Research Building, completed in 1986.

17. Oxford University, England

Donations to construction fund for three buildings in the Cecil H. Green College at Oxford University. (1978, 1980)

Funds to renovate the Radcliffe Observatory. (1978)

18. Suffolk University, Boston

A modest donation to help the school acquire additional space for expansion. (1982)

19. TAGER

Provided the funds for the TAGER headquarters building located on a small plot in the northwestern corner of the UTD campus.

20. Dallas Library (J. Erik Jonsson Library)

A donation to help fund construction of the new Dallas Library recently named for Cecil's longtime business partner. (1978–79)

21. The Salvation Army, Dallas

The Building Fund. (1973–74)

The principal donation for construction of the Salvation Army's headquarters building in Dallas. (1984)

22. The YWCA, Dallas

The principal donation for construction of the New Dallas Community YWCA. (1972–75)

An additional donation for construction of the swimming pool, later named the Ida Green Natatorium. Cecil also served as chairman of the fund-raising committee. (1972–75)

23. The YMCA, Dallas

A five-year commitment (1970–74) and a two-year commitment (1980–81) of funds for construction of a new building for the Dallas Metropolitan YMCA.

24. The Goodwill Industries, Dallas

In the late 1960s the Greens donated much-needed funds ($58,849) to help the organization get established in its new building on North Hampton Road. (1967–69)

25. The Dallas Symphony Orchestra

A major donation for the Dallas's new concert hall. (1983)

26. The Dallas Museum of Art

A major challenge donation for construction of new museum building. (1980)

27. The Dallas County Heritage Project

Sizable donations for building reconstruction. (1970s)

28. The Presbyterian North Village Retirement Home, Dallas. (1978–79)

A major donation for the building.

29. The National Academy of Sciences

Funding of the Cecil H. and Ida Green Building for NAS-NRC in Georgetown (Washington, D.C.) dedicated in 1987.

30. Cecil H. and Ida Green Suite in Baylor University Hospital, Dallas.

EDUCATIONAL AND MEDICAL CENTERS

1. Support of the Cecil H. and Ida Green Center for Reproductive Biology Sciences in UTHSCD.

2. A major donation for the Presbyterian Medical Center for Diagnostic Medicine in Dallas. (1982–83)

3. Ida's major 1987 bequest for the Children's Medical Center in Dallas for new space for the hematology/oncology clinic, to be named for Ida. (1987)

4. Ida's major 1987 bequest for the Ida M. Green Cancer Research Center of Scripps Clinic and Research Foundation.

5. Cecil's 1987 donation for MIT's Cecil H. Green Center for Physics.

FACILITIES

1. University of Western Australia, Perth, Australia

   An eighteen-volume set of *Geophysics*. (1954)

2. University of Sydney, Australia

   A KDF9 computer. (1964)

3. St. Mark's School of Texas

   Telescope, planetarium, and three-climate greenhouse.

4. Children's Medical Center, Dallas

   Lung and kidney machines for the Maggie Green Recovery Room. (1968–70)

5. TAGER

   The Cecil H. Green TV Network. (1970–78)

6. UTMBG

   One-half the cost of research vessel (R/V) *IDA GREEN*. (1973)

   Gravimeter for research vessel (R/V) *IDA GREEN*. (1975)

   NMR equipment for Dr. Andrew D. Suttle, Jr. (1983–84)

7. YWCA Dallas

   The Ida Green Natatorium. (1973)

8. UTD

   30,000 books on Central America. (1973)

9. UCSD

   IGPP—Accelerometers for IDA Network. (1975–87)

   SIO—Gear for research vessel (R/V) *WASHINGTON*. (1976)

10. Scripps Clinic

    Equipment for Radiology-Thermoscope. (1971)

    CAT body scanner. (1975)

11. SMU

    A special major donation for the purchase of teaching and research laboratory equipment for the School of Engineering and Applied Science. (1979)

12. Texas Instruments Inc.

    A set of KERA-TV "MD" audio-video tapes. (1980)

13. CSM

    An Allen electronic digital organ. (1980)

14. Austin College

    Communication equipment. (1980)

ENDOWED ACADEMIC POSITIONS

1. SCRF

    Green Associate Chair. (1961)

    Green Investigatorship in Medical Research. (1970)

2. St. Mark's School of Texas

    First Master Teacher. (1977)

    Second Master Teacher. (1983)

3. SMU

    Cecil H. Green Chair in Engineering. (1969)

    Cecil H. and Ida Green Chair in Electrical Engineering. (1979)

4. UBC

    Cecil H. and Ida Green Visiting Lectureships. (1972)

5. MIT

    Green Chair in Electrical Engineering. (1970)

    Robert R. Shrock Chair in Earth and Planetary Sciences. (1970)

    Green Chair in Education. (1974)

    Two Green Chairs in Physics. (1976, 1979)

    Two Green Chairs in Earth and Planetary Sciences. (1976)

    Two Green assistant professorships in Career Development. (1979)

6. TCU

    Cecil H. and Ida Green Honors Chair. (1969)

7. UCSD

   IGPP—Green Visiting Scientists. (1972)

   Partial funding of Harold C. Urey Chair. (1983)

8. UTMBG

   Cecil H. and Ida Green Chair in Marine Sciences. (1972)

   For visiting scientists program. (1974)

9. Stanford

   Green Chair in Geophysics. (1973)

   Ida M. Green Directorship of the Stanford University Libraries. (1987)

10. UTD

    Green Honors Chair in Natural Sciences. (1974)

11. Austin College

    Ida M. and Cecil H. Green Chair of Creative Educational Leadership. (1974)

12. UTHSCD-SMS

    Green Chair in Reproductive Biology Sciences. (1974)

13. CSM

    Charles Henry Green Chair in Exploration Geophysics. (1975)

14. NJCIRM

    Green Research and Investigator Chair in Basic Biological Research. (1975)

15. CIT

   Robert P. Sharp Chair in Geology (donation only). (1978)

   Charles H. Hewitt Dix Lectureship (donation only). (1980)

ENDOWED STUDENT AWARDS, SCHOLARSHIPS, FELLOWSHIPS, INTERNSHIPS, AND RESIDENCIES

1. St. Mark's School of Texas

   Greens' early donations to the Scholarship Fund and Cecil's services as a fund-raiser.

   Ida's 1987 bequest endowing five Ida Green Scholarships.

2. Suffolk University, Boston

   Green Undergraduate Scholarship Fund (3–4 students a term). (1980–88)

3. AAUW-sponsored Ida Green Fellowship for a Texas woman graduate student funded by Cecil. (1968, 1974, 1984)

4. CSM

   Cecil H. Green Gold Medal to a graduating senior in Geophysics. (1957–88)

5. MIT

   Ida Green Fellowships for beginning women graduate students (6–8 per year). (1974–88)

6. UTHSCD

   Green International Scholars' Program. (1975–88)

   Green Health Profession Career Training Program. (1975–88)

7. Stanford

   Green Graduate Student Fellowships in Geophysics (2–3 per year). (1976–88)

8. University of Sydney, Australia

   Harry Messel Research Fellowships in Physics (2 per year). (1979, 1983)

9. Oxford University, England

   The Joan and Richard Doll Fellowship in Green College. (1983–88)

10. SCRF

    Green three-year Residency in Internal Medicine. (1986–88)

11. TCU

    Ida's 1987 bequest endows a graduate fellowship in each of five departments: Chemistry, English, History, Physics, and Psychology.

OBJECTS OF ART, PAINTINGS, AND PHOTOGRAPHS

1. Dallas Museum of Art

   Two marble statues of classical antiquity; numerous smaller objects; Bybee collection of eighteenth-century American furniture.

2. A 66-foot long Mexican tapestry in CSM's Friedhoff Hall. (1975)

3. A miniature scale-model of a clipper ship in the Green Library Study Center of St. Mark's School of Texas.

4. "Spring Stirring," a black diorite sculpture by Donal Hord that stands near the main entrance of UCSD's IGPP laboratory in La Jolla.

5. A generous 1979 donation to the La Jolla Museum of Art.

6. Gittings-Neiman Marcus color portrait of Ida in MIT's Ida Green Room and in the Ida F. Green Hall for Graduate Women.

7. An Antony di Gesu photograph of Cecil and Ida hangs on the hallway wall of SCRF's Green Hospital. (1985)

8. An oil painting of Cecil and Ida by Anthony Wills hangs on the entry wall of SCRF's Green Hospital.

9. Enlarged photographs provided by the Greens hang in a number of other buildings that they funded.

SUPPORT OF SPECIAL PROGRAMS

1. SMU

School of Engineering. (1967–74)

Excellence in Engineering and Applied Science Education. (1981–85)

Music Therapy. (1975)

Special donation to pay off a large University debt. (1974)

2. GRCSW-SCAS-UTD

Excellence in Engineering Education. (1971)

3. MIT

MIT-GSI Student Cooperative Plan. (1951–65)

Excellence in Engineering and Applied Science Education. (1979)

Geophysical Analysis Group (GAG) Program. (1952–57)

4. CSM

Guy T. McBride, Jr., Honors Program in Public Affairs for Engineers. (1984)

Ida Green's 1987 bequest to be used to enhance CSM's program in Geophysics

5. Society of Exploration Geophysicists Education Fund.

6. National Academy of Sciences

Donation for Academia-Business-Government Cooperation. (1983)

Second International Conference on Biological Imaging. (1983)

7. SCRF

First International Conference on Biological Imaging. (1980)

8. UTSMSD

A program on family planning. (1969–70)

9. UTHSCD

Endowment of the Cecil H. and Ida Green Center for Reproductive Biology Sciences Training Program. (1975)

"M.D." the TV medical film on KERA-TV Channel 13, Dallas–Fort Worth. (1975–76)

A generous contribution to the Department of Cardiology by Cecil as a memorial to his longtime friend, A. Earl Cullum, Jr. (1985)

GENERAL OPERATING AND SUSTAINING SUPPORT
(All donations $10,000 or more)

Dallas Public Library (1942–87)

St. Mark's School of Texas (1950–87)

SMU—early general sustentation (1953)

St. Christopher's Church, Dallas (1955–56)

St. Luke's Church, Dallas (1955–58); Building Fund (1960)

Colorado School of Mines (1958–87)

Pilot School for the Blind, Dallas (1960)

Dallas Symphony Orchestra (1963–87)

Dallas Art Association (1965)

Goodwill Industries, Dallas (1967–69)

Scripps Memorial Hospital (1968)

Dallas County Heritage Project (1968)

Lamplighter School, Dallas (1968)

Greenhill School, Dallas (1968–70)

Jesuit College Preparatory School (1968–71)

SMU President's Council (1969–87)

TCU President's Council (1969–87)

Jarvis Christian College, Dallas (1969–71)

Thanksgiving Square, Dallas (1969–72)

TAGER—travel costs to Miami, Florida, etc. (1970, 1979–87)

Cistercian Preparatory College, Dallas (1971–84)

Judith Munk for her International House, San Diego (1971)

Bishop College, Dallas—cash donations (1971)

Trinity Valley School, Fort Worth (1972)

Ursuline Academy, Dallas (1972–84)

Besses O' Th' Barn Brass Band, England (1980)

Episcopal School of Dallas (1982–83)

Billie Press for work with handicapped children (1986)

Fund for Protection of the Peregrine Falcon—a generous donation (1977)

Member/Chairman—Cecil and Ida, together and separately have served on fund-raising committees for many organizations and institutions, including St. Mark's School of Texas, MIT, CSM, TCU, Bishop College, Stanford, UTHSCD, Children's Medical Center in Dallas, Dallas YWCA and YMCA, Dallas

Symphony and Dallas Museum of Art. From April 4-June 21,
1987, Cecil served as Honorary Chairman of the "Dr. Seuss from
THEN to NOW" exhibit in Dallas for the benefit of the Children's Medical Center of Dallas; and for the next few years,
starting on October 22, 1987, he will be Honorary Chairman of
MIT's *Campaign for the future*, which expects to raise $550 million
by 1992. In early 1988 he agreed to be Honorary Chairman of
UBC's fund-raising campaign. Ida's will, prepared in February,
1986, included a section titled *Residuary Bequests*, which dem-
onstrates that she remained to the very last the true philanthropist she desired most to be. The bequests, twenty-five in
number, included the following: Major bequests to:

Massachusetts Institute of Technology

Scripps Clinic and Research Foundation

Trustees of the Green Foundation, Dallas

Texas Christian University

University of British Columbia

St. Mark's School of Texas, Dallas

Leland Stanford, Jr., University

Substantial bequests to:

Episcopal School of Dallas

Austin College

Dallas Society for Crippled Children

The University of Texas at Dallas

The Dallas Museum of Fine Arts

Dallas Symphony Foundation

The Southwestern Medical Foundation, Dallas

Children's Medical Center, Dallas

Dallas Public Library

Colorado School of Mines, for Geophysical Department

La Jolla Foundation (now the Cecil H. and Ida M. Green
Foundation) for Earth Sciences

YMCA of Metropolitan Dallas

The Salvation Army, Dallas

Dallas Lighthouse for the Blind

The Timberlawn Foundation, Inc., Dallas

The Association for Graduate Education and Research (TAGER), Dallas

The Hockaday School, Dallas

The United Way of Metropolitan Dallas

SUMMARY

I have honored the request of the Greens not to indicate the dollar amount of either their joint or separate donations or the amount of Ida's bequests because such information might bring many more personal solicitations—like those of the past—that could not be granted. Instead, excepting those donations of which the dollar amounts have been made public, I have simply described them as principal or major, substantial, generous, or modest. It can now be recorded here, however, that Ida's bequests totaled in excess of $41 million and that, in addition, the joint and separate donations of the Greens before and since Ida's death now total at least $110 million, giving a grand total of more than $150 million that Cecil and Ida have thus far donated for the common good. And I have omitted most donations of less than $10,000, of which there have been many.

6
THE GREENS' SUPPORT OF TRADITIONAL EDUCATION

The Greens' support of traditional education was a natural outgrowth of their own attitudes and early experiences. During her girlhood in upstate New York, Ida became an avid reader, with the encouragement of her grandfather Flansburgh, and looked forward to a college education. But when she was unable to complete high school in Scotia, because of circumstances beyond her control, she concluded that she would have to educate herself through reading. She had not given up hope of further education, however, when she took her first job with General Electric in Schenectady.

Ida quickly demonstrated her intelligence and talent at GE, and had advanced to the responsible position of statistician by the time she met Cecil, but she still lacked a high school diploma. Marriage to Cecil, and the subsequent move to Boston, meant that Ida would have to give up her practical training at GE. In Boston, however, she was able to continue her education at the Brighton High School. By the time Cecil decided to seek a new job elsewhere, she had completed enough courses at Brighton High to qualify for college work. In the early 1940s, soon after she and Cecil became part-owners of Geophysical Service Inc. in Dallas, she began to study for a degree in industrial psychology at Southern Methodist University.

But once again, she had to suspend her studies after a few years, because of her time-consuming responsibilities as a corporate wife. When Cecil became vice president of GSI in 1941, Ida became even more deeply engaged in social affairs with employees of GSI. Afterward the relentless demands on her as the wife of GSI's president prevented Ida from completing her degree, though SMU proudly claims her as a distinguished alumna and has awarded her its highest honors. Her struggle to complete high school and her regret at having been unable to finish college gave Ida a special sympathy for students who have been frustrated in their efforts to continue their education. It is easy to understand why Ida has happily joined Cecil in funding various scholarships and fellowships, especially those for women.

Cecil, for his part, has been able to appreciate the value of education from three different perspectives: as student, teacher, and employer. In his earliest youth, urged on by his mother, he excelled in the rigorous Canadian primary and secondary program of study. His matriculation at MIT came at the price of sacrifices by his parents, who communicated the importance of education in their devotion to Cecil's future. When Cecil completed his formal training as an electrical engineer at MIT and took his first job at GE in Schenectady, he soon realized that he enjoyed teaching people more than designing electrical equipment. In colleagues and, later, in employees, he was quick to note the superior competence of those with advanced training. That he should concentrate on improving the quality of education wherever possible comes as no surprise. Over the last sixty years, he has steadily increased his interest in the whole educational process.

As early as the 1950s, Cecil and Ida took an interest in the primary and secondary schools of the Dallas community. Since then they have broadened that interest to include colleges and universities throughout the world—in Australia, Canada, and England, as well as in the United States. Their benefactions have gone to more than twenty-five different educational institutions. Varying in nature and amount, they have always been given to meet some critical need of the recipient.

The Greens began to enjoy some affluence as World War II ended and GSI's field operations increased under Cecil's leadership, first as vice president (1941–1950) and then as president (1950–1955). But the Greens could not think seriously about any major benefactions until several years after Cecil had advanced from president to board chairman, and the four owners of GSI—Green, Jonsson, McDermott, and Peacock—had decided to incorporate the manufacturing division of the company as Texas Instruments (TI) in 1951. The rise in the price of TI's common stock after it was first listed on the New York Stock Exchange on October 1, 1953, brought substantial wealth to each of the four founders, and by 1960 Cecil and Ida could consider large-scale philanthropy. Even as early as 1950, however, the Greens were already making modest donations to a few of the primary and secondary schools of the Dallas–Fort Worth area.

THE GREENS AND ST. MARK'S SCHOOL OF TEXAS

Cecil's interest in primary and secondary education was first aroused in the late 1940s following a 1943 fire at the Texas Country Day School, a small private preparatory school in Dallas in which Eugene McDermott had been involved since its founding in 1933. The school's main building, Davis Hall, was destroyed by the fire, and McDermott was engaged in raising funds for its replacement when Cecil first learned about the school from him.

At the same time, Cecil and McDermott, together with J. Erik Jonsson and numerous other Dallas businessmen, were deeply concerned about where they were going to find the scientists and engineers with doctorates who would surely be needed for the high-technology industries rapidly developing in the Dallas–Fort Worth area. One day while discussing the paucity of doctoral graduates being produced in the colleges and universities of North Texas, and what might be done about the problem, McDermott and Cecil suddenly asked themselves,

Hey! Haven't we been overlooking an important aspect of this "Ph.D.-gap" problem? Just what kind of education are the kids now in our public and private schools getting to prepare them to start work on a doctoral program ten or fifteen years from now? Maybe we'd better look into that problem too!

Cecil's questions about the private school in which McDermott was involved drew the response, "Why don't you come by sometime and see for yourself?" As Cecil tells it, "I did visit the school soon after and was hooked!"

Cecil visited the Texas Country Day School in 1950, the year when it merged with the Cathedral School of Dallas to establish the new St. Mark's School of Texas. Soon he and McDermott were collaborating in developing a program based on the principle that:

Since we can't do much with the public school system, why not build up St. Mark's to match some of those well-known prep schools in the Northeast, like Andover, Exeter and others?

Here was the kind of challenge that excited Cecil's entrepreneurial instincts. He and McDermott would build St. Mark's into one of the best prep schools in the nation, and he would make this a personal not just a financial investment in the school's future. McDermott didn't have to press his friend for financial support. On the contrary, Cecil no doubt took McDermott aback when he outlined a bold plan for an expanded St. Mark's. The school is too small, with only about 250 boys (in grades 1–12), suggested Cecil. Ultimately there should be around 650 to 700 boys, to gain optimum size and diversity, and a diverse faculty to match. New buildings would be needed as well as the most sophisticated instructional facilities and equipment. The faculty should design a course of study that would challenge each boy and at the same time give him the freedom to develop his special talents and interests to the fullest extent. To achieve these goals, millions would have to be raised from private donors. In the far-reaching and imaginative plan he outlined to McDermott, and to the Board of Trustees later, can be seen an early example of Cecil's outstanding ability as an educational statesman to persuade people with diverse opinions and interests to collaborate in a common enterprise.

In the mid-1950s the Greens helped to start a faculty salary endowment and a scholarship endowment. Cecil shared equally with McDermott the cost of the Science and Mathematics Quadrangle, dedicated in 1961 (figure 17). He early became a member of the Board of Trustees, and served as president of the board from 1964 to 1966. As president he joined with his fellow trustees and Headmaster Christopher Berrisford to formulate a far-sighted Plan for Development for the school which resulted in growth and expansion that even today, twenty years later, is not yet complete. During 1965, while board president, Cecil served as chairman of the successful Building Fund Drive, which made possible the major improvements envisioned in the board's development plan. Davis Hall was soon rebuilt, and in 1967 the Library Study Center, the Gymnasium, and the Natatorium were completed with money from the fund drive, much of it donated by the Greens themselves.

In the early 1970s, Cecil and Ida pledged a challenge grant that made possible construction of the H. Ben Decherd Art Building (figure 18), named in honor of another strong sup-

17   The Science and Mathematics Quadrangle of St. Mark's School of Texas, in Dallas, was started in 1958 and completed and dedicated in 1964. The Science Wing of the $1,250,000 quadrangle was initially funded by the Greens and McDermotts and was dedicated in 1958. Other donors, together with the Greens and McDermotts, later provided funds for additional wings and special facilities. Today the quadrangle is enclosed by a one-story, fire-resistant, air-conditioned brick complex consisting of special classrooms and laboratories for the sciences (astronomy, biology, chemistry, geology, and physics) and mathematics; a planetarium equipped with a $13,000 Spitz instrument and an adjacent observatory with a 12½-inch reflector telescope; a closed-circuit television system that includes a studio from which scientific experiments can be broadcast to every wing in the quadrangle; a spacious auditorium; and a climate-controlled three-zone greenhouse. Representatives of national organizations have characterized the building and its equipment as one of the best teaching facilities to be found in any preparatory school in the country. At the dedication of the Science Wing in 1958, Cecil remarked:

We look upon the quadrangle not as a gift but as an investment. The purpose, beyond that of bringing out the best in the boys and the teachers, is not to make mathematicians and scientists out of all the boys. We want to make sure they have a well-balanced education according to tomorrow's standards. [Quoted from the *Dallas Times-Herald,* November 14, 1961.]

Photo courtesy of St. Mark's School of Texas.

18   The H. Ben Decherd Center and the Green Library Study Center
at St. Mark's School of Texas. *Top:* The Decherd Center for the Arts
was partially funded by the Greens, together with other donors. The
Center contains 17,000 square feet and provides space for the studio
arts, ceramics, and theater workshops, a boys' choir, and a 400-seat
performance hall for concerts. *Bottom:* The Cecil and Ida Green Library Study Center was named in honor of the Greens in 1981, in recognition of "their benevolence and wise guidance." Photos courtesy of
St. Mark's School of Texas.

porter of St. Mark's. In 1977 they endowed a Cecil H. and Ida Green Master Teacher Chair—to which Evelyn Boone, a lower-school teacher, was appointed.

In the mid-1980s the Greens pledged a major donation to St. Mark's Fifty-Year Campaign, initiated to raise $6 million of endowment funds by the school's fiftieth anniversary in 1983. A portion of the Greens' challenge grant was designated to endow a second master teacher chair named in their honor like the first one, and another portion was to be added to the endowment for the financial aid program. These major donations have been critical to the development of the present-day St. Mark's. In addition to these, the Greens have made generous donations for teaching equipment (a planetarium, an observatory, and a three-zone greenhouse); for staff travel and visiting lecturers; for trees to improve the grounds; and for decorative and educational objects to make interiors more interesting (such as a clipper ship model in the library). Moreover they have worked tirelessly on behalf of the school. While Cecil served as a trustee, a fund-raiser, and a general advisor, Ida participated in the activities of the Mothers' Club, of which she was made an honorary member.

Proof that the Greens have endeared themselves to students and staff alike is eloquently recorded on the plaque that hangs in one of the major buildings they helped to fund, the Cecil H. and Ida Green Library Study Center (figure 18).

> For Their Loving Concern for the Students
> of St. Mark's School of Texas, Their
> Benevolence, and Their Wise Guidance
> this Library Study Center Is Gratefully
> Dedicated to
> Cecil and Ida Green
> November 3, 1981

THE GREENS AND OTHER PREPARATORY SCHOOLS OF TEXAS

Although Cecil and Ida have made their most substantial contributions of funds and personal effort to St. Mark's School of

Texas, they have not neglected other preparatory schools in the general Dallas–Fort Worth area. When they learned in 1968 that the Trinity Valley School in Fort Worth needed money to move to a better location, they contributed generously to finance the purchase of land and construction of new buildings. In Dallas they made modest donations to the Pilot School for the Blind (1960) and to the elementary (K-4) Lamplighter School (1968); and contributed substantially to building and maintaining the Greenhill School (K-12; 1968–70), Cistercian Preparatory School (1971–84), Ursuline Academy (1972–84), and Episcopal School (1982–83). Most recently, Ida left a generous bequest to the Hockaday School.

THE GREENS AND MASSACHUSETTS INSTITUTE OF TECHNOLOGY (MIT)

CECIL RETURNS TO MIT TWENTY-FIVE YEARS AFTER GRADUATION, AND A COOPERATIVE EDUCATIONAL PROGRAM IS BORN

After receiving his Bachelor and Master of Science degrees in Electrical Engineering at MIT's commencement in the spring of 1924 (as a member of the Class of 1923), Cecil did not visit the Institute again until the fall of 1950, more than twenty-five years after graduation. Nor did he keep in touch with the Alumni Association during that lapse of a quarter century. In 1950, however, Cecil and MIT once again crossed paths. In April of that year I first met Cecil and Ida Green in Chicago's Palmer House as fellow dinner guests of a mutual friend, William (Bill) Sherry, a Tulsa oilman and an MIT graduate (figure 19). Cecil discussed his difficulty in finding well-trained, interested recruits for GSI, and I suggested that he visit MIT when he was next in the Boston area.

In the late fall of 1950, Cecil came to my office in the geology building—I was then head of the department—and introduced himself as "Cecil Green of GSI," recalling our meeting in Chicago's Palmer House earlier in the year. He remarked that he had just visited his own department, Electrical Engineering, in

19   The author meets Cecil and Ida Green for the first time, as a guest of William J. Sherry, Oklahoma oilman, at a dinner in the Empire Room of Chicago's Palmer House on April 25, 1950. Guests (*from left to right*) are Mr. and Mrs. Neal Smith, Cecil H. Green, host William Sherry, Ida M. Green, and the author. Later that year Cecil and I organized the GSI Student Cooperative Plan. Photo by Palmer House photographer.

the hope of recruiting some upcoming graduates. Not a chance! he was rather brusquely told; several companies had been trying to hire the entire contingent of seniors. A bit chagrined at receiving such short shrift from his own department, Cecil had decided to extend his recruiting efforts to the geology department. Harking back to our dinner conversation in Chicago, he repeated that as the recently elected president of GSI, he felt that one of his major responsibilities was to recruit promising young scientists and engineers for his rapidly expanding company. He had decided to start recruiting at the institution, and the department, where he himself had been trained. It did not take us long to discover that we had a mutual interest. Cecil was looking for well-trained undergraduates to work as summer replacements for regular field-party crew members who wished to take vacation leaves, and hoped that some of his new recruits might be interested in later employment in the geophysical exploration industry. My problem, on the other hand, as head of the geology department, was to find summer employment for the fourteen geophysics majors who needed to earn some money while getting practical field experience. We quickly reached an understanding: Cecil would offer summer employment to every member of our junior class of geophysics majors, and I would encourage the students to accept his offer.

In our three-hour discussion that day, Cecil recalled the excellent training that he had received in the Cooperative Course in Electrical Engineering (Course VI-A) at MIT, and finally suggested that we organize a cooperative program of our own for the following summer (1951). From this evolved the GSI Student Cooperative Plan, which Cecil initiated in the summer of 1951, and directed for the next fifteen summers. He then turned over supervision of the Plan to one of GSI's vice presidents, Kenneth E. Burg, who continued the program for an additional two years until 1967, when it was replaced by a differently oriented training program more attuned to recent advancements in data processing.

This unique program had lasted seventeen years and had trained a total of 363 students from eighty-seven different colleges and universities in the United States and Canada.[1] The GSI Co-op Plan clearly accomplished its principal purpose as stated by Cecil on many occasions,

We want to give the student the opportunity to decide if he really wants to be a professional geophysicist or geologist; we are looking for the "naturals"—those who quickly discover that they like the kind of fieldwork done by exploration geophysicists.

Although not the first, this was by far the largest and most ambitious of the numerous cooperative and collaborative pro-

[1]The interested reader is referred to my booklet: *The G.S.I. Student Cooperative Plan: The First Fifteen Summer Programs, 1951–1965*, Dallas, Tex. (Published by Geophysical Service, Inc., 1966, 143 pp.)

grams, conferences, symposia, and the like that Cecil has organized and directed since his days as supervisor of GSI party chiefs. It illustrates well his innovative flair, and his strong belief in the importance of collaborative enterprises in training young scientists and engineers and in advancing technology.

CECIL BECOMES A MEMBER OF GEOLOGY'S VISITING COMMITTEE (1951) AND OF THE MIT CORPORATION (1958)

The success of the first summer of the GSI Co-op Plan heightened Cecil's interest in the curriculum of MIT's Department of Geology and in the quality of its geophysics majors. Consequently, when he was invited to become a member of the MIT Corporation's visiting committee for the department (renamed Geology and Geophysics in 1952), he immediately accepted and joined the committee for the school year (SY) 1951–52. Since then he has served continuously, except for SY 1968–69 when he withdrew temporarily to serve as president of the MIT Alumni Association. He was elected a term member of the Corporation in 1958; became a Life Member in 1961; and finally a Life Member Emeritus on his seventy-fifth birthday, August 6, 1975.

Service on consecutive MIT Geology and Geophysics visiting committees—on many as chairman—and as a Corporation member, gave Cecil repeated opportunities to learn about the department's most pressing needs—particularly for additional space and faculty members, and for more student aid. In due course Cecil and Ida responded generously to those needs, and they even motivated their longtime friends and business associates, Eugene and Margaret McDermott, to make their own contributions to some of the major needs of the Institute.

THE FIRST SUPPORTING FUNDS

The Greens' first response to needs of the department came in 1952, the year after initiation of the GSI Student Cooperative Plan. It was a generous contribution to be used as I saw fit to help students and faculty members meet small, unexpected expenses that could not be covered under the regular budget. Every departmental chairman who has been called upon for such funds will quickly understand my gratitude to the Greens and the appreciation of those who received help. From 1952 through 1960, Cecil arranged to have annual contributions of $1,000 to $2,000 sent to the department for my use in strengthening the geophysics and geology programs. The ultimate total of $13,000 was of inestimable assistance in meeting critical needs of the kind just mentioned.

In 1954, and again in 1956, Cecil gave the Institute 1,000 shares of Texas Instruments Inc. common stock valued at slightly more than $12 per share. These two gifts were saved and used later to create and endow the Cecil and Ida Green Fund which has provided a substantial portion of the costs of the daily afternoon tea and coffee social hour that has helped to bring together the students and faculty of all the branches of earth science.

THE CECIL H. AND IDA F. GREEN CENTER FOR THE EARTH SCIENCES

On April 12, 1959, the MIT earth scientists were thrilled by an announcement from President J. A. Stratton's office that,

. . . the Institute has received from Mr. and Mrs. Cecil H. Green of Dallas, Texas, a gift of $2,527,500. This represents its April [1959] market value of 30,000 shares of stock of Texas Instruments Incorporated in which form the gift was made. It will be used for the creation of a Center for Earth Sciences, a major building on the M.I.T. campus to house vitally important laboratories for work in geophysics, meteorology, oceanography and related fields.

Stratton went on to state,

We at M.I.T. feel that the earth sciences represent one of the nation's greatest technological challenges, comparable in importance in the next decade to the exploration of space, the search for new materials, and the communications. . . .

The gift . . . will enable M.I.T. to build a multi-story Center which will house the laboratories on its campus now actively exploring the physical environment. Geologists, chemists, physicists, meteorologists, and oceanographers will now be able to work side by side in a basic and applied scientific program which will have, I am certain, the greatest impact on our economy and society as a whole.

The announcement also included Cecil's reasons for their gift, as follows:

Within the all-encompassing field of science itself, the importance of the earth sciences has been increasing with almost explosive force. For our country to maintain leadership in these areas in competition with other nations, or even to keep abreast in the race for new knowledge, we believe it imperative that greater education and research effort be devoted [to] these fields, and that [that effort be directed] where it can bear fruit most quickly. [slightly revised by Cecil]

The events leading up to the Greens' decision to offer MIT the funds to construct a new building for the earth sciences are recorded in detail in chapter 4 ("The Green Building—How It Came to Be") of Volume 2 of my *Geology at M.I.T. 1865–1965*. Their intentions were first made known to senior MIT administrators on February 28, 1959, when the Greens hosted a dinner at the Ritz-Carlton Hotel in Boston for MIT President J. A. Stratton and Institute Secretary R. M. Kimball. In the course of the evening conversation, Cecil asked Stratton if he and Ida made the Institute a donation of about two and a half million dollars, what he would do with it. To this Stratton replied that he would assign it to the Department of Geology and Geophysics for a new building. According to Cecil, this evidence of President Stratton's full support of the earth sciences was exactly what he and Ida wanted to hear. Under such agreeable conditions the first major donation to MIT was thus pledged by Cecil and Ida; twelve more would come during the next twenty-five years.

On December 12, 1961, after almost two years of planning, construction began on an air-conditioned twenty-story concrete tower, designed by architect I. M. Pei. It was the first high-rise building to be constructed in Cambridge, and was officially completed on June 1, 1964 (figure 20). When the total cost of the

20   The twenty-story Cecil H. and Ida F. Green Building for Earth
Sciences at MIT houses the Department of Earth, Atmospheric, and
Planetary Sciences. The 277-foot-high, 128,910-square-foot tower,
dedicated October 2, 1964, was the first college building completely
funded by the Greens. It includes the Lindgren Library, with space for
50,000 books and maps; the 294-seat McDermott Hall; and the Ida
Green Room, a faculty-student lounge dedicated in 1974. Alexander
Calder's imposing stabile, *The Big Sail*, commissioned by the Mc-
Dermotts, stands in the court that bears their name. Photo courtesy of
The MIT Museum.

building soared to double the original estimate, the Greens generously agreed to increase their original gift, and even added a substantial fund for maintenance. It has long been characteristic of them that when they decide to support a particular project, they want it to be done right, and so it was with the funding of their first major college building.

An example of this attitude arose during the planning stage when Cecil, noting that there was no provision for air-conditioning, asserted that, "No modern building, like the one being planned here, should be without air-conditioning." He funded the additional cost.

The Greens had the same reaction when additions to the earliest plans called for a large lecture hall, a spacious library, and a multipurpose conference room for faculty-student use (later to be designated the the Ida Green Room). In short, because they regarded their gift as an investment in pleasure, they wanted to be assured that the building would bring pleasure to everyone who used it.

As an example of their generosity and of their regard for their longtime friends, Eugene and Margaret McDermott, the Greens suggested that the large lecture hall in the new building be named McDermott Hall, and they were deeply pleased when Corporation Chairman James R. Killian, Jr., announced the naming of the hall at the dedication of the Green Building on October 2, 1964 (figure 21).

REACTION OF THE MCDERMOTTS TO THE GREENS' PHILANTHROPY

It has often been said that the philanthropy of one person may induce that person's close friends to consider making similar gifts. It has been thus with the philanthropy of Cecil and Ida Green. Cecil once told me that when McDermott heard about the twenty-story tower that their gift would fund, he remarked, "Such a big building will need a lot of good students; maybe *we* should do something about that!" Soon thereafter, in May 1960, came the first of the gifts of the McDermotts, the endowment that established the Eugene and Margaret McDermott Scholar-

21 Dedication of MIT's Green Building for the Earth Sciences, October 2, 1964. *Top left:* Cecil and Ida Green, in step as usual, in front of their building on dedication day. *Top right:* Statement at west end of building. *Bottom:* Cecil speaking at the dedication. Dignitaries sharing the platform (*seated from left to right*): James R. Killian, Jr., chairman, MIT Corporation; R. R. Shrock, chairman, Department of Geology and Geophysics; I. M. Pei, the building's architect; E. A. Crane, mayor of Cambridge; and Ida F. Green. Photos courtesy of The MIT Museum.

ship Fund, to assist undergraduate students from the Southwest to attend MIT.

Although he was not a graduate of the Institute, Eugene McDermott believed that MIT was one of the greatest educational institutions in the world. He also believed that the Southwest offered the best career opportunities in the United States for young people, once they had thorough training at the college level. At the time they established their Scholarship Fund with a munificent endowment of $1,218,000, the McDermotts said:

We are motivated in making this gift by a desire to give young men and women of Texas and the Southwest opportunities to receive an education of the highest standards, together with a broadening of experience and perspective, such as attendance at MIT affords. Upon returning after graduation to the Southwest, either as educators or participants in industry, they thus will contribute to the enrichment of this region's economic and cultural life.

The income from the Fund, the largest single gift for scholarship aid ever made to MIT, now annually assists some one hundred undergraduate students from Texas and the Southwest. Thus far, through June 1988, more than 700 McDermott Scholars have attended MIT.

Soon after the endowment of the scholarships came the announcement that the McDermotts would provide the funds to commission Eugene's Stevens Tech schoolmate, famed sculptor Alexander Calder, to create a suitable stabile (*The Big Sail*) for the area fronting the Green Building. The announcement further stated that their donation would cover the cost of erecting the stabile and beautifying the area around it, now named McDermott Court in their honor (See figure 20).

The continued interest of the McDermott family and friends in art at MIT is shown by a recent (1976) contribution to the gift of Henry Moore's bronze sculpture—*Three Piece Reclining Figure, Draped*—that graces Killian Court, and endowment support of the Eugene McDermott Award, initially established by the Council for the Arts at MIT and presented annually for "major contributions to the Arts as a means of fulfillment." More recently has come endowment of the Eugene McDermott professorship in the Brain Sciences and Human Behavior (1980), and still more recently, generous donations to the Gyorgy Kepes Fellowship Prize, to the President's Discretionary Fund, and toward construction of the Wiesner Building.

The Green Building at MIT was the first major academic structure outside Texas that the Greens completely financed. The only other college building they had provided for—together with McDermott and Jonsson—was the Founders Building on the newly created campus of the University of Texas at Dallas; the events leading to the opening of UTD are discussed farther on in this section.

Gratified by the instant benefits that the Green Building had for the development of MIT's Department of Geology and Geophysics, and increasingly aware of the Institute's great need for endowed chairs to attract the finest possible faculty in engineering and science, the Greens next turned their philanthropic attention from bricks and mortar to endowed professorships.

DISTINGUISHED CHAIRS ENDOWED BY CECIL AND IDA GREEN

The importance of endowed chairs was clearly described by Hans J. Hillerbrand, provost of Southern Methodist University, in the introduction of an 1983 in-house pamphlet titled *Endowed and Distinguished Professorships:*

Like all philanthropy, the endowment of a professional chair is an act of courage and faith. For what is implied is that the academic discipline is worthy and important, that the institution will chose the incumbent wisely, and the chairholder, in turn, will fulfill the University's expectations for excellent teaching and scholarship.

Endowed professorships affirm that a university has no greater asset than its faculty . . . the scholars who perpetuate the accumulated knowledge of mankind and who create new knowledge. In so doing, they serve both the institution with which they are affiliated and the society of which they are a part. Without them, the young would not be educated and society would be impoverished.

Once they have decided on a course of action, the Greens characteristically do things in a big way. The magnitude and diversity of their generosity has nowhere found fuller expression than in their endowments of distinguished chairs. To the institutions receiving their benefactions, they have emphasized that they share Hillerbrand's view of the importance and expectations of endowed professorships.

The funding of endowed chairs, starting in 1969, marked a conspicuous change in the direction of the Greens' support of higher education. Previously, they had provided generous funding primarily for buildings and facilities, such as UTD's campus and Founders Building (1951), MIT's Earth Sciences Building (1964), TAGER (1965), and UBC's Cecil Green Park (1967). Since January 1970, they have steadily added endowed chairs to their other projects, with the result that by 1988 they will have endowed thirty-one such chairs, which together have received a greater portion of the Greens' total philanthropy than any other category of educational support.

At MIT, Cecil's American alma mater, two chairs were announced in January 1970. The Cecil H. Green Professorship in Electrical Engineering carried the following citation:

The purpose of the chair is to help individual members of the electrical engineering and computer science faculty move into new areas of research. Appointment to the chair is for two to three years.

The second, the Robert R. Shrock Professorship in Earth and Planetary Sciences, was

. . . established in 1970 . . . to honor Professor Shrock, former head of the Department of Earth and Planetary Sciences, who retired that same year.

In 1974, the Cecil and Ida Green Professorship in Education was established. Generally carrying a three-year appointment, it

is intended to promote innovation in education and to encourage study and research in areas related to the techniques of learning and teaching.

Additional distinguished chairs were established at MIT in 1976 when the Greens endowed two Cecil and Ida Green Professorships in Earth Sciences, to strengthen further the Department of Earth and Planetary Sciences, and for a Cecil and Ida Green Professorship in Physics, the fundamental science on which Cecil's professional career has been based.

In 1979, the Greens endowed a second Cecil and Ida Green Professorship in Physics, and simultaneously established a Fund for Excellence in Engineering and Applied Science Education. In the same year they provided most generously for two typically innovative positions—Endowed Career Development Professorships—appointments which can be awarded to unusually promising young scientists and engineers embarking upon an academic career of their choice, \

the purpose of the appointments being to encourage some of the most promising MIT doctorates to prepare for an academic career rather than for one in business, industry, or government service.

The latter two positions clearly speak for the Greens' continuing concern about student education and career development.

In all, Cecil and Ida have created nine endowed chairs at MIT alone—an extraordinary expression of the Greens' trust in the excellence of MIT, and their commitment to fostering excellence by their philanthropy.

THE IDA GREEN FELLOWSHIPS FOR MIT WOMEN GRADUATE STUDENTS

Knowing Ida's long-standing interest in furthering educational opportunities for women in science, engineering, and related fields, Cecil gave money to the American Association of University Women in 1968 to establish the first Ida Green Fellowship. The Greens' satisfaction with this AAUW-administered award, and especially Ida's desire to see outstanding women graduate students encouraged to attend MIT, led her and Cecil to establish at the Institute a million dollar endowment fund annually providing financial support for six to eight first-year MIT women graduate students who are designated Ida Green Fellows.

The 1986–87 record shows that a total of ninety-four women, from more than fifty undergraduate institutions, have won these coveted one-year fellowships since the program began in 1974. More is written about them in the later section, headed "Awards to Students." The students' names are listed in appendix G.

THE IDA GREEN ROOM IN THE GREEN BUILDING
In view of the magnitude of the endowment provided by the Greens for the Ida Green Fellowship program, it seemed altogether appropriate to honor Ida further for her longtime dedication to women students' success. Income from the Cecil and Ida Green Fund, endowed by two gifts of TI stock in the mid-1950s, has been used to help meet the expenses of the daily tea and coffee hour held in the ninth floor conference room of the Green Building since the building was completed in 1964. Cecil enthusiastically approved the suggestion that the room in which students and staff members gathered to socialize be named for Ida. It took vigorous persuasion to win Ida's consent, however. Greatly pleased over the naming of the fellowship program in her honor, she didn't think she needed further recognition from MIT. Nevertheless, the name was approved by the Corporation, and the room was formally dedicated on December 4, 1974, as the last event of the two-day conference celebrating the tenth anniversary of the dedication of the Green Building. The Ida Green Room (figure 22) has long since taken an honored place among the memorial rooms of MIT. The dedicatory plaque reads:

She has shared equally and creatively with her husband, Cecil, Class of 1923, in providing extraordinary benefactions for improving education for men and women at M.I.T.

THE IDA FLANSBURGH GREEN HALL FOR GRADUATE WOMEN—NAMING AND DEDICATION
On the warm, sunny Friday afternoon of June 10, 1983, several hundred students, faculty, and friends of the Institute gathered in the dining room of McCormick Hall to attend the ceremony

22   MIT's Ida Green Room. *Top:* Ida F. Green. Gittings photo (Nei-man-Marcus). *Bottom:* The Ida Green Room (54-915) at the time of dedication, December 4, 1974. Photo courtesy of The MIT Museum: Russell Clarke.

dedicating the next-door building (at 350 Memorial Drive) as the Ida Flansburgh Green Hall (figure 23).

This architecturally distinctive eighty-year-old brick-and-stone building, which had served successively as a residence and a hospital under several different organizations, the last being MIT, had been completely renovated and altered to become a well-appointed dormitory for MIT women graduate students. The story of Ida Green's life and accomplishments, and of her longtime devotion to the task of improving women's lot in the modern world, is set forth in a booklet specially prepared for the dedication by Institute Secretary Vincent A. Fulmer.

The ceremonies closed with a glowing tribute to Ida from MIT President Paul E. Gray, and as the applause subsided, Cecil was heard to remark, "Now I know how Mr. Thatcher must feel!"

SOME OTHER BENEFACTIONS
In addition to their large-scale donations to MIT, the Greens have made other smaller, but nevertheless much-appreciated, gifts to the Institute, including the two Gittings-Neiman Marcus color photographs of Ida that hang in the Hall and in the Room bearing her name, funds provided for refurnishing the Ida Green Room in 1974, and modest donations to establishment of the Vannevar Bush Room and for designation of a numbered seat in recently refurnished Lecture Room 10-250.

Then finally, in late 1987, when the typescript of this biography was undergoing final revision, the MIT Corporation issued two stirring announcements: the first was that the Institute had received a major bequest from the estate of the late Ida M. Green, who died the day after Christmas 1986. The second announcement was that Cecil had agreed to be the Honorary Chairman of MIT's *Campaign for the Future* and had made one of the major personal pledges toward the Campaign goal of $550 million by June 1992. Cecil's commitment is to be used for the renovation of Building 16 (Dorrance), which after renovation will house "The Green Center for Physics, named for Cecil H. and Ida F. Green, MIT's great friends and benefactors."

MIT has recognized both Cecil and Ida by making them Hon-

23   Ida Flansburgh Green Hall. *Top:* The Ida Green Hall for MIT
women graduate students was dedicated on June 10, 1983. *Bottom:* Ida
F. Green on dedication day. Photos courtesy of The MIT Museum.

orary Members of the Corporation, by electing Ida an Honorary Alumna, and by bestowing on Cecil in 1973 its highest and rarest award, Honorary Lecturer, thus far given only to Britain's former Prime Minister Winston Churchill, and to Eugene McDermott, like Cecil a co-founder of TI and one of MIT's generous benefactors.

THE GREENS AND COLORADO SCHOOL OF MINES (CSM)

Cecil's interest in the Colorado School of Mines was first aroused in his early college years at UBC when he became acquainted with a Mines graduate while on a summer job at a copper smelter in northern British Columbia. The graduate, a Mr. Speight, who superintended the smelter, made the profession of mining engineering sound most exciting and CSM an excellent place to go, if he, Cecil, should leave UBC for further training in engineering. Although Cecil ultimately followed his first interest, electrical engineering, and chose MIT to complete his training, he never forgot his good impression of CSM as an engineering school. So when he became president of GSI and concluded that he must actively recruit promising young college graduates for his expanding company, Mines was one of the first colleges that he visited.

John Hollister had just succeeded C. A. Heiland, who had founded the Department of Geophysics, and during several visits in the 1950s, Cecil and John struck up a warm friendship that has sustained Cecil's and Ida's interest in CSM. Having assured himself that CSM graduates would be excellent future employees for GSI, Cecil invited three students to participate in the second summer program of the GSI Student Cooperative Plan in 1952. Thereafter, Mines students participated regularly in successive summers, and a total of twenty-four participants (second only to MIT's sixty) had attended through 1965, by the end of the fifteenth session and of Cecil's term as director.

THE CECIL H. GREEN GOLD MEDAL

Cecil further signified the value he placed on thorough training and his confidence in the quality of CSM's graduates in geophysics by providing for a solid gold medal to be given annually to "the graduating senior who has attained the highest rating in the combination of scholastic achievement, personality and integrity." Recipients are chosen by the Mines geophysics faculty and administration.

The Cecil H. Green Gold Medal was first awarded in 1957. Up to 1987 thirty-one medals had been awarded—convincing evidence that Cecil continues to encourage excellence in academic performance coupled with development of personal character and integrity, and that he himself remains devoted to the advancement of training in exploration geophysics. The names of the recipients, and of their home towns, are listed in appendix G (see figure 24).

THE CECIL H. AND IDA GREEN GRADUATE AND PROFESSIONAL CENTER

Repeated visits to CSM in the later 1950s and the 1960s, during which Cecil became better acquainted with the professors and students in the geosciences, convinced him that the school was conducting the kind of training program that he strongly favored—one based on practical instruction in laboratory and fieldwork and oriented toward professional careers in industry.

Predictably, when Mines' President Orlo E. Childs, with whom Cecil had become acquainted meanwhile, brought to his and Ida's attention the major needs of the school, and sought their financial support, they responded generously. Their large initial pledge stimulated other individuals, as well as foundations, corporations, and the state of Colorado, to complete the funding for a building that could serve both the school and the town of Golden as a common center. Then when Cecil learned that the completed funding target had not included the cost of equipping and installing the specialized laboratories planned for the interdisciplinary geoscience research area in the basement, he commented, "It doesn't make sense to leave the basement unfinished, particularly since that is the area where much of the future research in the geosciences is to be done." Accordingly he and Ida pledged the additional funds for the desired basement laboratories—a substantial contribution.

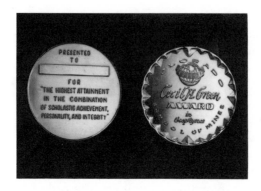

24   The Greens and Colorado School of Mines. *Top left:* The Cecil H.
Green Gold Medal. *Top right:* Cecil's father (ca. 1915), in whose honor
he endowed CSM's Charles Henry Green Professorship in Exploration
Geophysics. *Bottom:* The Cecil H. and Ida Green Graduate and Professional Center, jointly funded by the Greens and the State of Colorado,
was dedicated on April 17, 1972. The three-story, 157,000-square-foot
building houses the Department of Geophysics and the Department
of Mineral Economics, as well as conference, food service, and auditorium facilities. A magnificent tapestry and a modern Allen organ, also
donated by the Greens, make the building more pleasant for town-
and-gown activities. Top left and bottom photos courtesy of CSM; top
right photo provided by Cecil.

Today, the Cecil H. and Ida Green Graduate and Professional Center (figure 24) commands the admiration of every visitor to CSM's campus. The $3.5 million steel-and-concrete building, with three floors and 157,000 square feet, was dedicated on April 14, 1972. Guy T. McBride, Jr., CSM's president, in his introductory statement in the dedication booklet, wrote:

Some new buildings merely enlarge a campus, while others change it in a fundamental way . . . surely the Green Center is one of the latter kind. It was in fact envisioned to be such from its early concept in the minds of its donors and of prior School administrations, for it was planned to include, as it now does include, not only classrooms and laboratories of Geophysics and Mineral Economics Departments and the Brown Computing Center, but auditoria and meeting rooms to enable its use as a campus and community center, and as a site for state and national professional conferences.

In a letter to an old-time friend, who wrote complimenting the Greens on their donations for the Green Center, Cecil replied:

It is wonderful to receive such an understanding, contemplative, and, may I say, sentimental letter from those who knew us when—who liked us then and still do now! Ida and I feel fortunate and blessed that we were able to help make possible the new nerve center for the Colorado School of Mines. We liken it to a real investment in pleasure and satisfaction, and now along comes your wonderful letter as a valuable dividend resulting from this investment.

Again and again on later pages, Cecil speaks of their gifts as "investments in pleasure and satisfaction" and the laudatory remarks from close friends as "dividends" from those investments. In his remarks at the dedication of the Center, Cecil reiterated some of his strongly held views about the importance of cooperation and collaboration in bringing a funding drive to a successful conclusion:

. . . I think it worth emphasizing that state and private resources can indeed work together very effectively in a worthwhile common cause or project . . . all of us agree that this new Center represents a good example of what can be accomplished by cooperative planning, joint effort, and collective support.

Then he went on to reveal how he and his associates approached the fundamental question: Will the Center serve a really useful set of functions?

This was the principal question which had to be answered affirmatively, when, as a joint state-private enterprise, this project was first launched some seven years ago [1965] under the capable promotional guidance of Dr. Orlo Childs, then president of Mines. The future needs of Colorado School of Mines were analyzed carefully, and in consequence, this Center was envisioned as a logical, practical means for solving a well-thought-out need. So today, at this time of dedication, I believe that I speak for all donors when I say that the finally completed Center is a real tribute to the visionary and continuously persistent and combined efforts of administration, faculty, trustees and state representatives.

In the foregoing comments one can discern Cecil Green's guidelines for his own personal philanthropy, and his strong belief that "success comes from seeking the help of people, rather than in spite of them."

One of the outstanding features of the Greens' philanthropy is the fact that once they have funded a specific undertaking, they continue to keep informed about it—be it a new building; a new facility; an endowed chair; a scholarship, fellowship, or other student award (a medal); or a special research or training project. In thus keeping informed, they not only enjoy the results of their sharing; they also find out if additional donations would be merited, even if not requested.

Having learned how well the building that he and Ida had made possible was serving its intended purposes, Cecil asked himself, "How can we make it even better?" A color portrait of the Greens, provided by them, already hung in the main corridor (figure 25). Then in 1975, when they learned that a Mexican tapestry of magnificent proportions (16 feet wide by 66 feet long, and weighing 1,500 pounds), was for sale, they donated the money for its purchase. Today this crescent-shaped tapestry adorns the high back wall of Friedhoff Hall in the Center.

Still later, in 1980, the Greens contributed a modern Allen digital elecronic organ for the main auditorium (the Bunker Memorial Auditorium) of the Green Center, where commencement exercises are held each year, and where the Golden community enjoys organ recitals.

That the Green Center was functioning well as a town-and-gown facility by 1980 is evidenced by the prediction that the Center's varied program of activities for that year would involve a total of some 50,000 participants. Thus the Center has been a continuing reminder of the Greens' sensitivity to the need for a place where town and gown can come together in cooperation rather than in confrontation and conflict. During April 7–8, 1981, Cecil himself used some of the Center's facilities to conduct his first CSM-GSI Conference on geophysical matters.

THE CHARLES HENRY GREEN CHAIR IN EXPLORATION GEOPHYSICS

Obviously pleased with the several philanthropic investments they had made at CSM, and wishing to encourage further the school's program of instruction and research in applied geophysics, Cecil and Ida generously endowed a chair in that discipline in mid-1975. Cecil requested that it be named the Charles Henry Green Chair in Exploration Geophysics in memory and in honor of his father, whose work as an industrial electrician in British Columbia, and later in California, had inspired Cecil to prepare for a professional career in electrical engineering. The elder Green (figure 24) had first worked on Vancouver Island, then moved to a copper smelter at Anyox near Prince Rupert, B.C. Cecil worked at the same Anyox smelter in the summer of 1919, as an eighteen-year-old college student. This opportunity to learn firsthand about the work of electric technicians led to his decision to enter MIT in 1921 as a transfer student from the University of British Columbia.

Cecil owed much to his parents for paying the costs of his years at MIT, and he was mindful of that debt when he spoke at the investiture of Dr. James Edward White, the first Charles Henry Green Professor of Exploration Geophysics, on November 19, 1976:

25   The Greens and Colorado School of Mines. *Opposite page, top:* Cecil and Ida in the main lobby of CSM's Cecil H. and Ida Green Graduate and Professional Center, following the unveiling of the portrait that hangs on the wall behind them as a permanent feature. This picture of the Greens with their color portrait in the background was printed on page 12 of the May 1972 *Mines Magazine. Opposite, bottom:* Dr. John W. Vanderwilt, president of Colorado School of Mines, hands Cecil his first honorary degree, Doctor of Engineering, in 1953. This degree was awarded to Cecil in recognition of his contributions to the profession of engineering and interest in the training of student engineers. *Above:* CSM's Cecil H. Green Geophysical Observatory, one of five similar observatories developed with government funds, was named for Cecil in recognition of his advancing the art of geophysical exploration by seismic means. The actual recording instruments are in a quarter-mile-long abandoned mining tunnel driven horizontally into the mountain directly behind the office building to which data are telemetered. The observatory, located near Evergreen, a few miles west of Golden, is in an ideal location for making precise measurements because the environment in the tunnel undergoes little change of temperature during the year. Top left photo courtesy of CSM; bottom left photo courtesy of CSM: William Smyth; above photo taken by Daniel Williams.

Happily, my adolescent and impressionable years—that is, from age five to twenty-one—were spent in British Columbia, where I also profited from the inspired influence of two wonderful individuals—my mother and father! In the newly emerged field of electrical power and its use, my father, as an energetic and enterprising man, became a maintenance electrician.

Then, thanks to the continued backing of my father, still at Anyox . . . I was able to complete my education in 1924 with a Master's Degree in Electrical Engineering.

I tried to repay my mother and dad through the subsequent years for their guidance, inspiration and material assistance . . .

and this he did, in full measure, as long as they lived. His father died in 1949, at the age of seventy-one; his mother in 1973, on the day before her ninety-fifth birthday.

The Charles Henry Green Chair, left vacant in 1986 by the retirement of James Edward White, will be reoccupied in September 1988 by Kenneth L. Larner, an alumnus of CSM (1960) and MIT (1970), and president-elect of the Society of Exploration Geophysicists (SEG's *Geophysics: The Leading Edge of Exploration*, March 1988, p. 45).

THE GUY T. MCBRIDE, JR., HONORS PROGRAM IN PUBLIC AFFAIRS FOR ENGINEERS

Maintaining their longtime interest in CSM, the Greens recently (1984) pledged a substantial donation to a permanent endowment for an honors program in public affairs for engineers, named for Guy T. McBride, Jr., recently retired president of CSM. The program, according to a letter from William K. Coors to Cecil, dated February 27, 1985, " . . . seeks to introduce students to the complex social, political and economic issues confronting today's engineer."

Finally, the 1987 bequest that Ida left CSM has been named the Ida M. Green Endowment for the Enrichment of Teaching and Research in Geophysics. According to a news item in the summer 1987 issue of *Mines Today*, "Income from the endowment will be utilized to augment the Green Professorship in Geophysics and also to address special needs and opportunities within the department."

The philanthropy of the Greens to Colorado School of Mines has not gone unrewarded. In 1953 CSM awarded Cecil the first of his twelve honorary degrees—an Honorary Doctor of Engineering (figure 25). Eight years later, the newly established geophysical observatory at nearby Bergen Park was named for him—the Cecil H. Green Geophysical Observatory (figure 25).

THE GREENS AND SOUTHERN METHODIST UNIVERSITY (SMU)

As mentioned previously, SMU was by far the most important academic participant in the earliest development stages of the TAGER Network, thanks to Dean Martin's technical assistance and President Tate's counsel and guidance in developing the courses that were to be offered by his university. And it should be recalled that not only did SMU offer 90 percent of the academic courses in the beginning years, it continues to offer more than half of the courses today. Little wonder, then, that Cecil and Ida have had such a long and productive relationship with SMU.

The Greens' relationship with SMU actually dates back to their earliest years with GSI when they were moving about frequently in Oklahoma, Texas, Louisiana, and California, while Cecil was a GSI party chief and later regional supervisor. Cecil kept a sharp eye open for likely recruits among the SMU's geology graduates, and more than once discovered one of those "naturals" he was always seeking, like Robert C. Dunlap, Jr., who signed on as a field hand, advanced to party chief, and ultimately became president of GSI.

The relationship became stronger when he and Ida bought their one-quarter share of GSI, moved to Dallas, and purchased their first home, which was near the campus. Cecil had become acquainted with SMU President Willis M. Tate in the 1930s, and quickly developed respect for him as a person and a scholar. As he and Ida learned more about Tate and SMU, they began to consider how they could help the university most effectively. Their first donation, as Cecil recalls, was in 1953 when a share of TI common stock was listed at $5.56. The one hundred-share donation was for the university's general support. Later in the

1950s, they joined their GSI-TI business partners, the Erik Jonssons and Eugene McDermotts, in providing their share of the construction costs of SMU's Science Information Center (figure 26), dedicated in November 1961. Early in 1961 they had also agreed to join their GS-TI partners in sharing the expenses of renovating the university's McFarlin Memorial Auditorium so that the Dallas Symphony Orchestra could play there.

Now in the early 1960s, when Cecil became deeply involved in developing the TAGER program, he naturally turned to his longtime and respected friend Tate for advice and assistance, knowing from public reports that "between 1963 and 1970 the six counties composing the Dallas–Fort Worth metropolitan areas would require about 2000 Ph.D.'s in science and engineering alone." This was the Ph.D. gap that first GRCSW and then TAGER were chartered to help fill, and Cecil and Tate would spend many hours working together on the problem. During this time the TAGER headquarters occuped 20,000 square feet in SMU's Science Information Center. Cecil well knew that to be successful, the TAGER program would have to depend almost entirely, at least during its early years, on the courses to be offered by SMU's School of Engineering, under the leadership of its new dean, Thomas L. Martin, Jr.

In addition to his personal efforts as board chairman of TAGER, Cecil also made three substantial commitments of time and money in support of Tate and SMU's School of Engineering—a three-year support program of $20,000 annually, starting in 1967; a four-year membership on Tate's President's Council (1969–1972), which called for a donation of $10,000 annually; and a five-year program (1970–1974), which involved a donation of $50,000 annually.

After making the principal pledge for construction of the TAGER TV network, which brought Martin to SMU, the Greens made another major donation to the university's School of Engineering and Applied Science to endow an honors chair, the Cecil H. Green Professorship in Engineering, to which Martin was appointed immediately after its establishment in 1969.

In order to encourage further improvement in general engi-

26   The Science Information Center of Southern Methodist University, dedicated on November 31, 1961, is a four-level, 77,000-square-
foot building designed with pillars to support an additional floor. It
was made possible by gifts from Cecil and Ida Green and from the
Texas Instruments Foundation, the McDermott Foundation, and the
Jonsson Foundation. More than half of the space is used for the Science Library. In the early 1960s the Graduate Research Center of the
Southwest occupied 20,000 square feet according to Berkner's statement on page 6 of GRCSW's 1963 *Annual Report*. Courtesy of Southern
Methodist University.

neering, the Greens made a major commitment to donate a total of three separate funds over a five-year period, starting in 1979:

1. The first to be used for the purchase of laboratory equipment for the various departments in the School of Engineering and Applied Science.

2. The second to endow a second distinguished chair—the Cecil H. and Ida Green Professorship in Electrical Engineering.

3. The third to endow the Cecil and Ida Green Fund for Excellence in Engineering and Applied Science Education.

Two additional donations by the Greens deserve mention because of their unusual nature. In 1974 they gave a single gift to the school of Engineering to offset a debt of $207,200, and in 1975 they donated $25,000 to support a program in music therapy.

In recognition of their service to and support of the university, SMU has repeatedly honored both Ida and Cecil. Ida was elected an Honorary Member of Mortar Board, SMU's highest honor, in 1973 and was cited as a Distinguished Alumna in 1977. Cecil was made an Honorary Alumnus in 1962 and was given an Honorary Doctor of Science degree in 1967. On October 10, 1981, in his eighty-second year, Cecil was recognized by SMU as "father" of the TAGER TV Network, and commended "for his pioneering efforts in advanced engineering education in the [Dallas–Fort Worth] Metroplex." (*Park Cities People*, Dallas, October 23, 1981.) That the Greens still maintain a lively interest in SMU is shown by Cecil's recent membership on the Geology Department's visiting committee and Ida's service of twelve years on the Board of Trustees.

THE GREENS AND AUSTIN COLLEGE (SHERMAN, TEXAS)

The Greens became interested in Austin College in the 1960s when Cecil made the acquaintance of Austin's president, John D. Moseley. Under Moseley's leadership and at Cecil's invitation, the college subsequently joined the TAGER program, and continues to participate actively in the expanded AHE program of 1987. Cecil, in turn, was named a trustee and member of the Executive Committee of the college.

Both Cecil and Ida were much impressed by the innovative educational program that Moseley was directing. Cecil liked particularly the strict discipline that Moseley and his staff required students and their parents to accept—especially because at the time, college students across the country were rebelling against authority. Both he and Ida approved the innovative curriculum with its strong emphasis on the liberal arts and notably on training for careers in both communication (radio and television) and the performing arts. When Ida learned about the curriculum, she enthusiastically agreed with Cecil's proposition that they fund the construction of a much-needed building designed specifically as a Communication Center. For she still retained fond memories of her youthful performances in "Pollyanna" and "Black Velvet," the first radio programs broadcast by the General Electric Station WGY Players in Schenectady in the early 1920s. The proposed new building was for Cecil, too, an especially promising investment in pleasure because it "could help young people find themselves in terms of their natural and individualistic inclinations," as Cecil put it in his comments at the dedication (figure 27).

The Greens were justly honored at the dedication on April 28, 1972, when the Board of Trustees announced its unanimous decision to name the new building the Ida Green Communication Center, and to dedicate it "to the Advancement of Human Understanding and Appreciation through the Arts and the Science of Communication [inscription by the entrance to the Center]." In addition, the space equipped for theatrical instruction was designated the Ida Green Theater, and another specially furnished space was named the Cecil Green Conference Suite.

Following their donation for construction of the Communication Center, the largest gift ever received by Austin College, the Greens generously endowed a distinguished professorship, the Ida M. and Cecil H. Green Chair of Creative Educational Leadership. This innovative position permits Austin's administrators to bring distinguished scholars to Sherman as visiting

27 Front (*top*) and rear (*bottom*) views of the Ida Green Communication Center at Austin College (Sherman, Texas). The building, funded by the Greens and dedicated on April 28, 1972, is a multipurpose structure containing more than 50,000 square feet of floor space. The Center is named for Ida in recognition of " . . . her interest in the arts and for her responsible leadership as a trustee in educational and cultural organizations." It includes the Ida Green Theater; the Cecil Green Conference Suite; numerous other conference rooms; an art gallery; the Paul Beardsley Arena, a flexible experimental theater; the Studio Theater; shops for scenery and costume construction; and a master control room that is the center of the entire building. The Center was designed specifically to encourage student training in the communication media and performing arts, and student participation in the college's continuing education program. Photos courtesy of Austin College.

lecturers, to be heard by both townspeople and members of the university, thus adding important support to Austin's special program entitled Individual Development: Encounter with the Arts and Sciences (IDEAS).

To show their gratitude to the Greens for their philanthropy, Austin College's Board of Trustees awarded an Honorary Doctor of Laws to Cecil in 1966 and an Honorary Doctor of Humanities to Ida in 1977.

THE GREENS AND TEXAS CHRISTIAN UNIVERSITY (TCU)

Cecil first became acquainted with TCU's Chancellor James M. Moudy in the 1960s, when the two of them became involved in the deliberations that led to the chartering of TAGER to TCU's becoming one of the seven original active prticipants in the consortium. During visits to the TCU campus, Cecil and Ida had numerous opportunities to learn firsthand about the university's educational program, and they admired its excellent quality. When informed of the institution's plans for improvement and expansion, and for raising needed funds, they decided to take a more active role in the school's affairs.

Cecil first served as a member of the President's Council and of the Board of Trustees, then as an Honorary Trustee, and also as a member of the Board of Directors of the TCU Research Foundation. These memberships gave him further information about TCU's financial needs and motivated him to consider how best he and Ida might help the university.

In 1969 they endowed a special chair, the Cecil H. and Ida Green Honors Chair, which provides for internationally recognized scholars in various disciplines to hold limited-time appointments as Visiting Green Professors on the TCU faculty. The visiting scholar may come for only a few lectures, for a month, or for an entire semester—an innovative idea on Cecil's part that is an excellent arrangement because of the flexibility it gives the TCU administration. As an indication of the high value of Cecil's suggestion regarding length of appointment, ten specialists were scheduled to come to TCU as Visiting Green

Professors during the 1984–85 academic year. The first of these was Ellen Johnson, a historian of contemporary art. Three of her lectures during her brief visit, May 4–8, were open to the public. In fall 1987 Hans Link of West Germany's Karlsruhe, world renowned in the field of the philosophy of technology, held the Green Honors chair.

Since the Greens endowed the Cecil H. and Ida Green Honors Chair, fifteen scholars have held the prestigious post. In addition, more than a hundred writers, scientists, educators, and other notable persons have spent short periods of time at TCU as Visiting Green Professors.

As early as 1973, TCU's Mary Couts Burnett Library was proving to be too small to provide urgently needed shelving, special rooms, and storage space. When these library needs became critical, the university initiated a $10 million fund-raising campaign. The Greens elected to help the campaign off to a good start by pledging one of the major challenge gifts. Accordingly, when the refurbished and expanded library was dedicated on March 25, 1983, the main reading room was named the Cecil and Ida Green Reading Room in their honor.

In 1987 came the exciting announcement that Ida had left TCU a major bequest, which the Executive Committee of the Board of Trustees immediately and unanimously voted to use to establish the Ida M. Green Graduate Fellows Program. In a letter dated August 26, 1987, Chancellor William E. Tucker advised Cecil that "An Ida M. Green Fellowship has been assigned to each of TCU's [five] doctoral-granting departments . . . Chemistry, English, History, Physics, and Psychology." The university expects the first Ida M. Green Fellows to begin residence in the fall of 1988.

A grateful TCU had previously awarded honorary degrees to the Greens: a Doctor of Science to Cecil in 1974, and a Doctor of Humanities to Ida in 1977.

THE GREENS AND BISHOP COLLEGE

Bishop College, in Dallas, was one of the seven original academic institutions to join the TAGER consortium. When Cecil first visited the college, he quickly learned from its president, Dr. Milton K. Curry, Jr., about the problems that the college faced in trying to develop a suitable curriculum and a stimulating environment for students. Dr. Curry was making great efforts to improve his school, but he was having an uphill struggle. Cecil decided to help Curry in some way. He and Ida at first made modest donations in 1969 and 1971. Soon thereafter, when he was asked to serve as chairman of the 1974 Capital Fund Drive, with a goal of more than $5 million, Cecil consented to do so. While inducing others to contribute to the drive, he and Ida pledged a major challenge gift, making certain thereby that the college could develop the arts center that Curry so much wanted. Additional donations that Cecil was successful in soliciting from a variety of contributors helped the college to attain its large financial goal.

THE GREENS AND THE UNIVERSITY OF DALLAS

When the University of Dallas, a small Catholic college, started a campaign in 1983 to raise funds for construction of a two to three million dollar science building (figure 28) to be named for Patrick E. Haggerty, one of the four founders of Texas Instruments Inc., the Greens, Jonssons, and McDermotts pledged major challenge gifts in honor of the co-partner who had played such a critical role in the early development of Texas Instruments. The Patrick E. Haggerty Science Center was dedicated in 1985.

The Greens' further connection with the University of Dallas derives from the fact that it was one of the seven original educational institutions to join the TAGER consortium. And like Austin College and Bishop College, among others, it continues today, twenty years later, as an active participant in the greatly expanded AHE-TAGER program, in which Cecil still maintains a lively interest.

The university for its part recommended Cecil for the Papal Gold Medal, which he received from Pope Paul VI in 1965, and awarded him an Honorary Doctor of Civil Jurisprudence in 1976.

28   The Patrick E. Haggerty Center at the University of Dallas, Irving, Texas, was funded principally by the three original founders of Texas Instruments Inc.—Cecil H. Green, J. Erik Jonsson, and Eugene McDermott. The building, formally opened in March 1984, contains the offices, laboratories, and support areas for the Departments of Biology, Chemistry, Mathematics, and Physics. Photo courtesy of the University of Dallas.

THE GREENS AND THE UNIVERSITY OF TEXAS MEDICAL BRANCH AT GALVESTON

Cecil's leadership role in founding and developing GRCSW-SCAS, which became UTD, and in creating TAGER, quickly became widely known throughout Texas, with the result that his services were soon sought by many institutions. One of these, the University of Texas Medical Branch at Galveston, invited him to chair an advisory committee appointed to select a director for the Earth Sciences Division of its Marine Biomedical Institute. Upon joining the committee, Cecil learned that the whole University of Texas System wanted to offer the directorship to Columbia University's famed geophysical oceanographer, W. Maurice Ewing, Higgins Professor of Geology and Founder-Director of the Lamont-Doherty Geological Observatory. But would Ewing leave the observatory, which he had founded and made known throughout the scientific world, and accept the directorship of a less well-known laboratory in Galveston?

As it happened, Cecil and Ewing knew each other quite well (figure 29). In the late 1960s Maurice had approached Cecil with a request for financial support, but without success, although Cecil greatly admired him as a person and a scientist. Now it was Cecil's turn to make the approach. Would Ewing accept an appointment as professor of geophysics in the UT's Medical Branch in Galveston, and establish a new geophysical laboratory? Ewing asked if he could bring along several of his talented associates from Columbia's Lamont-Doherty Geological Observatory—a request that was granted to the delight of everyone concerned, when he accepted the appointment. There were some aspects of the offer that made it particularly attractive at the time. Ewing was nearing sixty-five years of age and was hoping to free himself of some of his more onerous and often frustrating responsibilities, so as to devote more time to marine research, some of which he could continue from Galveston as a base. He was undaunted by the Texas weather, for he had known the "red-dusters" of the Panhandle region of West Texas, having been born and reared in the small town of Lockney, near

29   Cecil (*right*) and the late W. Maurice Ewing, renowned geophysicist and recipient of the prestigious Vetlesen Award. Cecil induced Ewing to accept appointment as Professor of Geophysics in the University of Texas Medical Branch in Galveston, with the responsibility of creating a new geophysical laboratory in the Earth Sciences Division of the Branch. Cecil stands beside some of the equipment purchased for Ewing's laboratory. Photo courtesy of the University of Texas Medical Branch in Galveston.

Lubbock. Furthermore he had done some of his more important marine seismic work in the Gulf of Mexico, and hence was familiar with the Gulf Coast. He also felt that Texas ought to have a first-class geophysical training program somewhere—there was none at the time. Finally, and probably most important, he would be coming back home after a long and sometimes turbulent career in crowded New York City.

Soon after Ewing's appointment in early 1972, as professor and founding chief of the Earth and Planetary Sciences Division of the Marine Biomedical Institute, the Greens made good Cecil's promise of support by generously endowing an honors chair which he immediately occupied as Cecil H. and Ida Green Professor of Marine Sciences.

Sadly for his many friends and the science that he loved, Ewing had only a scant two years in his new position before he died on April 28, 1974. Even though his tenure was short, he had established a sound research program in geophysical oceanography with the assistance of the three talented associates he had brought with him from Columbia—J. Lamar Worzel, H. James Dorman, and Gary V. Latham.

As further evidence of their confidence in Ewing and his young associates, the Greens pledged a challenge gift of approximately half the cost of a new ship designed for research at sea. The 135-foot, 198-ton ship, named the R/V *IDA GREEN* in Ida's honor, was commissioned and christened at the Galveston Yacht Basin on May 31, 1973, and was soon conducting research in the Gulf.

As Worzel wrote in his 1974 GSA *Memorial* to W. Maurice Ewing, 1906–1974 (p. 734),

Characteristically, in his stay at Galveston he was addressing himself and his colleagues to applying 24-channel common-depth-point seismology developed in the oil industry to problems of the continent-ocean transition zone. He saw the exciting results from the first line of data just before he was stricken.

A year after Ewing's death the Greens made several additional donations to the Medical Branch, one of which was used to purchase an improved gravimeter for the R/V *IDA GREEN.*

Worzel followed Ewing as director of the geophysical laboratory of the Marine Biomedical Institute, and in this capacity directed the oceanographic research of the R/V *IDA GREEN*, much of which involved ocean-bottom coring. However, just before acquiring the R/V *IDA GREEN*, Worzel and his crew had used two other ships to conduct seismic work in the north-central section of the Gulf of Mexico, as shown on the accompanying chart (figure 30) and reported on in the *Bulletin of the AAPG* (Shih, Worzel, and Watkins, 1977), from which the chart was taken. One of the results of this seismic work, the first major sea project undertaken from Galveston, was the informal naming of two closely associated features in the Greens' honor: (Cecil) Green Seamount and (Ida) Green Canyon, both of which are discussed more fully in the chapter titled *Awards and Honors*.

Upon Ewing's untimely death in 1974, it was decided to use the endowment to bring to the Galveston campus each school year a few outstanding scholars who would give one or two lectures in their particular field of specialization. These visiting scholars were chosen for their diverse areas of interest, with the hope that their lectures, seminars, and informal discussions would produce a coupling of the basic sciences, particularly the physical sciences and engineering, with the biological sciences and clinical disciplines.

Such a program was exactly the kind of collaborative exchange of ideas that Cecil strongly favored and that he and Ida had already established elsewhere. He was convinced that the cross-fertilization resulting from the exchange of ideas and information between diverse disciplines was bound to be mutually advantageous. In the ten years (1978–1987) since the inception of the Visiting Scientists program, thirty-eight distinguished scholars from North America and abroad have spent short periods of time at the Marine Biomedical Institute. Among these Green Visiting Scientists (a complete list can be found in appendix F) have been Philip H. Abelson (chemist), David H. Baltimore (molecular biologist), Sir Richard Doll (epidemiologist), Patrick McGeer (neuroscientist), Harry J. Buncke (plastic surgeon), Frank Press (geophysicist), Glenn T. Seaborg (nuclear chemist), Frederick Seitz (solid state physicist), and Ruth Davis (computer scientist).

Andrew D. Suttle, Jr., professor of nuclear physics and radiochemistry and special assistant to the director of the Marine Biomedical Institute at Galveston, wrote me on February 11, 1985 that the income from the Greens' donation has contributed " . . . materially to stimulating new ideas and programs and to exchanging information between our faculty and others who are interested in a particular discipline. Interactions between the physical sciences and engineering, on the one hand, and the biological disciplines and the healing arts, on the other, have been among the most beneficial."

The Greens made one further donation to the Medical Branch at Galveston, to help fund a special NMR laboratory for another of their very close friends, the aforementioned Andrew D. Suttle, Jr. That donation is also mentioned farther on in the section of the Greens' support of medical research.

Probably no person could evaluate the effects of the Greens' benefactions to the UT Medical Branch at Galveston better than Professor Suttle, who wrote them the following expression of appreciation on March 6, 1985:

Cecil and Ida, you do not know how significantly you have shaped and ordered the development of the Marine Biological Institute and the various academic departments with which its members are associated. Without doubt, you first added significantly to our breadth and our abilities when you made it possible for us to retain the late Maurice Ewing, Ph.D., as the leader of our group in the field of physical oceanography. The addition of Dr. Ewing certainly enhanced our reputation and strengthened our ability to conduct a broad spectrum of research. Subsequently, you graciously and generously supported him by procuring the research vessel *Ida Green*, which was used very successfully by him and his immediate successors, . . . and I understand has been used effectively by the University at Austin in subsequent studies.

And now, after that generous gift, you further enhance our capabilities and our scope by making it possible for us to have this excellent imaging instrument. You have indeed been good to us, and we do appreciate it.

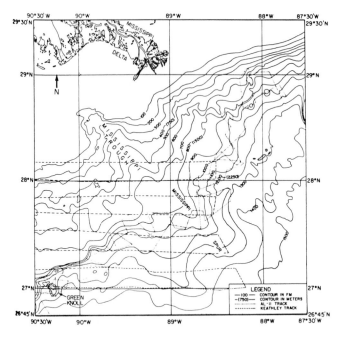

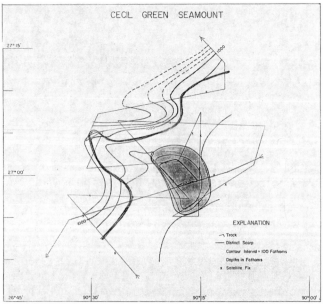

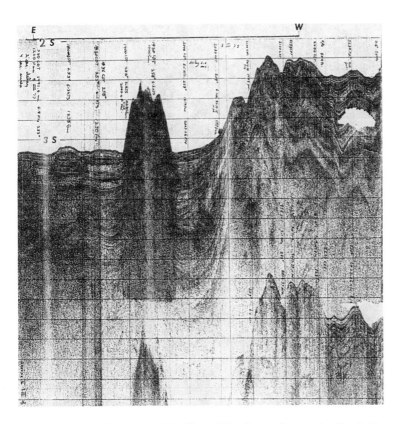

30 Cecil Green Seamount (Knoll) and Ida Green Canyon in the Gulf of Mexico. *Top left:* Bathymetric chart of the north-central Gulf of Mexico. Note Green Knoll in the far southwest corner of chart (Shih, Worzel, and Watkins, reprinted by permission of American Association of Petroleum Geologists). *Bottom left:* Original sketch of Cecil Green Seamount (courtesy of J. L. Worzel). *Above:* Submarine cross section showing Green Seamount (Knoll) rising prominently above the Gulf bottom (courtesy of Earth & Planetary Sciences Division, University of Texas, Marine Biomedical Institute: J. L. Worzel et al.).

The great potential of a challenge gift is demonstrated by the result of the Greens' 1983 donation to UTMBG to help Suttle develop a laboratory for biological imaging. UTMB President William C. Levin is reported, in the Branch's *Impact* for September 23, 1983, to have " . . . expressed gratitude to the Greens for their farsightedness in helping pave the way for operational support for the initial period of development of this new medical tool [nuclear magnetic resonance imaging] with their generous gift.'" Levin was reporting that a total of more than $2.8 million had been presented to UTMB for the purchase and installation of an NMR imaging unit: " . . . $2.3 million from the Galveston-based Sealy and Smith Foundation for the John Sealy Hospital and a $500,000 challenge operational grant from Cecil H. and Ida M. Green of Dallas."

Cecil still maintains a lively interest in the geophysical program that Ewing initiated and developed at Galveston by serving on the Visiting Committee of the UT Institute for Geophysics at Austin, with which Ewing's original geophysical laboratory was merged in 1982.

THE IDA AND CECIL GREEN EDUCATIONAL FUND FOR THE DALLAS EASTER SEAL SOCIETY
In June 1987, Lyda Hill, treasurer of the Dallas Easter Seal Society, informed Cecil that one of Ida's 1987 bequests—not included in the long list on an earlier page—will be used to create the Ida and Cecil Green Educational Endowment Fund. The society plans to use the income from this most welcome bequest to help provide the educational services for handicapped children who could not otherwise afford them.

SUPPORT OF TRADITIONAL EDUCATION IN CALIFORNIA
When in a jocular mood before an appropriate audience, Cecil likes to inform his listeners that in the early morning of the 18th of April, 1906, he was awakened by plaster from the ceiling falling on his face as the result of the great San Francisco earthquake. He was then not yet six years old, but he vividly remembers many aspects of that awesome phenomenon as he

and his mother wrenched open the door of his bedroom and rushed out of their Jackson Street apartment near Van Ness Avenue. Although Cecil likes to jest that this event may have foretold that some day he would be profiting from the use of deliberately produced mini-earthquakes, which he in fact did, he quickly adds that upon serious consideration of possible professions, he decided to be an electrical engineer.

Such was Cecil's introduction to California, to which he and Ida returned in the late 1930s, when he became the regional supervisor of all GSI field parties in the state. In 1940 Cecil moved his parents from Oakland to a new home in San Diego, to join other Green relatives. Frequent family visits familiarized the Greens with the San Diego–La Jolla area, a region so much to their liking that they decided ultimately to establish a second home there. They made their first California investment by purchasing a house and garden at 338 Via del Norte in La Jolla in the spring of 1959.

First as vacationers, then as frequent visitors, and finally as part-time residents, they developed a wide circle of friends, particularly academicians in the San Diego branch of the University of California (UCSD) and medical personnel at Scripps Clinic and Research Foundation (SCRF) in La Jolla.

THE GREENS AND THE MUNKS

Judith ("Judy") Horton Munk, a talented architect and wife of UCSD's famed geophysicist, Walter H. Munk, told the Greens about her interest in the International House Foundation. The foundation, though located on the campus of the University of California at San Diego, is nevertheless an independent communal center that obtains its support from private donors. Its humanitarian objectives soon attracted a generous donation from the Greens.

At about the same time, 1964, Cecil learned that the Munks greatly admired a striking piece of sculpture recently completed by the late Donal Hord, which he had named *Spring Stirring* (figure 31). Today, thanks to the Greens' generosity, and to their affection for Judy and Walter, the mantled statue of black diorite faces the main entrance of UCSD's Institute of Geophysics and

31 *Spring Stirring* (1948), a black diorite sculpture by the late Donal Hord (*left*), now sits in a large shallow diorite bowl in front of the laboratory of the Institute of Geophysics and Planetary Physics, a division of the University of California at San Diego, located on the La Jolla bluffs. The local story about the sculpture is that when Cecil and Ida first visited the new IGPP laboratory, which was directed by a close friend, famed geophysicist Walter Munk, they were much impressed by it. When they asked Walter if they could help the laboratory in any way, Judith, Walter's wife, who deeply admired Hord's sculpture, excitedly answered "yes"—"they could buy *Spring Stirring* for the IGPP's grounds." Being fond of "Judy," the Greens promptly bought the sculpture, then donated it to UCSD with the understanding that it would be placed where she desired. It was another example of Cecil's and Ida's innate generosity—matching their means with a close friend's desires. They were wont to call such an action "an investment in pleasure." Photo courtesy of UCSD's IGPP.

Planetary Physics (IGPP) building, where Munk and his associates carry on their worldwide geophysical research.

True to form, Cecil lost no time determining the nature of Munk's research and soon indicated his willingness to help. Opportunities for providing financial support were not long in coming. Even before they came, however, he made his own move.

THE CECIL H. AND IDA M. GREEN FOUNDATION FOR EARTH SCIENCES (ORIGINALLY THE LA JOLLA FOUNDATION FOR EARTH SCIENCES)

Highly pleased with the results of the visiting scholar/lecturer programs that he and Ida had initiated in several other universities, Cecil was naturally delighted by Munk's enthusiastic response to his offer to support a similar program at IGPP. Munk is reported, in the *San Diego Union* of September 20, 1972, to have said:

When Cecil Green suggested that he wanted to do something to help the institute, I told him the establishment of a visiting scholar fund would be the best thing. . . .

The idea of the foundation was to provide funds to bring an outstanding scholar to the institute to stimulate our thinking. What we want is somebody to shake us up.

The result of the 1971 and earlier conversations between Cecil and Walter was that on January 19, 1972, the Greens pledged the funds for establishment of the La Jolla Foundation for Earth Sciences as a nonprofit charitable California corporation (foundation). In the words of the donors:

This gift is in trust for the benefit of the Institute of Geophysics and Planetary Physics, La Jolla Laboratories of the University of California, San Diego (hereinafter called the "Institute," . . .) for the benefit of the entire Earth Sciences community at Scripps Institution of Oceanography, La Jolla, California. . . .

The purpose of the gift is to provide financial assistance to the Institute for the creation and support of one or more lectureships. Each recipient of a lectureship shall be a visiting scholar in the Earth Sciences on short term appointment, who engages in one or more of the following activities: research, teaching, conferences and lectures.

The foregoing two paragraphs, excerpted from Cecil's letter of January 19, 1972, show how thoughtfully and clearly Cecil and Ida customarily outline the purpose of any major gift they make.

It should be noted that this particular gift by the Greens is unique in that it was used to establish a nonprofit charitable California corporation (foundation) independent of the University of California's financial management, yet a part of the financial support of the university's Institute of Geophysics and Planetary Physics. The Greens chose the foregoing legal form, with its own board of directors, so that some or all of the endowment funds in the foundation could be invested and reinvested with fewer restrictions than those governing the financial transactions of the university.

The endowment of this innovative foundation now exceeds more than $2 million, thanks both to the vigor and imagination of the late William Schofield, a director of the foundation, and to a generous bequest left by Ida.

The Green Scholars are selected and invited by Munk and his co-directors. They typically spend three or four months at IGPP's La Jolla laboratories. Ten years after its inception, Munk could report, "The program has been international in scope. I count eleven Americans, ten British, two Canadians, one Frenchman and one Italian. . . . The Foundation has been a very valuable asset to IGPP." In a 1984 letter to IGPP, Cecil wrote:

As a devout collaborationist, I find it easy to say that the operation of this Foundation provides an excellent and indeed an outstanding example of how much more can be accomplished and in quicker time by a team comprising selected lay people with academic individuals and to the point where everyone thoroughly enjoys that grand feeling of accomplishment, and this is evidenced by all of the testimonial letters which you have collected in this volume from the entire roster of visiting scholars.

The foregoing statements were taken from an exemplary in-house history of the Cecil H. and Ida M. Green Foundation for Earth Sciences prepared in 1984 by the staff of the foundation. The history is complete to 1982, with pictures of the Greens, the Green Scholars, and the foundation directors, and with brief comments by or about most of those involved in the development and activities of the foundation.

Early in 1983 the directors of the foundation voted to change the name to the Cecil H. and Ida M. Green Foundation for Earth Sciences in recognition of the outstanding role the Greens have played worldwide in advancing the earth sciences.

THE CECIL AND IDA GREEN PIÑON FLAT GEOPHYSICAL OBSERVATORY OF UCSD'S IGPP

Cecil was particularly interested in the sophisticated instrumentation that the geophysicists of IGPP were using at home and abroad to measure the fundamental physical properties and behavior of the earth. Being experienced in producing man-made earthquakes by virtue of his years as an exploration geophysicist with Geophysical Service Inc., he naturally wanted to become involved in IGPP's worldwide research. So when he learned that the institute needed additional money to acquire ownership of the land occupied by its geophysical observatory on Piñon Flat, he and Ida made a generous 1978 donation to supplement the funds available for purchase of the desired land.

The Piñon Flat Geophysical Observatory, now named for the Greens, is located on a tract of some 480 acres of land about a hundred miles northeast of La Jolla (figure 32). Freeman Gilbert, then associate director of IGPP, wrote me on August 24, 1983,

At the Observatory . . . there are several state-of-the-art instruments for measuring the earth's acceleration, strain and tilt. It is one of the world's best instrumented observatories and it exists largely due to the Greens' generosity.

PROJECT *IDA*

In his letter, Gilbert also added information about a second important IGPP research project involving the Greens:

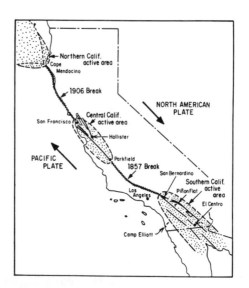

32   The Cecil and Ida Green Piñon (Pinyon) Flat Observatory of
UCSD's IGPP. *Top:* Map (*by SIO's Wyatt and Agnew*) showing location
of the Piñon Flat Observatory in the active earthquake area of southern California. The San Andreas fault is indicated by the heavy zigzag
line. *Bottom:* Cecil and Ida (in front of sign) with UCSD administrators
and scientists (from left to right): William A. Nierenberg, SIO director;
the Greens; Frank Wyatt, SIO; Richard C. Atkinson, UCSD chancellor;
J. Freeman Gilbert, IGPP associate director; and in front row, Walter
Munk, IGPP associate director; and Jonathan Berger, SIO and observatory director. The group assembled at the recent ceremonies officially naming the observatory after the Greens. The observatory is
ideally located to measure movements along local faults that can be
useful in earthquake prediction. Photo courtesy of UCSD–SIO–IGPP.

Finally, you may have heard of our Project IDA [see figure 33]. The letters stand for International Deployment of Accelerometers, a project that we have operated for almost a decade. It has led to a global network of low frequency seismographs, there are 17 now, and the digital data, IDA data, are used by scientists in several institutions to study the structure of the earth and earthquake source mechanisms. Ida and Cecil provided the funds to obtain the seismometers (accelerometers). Without their help Project IDA would not have been a success, and our knowledge about the earth would have been poorer.

To substantiate his continued interest in IGPP's worldwide seismological research, Cecil, in a July 6, 1987 letter to Gilbert, pledged a major donation to UCSD's Green Foundation for Earth Sciences, to be paid in three annual installments, starting in 1987,

. . . for the special purpose of completing the modernization of every recording station in [the] project IDA worldwide network with the installation of horizontal seismometers as a desired supplement to the present-day vertical units.

GEAR FOR THE R/V WASHINGTON

Being a part-time resident in La Jolla gave Cecil opportunities to learn about the research program of the renowned Scripps Institution of Oceanography; hence he was always alert to SIO's problems and needs, particularly those involving ships. His interest in ships stemmed from his knowledge of the operations of the two M/V CECIL H. GREEN's belonging to GSI, the company he formerly headed. Accordingly, when he learned from friends at SIO in 1976 that one of the Institution's ships, the R/V WASHINGTON, urgently needed some expensive gear in order to continue operations, he responded immediately by donating the requisite funds.

OTHER DONATIONS TO THE UNIVERSITY OF CALIFORNIA

When the Greens were recently solicited in 1983 for a donation to an endowed chair honoring University of California's distin-

268
Chapter 6

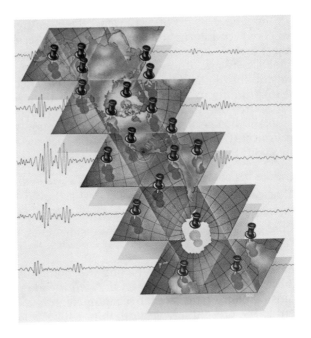

33   Project IDA (International Deployment of Accelerometers) is a
global digital seismic network that collects data for low-frequency
seismology (DC Agnew et al., *EOS*, April 22, 1986). In 1973 a small
group of geophysicists at UCSD's IGPP developed the concept of a
global network of accelerometers and approached Cecil and Ida Green
for financial support. Cecil, being an experienced exploration geophysicist, immediately understood the purpose of the proposed network
and together with Ida agreed to underwrite the cost of the accelerometers if the IGPP group could find operational support. Fortunately
for science, the National Science Foundation agreed to fund the cost
of deployment and operation of the instruments, so that installation
of them could begin in late 1974, under the name Project IDA (with
the acronym honoring Ida Green). By 1986 the IDA Network had the
20 stations shown in the diagram, and was planning for the next decade. Again the Greens participated in the future operation of the network by additional support. The accompanying diagram, prepared by
SIO's Steven Cook, and published in a slightly modified form on the
cover of AGU's *EOS* for April 22, 1986, shows the current distribution
of the IDA stations. Diagram courtesy of SIO (Cook) and *EOS, Transactions of the American Geophysical Union*, 1986, copyright by the American Geophysical Union.

guished chemist, the late Harold Clayton Urey, they responded generously.

As Cecil became more directly involved with the research activities of the Institute of Geophysics and Planetary Physics and of Scripps Institution of Oceanography, and made close friends in those and other divisions of the University of California at San Diego, he discovered that UCSD had no faculty club building. Having seen the great value of faculty clubs at other academic institutions, he concluded that such a facility was needed at UCSD to encourage more frequent communication between members of the different faculties. So in 1985, following one of their traditional patterns of giving, Cecil and Ida expressed to UCSD's Chancellor Richard C. Atkinson their willingness to pledge a major donation for a completely new building to be a faculty club, provided the University of California would fund the remaining construction costs. Chancellor Atkinson, who also deplored the lack of a faculty club at UCSD, seconded the Greens' perception:

"There is a great need at the university for a place where faculty, staff and guests can meet informally over lunch or at other times for social exchange, intellectual stimulation or for discussion of their work" (quoted from *La Jolla Light,* July 17, 1986).

The university met the Greens' challenge, and the new building is now a reality. The $2.5 million Cecil and Ida Green Faculty Club opened informally in mid-March, 1988, and was dedicated the following April 14 (figure 34).

THE GREENS AND CALIFORNIA INSTITUTE OF TECHNOLOGY (CALTECH)

As might be expected from Cecil's interest in seismology, he occasionally visited Pasadena to confer with the famous seismologists at the California Institute of Technology. Although his relations with Caltech have never been close or extensive, he did make a substantial challenge donation in 1978 to an honors chair named for, and first occupied by, Robert P. Sharp, Caltech's distinguished geologist, and a nominal donation toward a special lectureship honoring C. Hewitt Dix in 1980.

THE UNIVERSITY OF CALIFORNIA SAN DIEGO FACULTY CLUB   ARCHITECTS MOSHER / DREW / WATSON / FERGUSON

34   The artists' rendering of the Cecil and Ida Green Faculty Club of the University of California, San Diego. The building was dedicated on April 14, 1988. It is an ever-present reminder of the Greens' thoughtful and timely support of some of UCSD's most urgent needs. Courtesy of UCSD and Mosher, Drew, Watson & Ferguson, architects.

THE GREENS AND STANFORD UNIVERSITY

Cecil's longtime relationship with Stanford University started in the late 1920s when, as a recently graduated electrical engineer, he began working for the Federal Telegraph Company in a small plant located in Palo Alto at the corner of University Avenue and El Camino Real on the edge of the Stanford campus. He was fortunate to be assigned to work side by side with another young electrical engineer, Charles V. Litton, a Stanford alumnus ('24), whose inventive genius was based on an excellent education. And from friendships with other outstanding Stanford alumni—Frederick E. Terman ('20), Wallace Sterling ('38), Max Steineke ('21), Herbert Hoover, Jr. ('25), Kenneth Crandall ('24), and George Knox ('22)—Cecil developed a deep respect for the university and its graduates. Years later, when he became president of GSI, he established working relationships with Stanford's geology and geophysics departments, in an effort to find well-trained young scientists and engineers for his company.

His quest for college-trained employees led Cecil to become a member of visiting or advisory committees to leading earth sciences departments. In such a relationship with Stanford, Cecil soon became acquainted with the more pressing needs of both the earth science departments and of the university as a whole. Thus, when he was solicited for a critical donation to supplement an anticipated NSF grant to cover construction costs of an earth sciences building, he responded with alacrity. Imagine Cecil's astonishment when Kenneth Cuthbertson, vice president for development, advised him that the NSF grant had been approved for an amount somewhat larger than anticipated, so that only part of his donation would be needed, and hence the remainder would be returned to him with thanks! Such integrity on the part of the Stanford administration "warmed the cockles of my heart," as he puts it—and enhanced his already favorable opinion of the university's policies and personnel.

Impressed by the outstanding quality of the several different departments in the School of Science, which had been developed in the 1960s and early 1970s, the Greens in 1973 pledged a major donation for an endowed honors chair, to be designated the Cecil H. and Ida M. Green Professorship in Geophysics. Following this gift came another major donation from the Greens, in 1976, this one for an endowed fellowship program, again for the Department of Geophysics. Thus far, through SY 1986–87, some twenty-five students have been awarded Cecil and Ida Green Graduate Student Fellowships in Geophysics. The Greens established the prestigious fellowships "to provide special recognition for exceptional students in the broad field of geophysics, including (but not restricted to) applied geophysics."

Cecil's increasing involvement in the affairs of Stanford, outside those of the School of Science, brought to his attention the farsighted plan to enlarge the main library in anticipation of the university's needs in the coming decades. The bold plan was the very kind of educational enterprise that excited Cecil's interest, so in September 1976, with Ida's enthusiastic approval, he pledged a munificent donation to be paid in six annual installments while the old part of the library complex was being renovated and a spacious new wing constructed. In view of the magnitude of the challenge gift—some 40 percent of the total cost of the renovation and new construction—the university voted to name the completed library center the Cecil H. Green Library (figure 35).

At the dedication of the library complex on April 11, 1980, it was announced that a special area in the new wing had been named the Ida M. Green Room. Since 1980 it has been used extensively for conferences, committee meetings and other gatherings. A color photo of Ida graces one wall of the room. Two co-founders of Texas Instruments are also represented in the list of donors, no doubt as a result of Cecil's solicitation—J. Erik Jonsson donated funds for a government document area; and Margaret Milam (Mrs. Eugene) McDermott, Mary McDermott, and the Eugene McDermott Foundation contributed funds for the Eugene McDermott Room now devoted to the teaching of computer searching of bibliographic and other data files. In addition the Greens provided the portrait of Cecil that adorns one of the walls of the East Wing.

At the dedication, Cecil nostalgically recalled his longtime association with many Stanford friends and former associates, then raised the question no doubt in the minds of many in the audience:

As a retired engineer and industrialist, I have sometimes been asked, Why this support for Stanford, and its library system in particular, when neither Ida nor I were students at this University? The answer relates largely to my long professional career in a free enterprise system within a highly competitive environment. I soon learned that ultimate industrial success was directly dependent upon the careful selection of young people, and especially in regard to such human factors as quality of educational background, inherent motivation, plus natural innovative ability.

After commenting on how he and Ida first began to direct their philanthropy toward support of educational institutions, then in later years to support of medical research and health care, he continued:

All of these collaborative, or interdisciplinary, concepts just naturally led to a common crossroads, or nerve center; and, of

35 The Cecil H. Green Library of Stanford University. The complete library combines the old Main Library, built in 1918, and the new major addition on the right, which doubles the capacity of the library complex. The main entrance to the new wing is in the left-hand section. The tower of the Hoover Institution rises in the background. The complete library, named for Cecil, was dedicated on April 11, 1980. Photo courtesy of Stanford University Archives: Souza.

course, the name Library is virtually a synonym for such a meeting place.

So, from the point of view of a specialist who has finally come around to being also a generalist, I would say that a library literally illuminates future frontiers of thought, as it also preserves the accumulated knowledge of the past! . . .

So it can be said that no other single asset is so deeply involved in the academic activities of all members of the university community. Thus, I feel that this Stanford Library, in representing humanity's temple, is consequently a truly sophisticated and effective instrument for teaching and learning.

The Greens have made two additional major gifts to Stanford since the dedication of the Cecil H. Green Library. A large bequest from Ida's estate, received in early 1987, was designated by Stanford's Board of Trustees at their May meeting for the establishment of the Ida M. Green Directorship of the Stanford University Libraries. David C. Weber, director of Stanford University Libraries since 1969, and widely recognized as one of America's leading academic librarians, was elected by the board to be the first occupant of this prestigious position. In late June 1987, Cecil informed me that he had pledged another major donation to Stanford toward the cost of a new building to be named the Cecil H. and Ida M. Green Earth Sciences Research Building.

Stanford awarded its most prestigious academic honor to Cecil in 1982 by appointing him Consulting Professor of Geophysics, an honor, like MIT's Honorary Lecturer, which takes the place of the traditional honorary doctoral degree of other educational institutions. Early in 1988 Cecil learned that he and Ida would be co-recipients of yet another Stanford honor—the Degree of the Uncommon Man. This award by the Board of Governors of Stanford Associates, and previously granted to only fourteen Stanford alumni and alumnae, recognizes outstanding past service to the university. The award to Cecil H. Green and Ida M. Green (posthumously) is the first to nonalumni of Stanford and the first to co-recipients, and will be presented at two special dinners, the first at Dallas on November 15, 1988, the second at San Diego on March 17, 1989.

THE GREENS AND HAWTHORNE COLLEGE

Cecil's interest in Hawthorne College, located near Antrim, New Hampshire, dates from October 18, 1986, when his old MIT friend, Vincent A. Fulmer, became president of the college. When Cecil learned that the college was especially noted for its excellent program of training airplane and helicopter pilots and business managers for the aviation industry, and that it was laboring under a heavy debt that required substantial borrowing to keep it in operation and to realize the hopes of its cooperative program, his concern and sympathy were aroused. Admiring Vincent and Alma for the courage that they exhibited in accepting the presidency, when they well knew that the debt situation

urgently demanded resolution, Cecil pledged a substantial donation to help the college—a most timely act.

Fulmer was well aware of Cecil's career as an entrepreneur
and of his and Ida's fame as philanthropists. As board chairman
of Suffolk University he had bestowed upon Cecil the honorary
degree of Doctor of Commercial Science on June 11, 1978. Now
as the new president of Hawthorne College, he conceived the
doubly brilliant idea of awarding the first-ever honorary degree
Doctor of Philanthropy *jointly* to Cecil and Ida, the honor to be
bestowed at commencement, scheduled for Sunday, May 1,
1987. This was not to be the ordinary commencement, however;
it was to be held in conjunction with the celebration of Hawthorne College's Silver Jubilee (1962–1987). Furthermore, the
name of the special degree, and the fact that it was to be jointly
bestowed, would set two precedents.

Then came a devastating wet snowstorm that disrupted electric power and telephone lines on the campus, forcing Fulmer
to close the college and postpone commencement until the 24th
of May. Having come all the way to Boston from Southern California for the first scheduled commencement, Cecil nevertheless decided to return on May 24th because he did not want to
disappoint Vincent and Alma Fulmer.

Fulmer, as board chairman and president, bestowed the honorary degree, prefacing the award as follows:

Ladies and gentlemen, we come now to an historic moment in
American higher education—the awarding of this country's first
Joint Honorary Degree to a distinguished couple and the designation of their honorary degree in a field in which no college
heretofore has ever awarded such a degree. We are breaking
with tradition twice in what we are about to do.

In a moment I am going to ask Dean Kenick to bring Cecil H.
Green forward to receive for himself and his beloved wife, Ida
M. Green, a joint degree of Doctor of Philanthropy . . .

Thus, Cecil for the twelfth time, and Ida for the fourth, were
recognized for their philanthropy by academia's most prestigious award.

On the day following commencement, a vigorous oak sapling, to be a symbol of the esteem in which the Greens were held, was planted about fifty yards southeast of the president's house along the easterly border of the walkway leading south to Peabody Hall (figure 36).

THE CECIL H. AND IDA GREEN BUILDING OF THE NATIONAL ACADEMY OF SCIENCES—A NEW HOME FOR THE ACADEMY'S NATIONAL RESEARCH COUNCIL

The Greens, in 1986, made a major gift to the National Academy of Sciences for purchase and renovation of a building in the Georgetown section of Washington, D.C., desired as a new home for the Academy's National Research Council. This building, named the Cecil H. and Ida Green Building of the National Academy of Sciences in honor of the donors, was dedicated at a special international celebration September 21, 1987, which was attended by a large group of distinguished scientists from around the world. It will become the principal building of the National Research Council complex in the Georgetown location (figure 37).

THE AUSTRALIAN CONNECTION: HARRY MESSEL AND THE UNIVERSITY OF SYDNEY

Cecil first met Harry Messel in 1955 during a visit to Australia for GSI. The president of WAPET (Western Australia Petroleum Company), upon learning of Cecil's interest in education, told him of a young physics professor in the University of Sydney who was a real "go-getter," Harry Messel by name, and suggested to Cecil that he ought to meet him.

Canadian-born Messel had arrived in Australia in 1951 to receive an appointment at the University of Adelaide. Within less than a year, however, he was appointed to the then sole chair of physics at the University of Sydney, given the authority to expand Sydney's School of Physics into five departments, each headed by a professorial chair, and assured by the government of financial support for his formidable task.

36 Tree planting at Hawthorne College, Antrim, New Hampshire, May 25, 1987. The young oak tree was planted as a memorial to Cecil and Ida Green, generous benefactors of Hawthorne in a time of great need (*top left*). Against the background of Peabody Hall, from left to right, stand Hawthorne President Vincent A. Fulmer, Mrs. Theodora W. Shrock, Dr. Cecil H. Green, Mrs. Alma Fulmer, and Prof. Robert R. Shrock. The two planters, Fulmer and Green, stand ready (*top right*); they shovel (*bottom left*); the tree is planted, and Fulmer is pleased (*bottom right*). Photos courtesy of Hawthorne College photographers.

37   The Cecil H. and Ida Green Building of the National Academy of
Sciences, located in Georgetown, Washington, D.C., is the long rectangular structure in the immediate foreground facing Wisconsin Avenue and marked at one end by a large white dot with two short rays.
The Greens, for whom the building was named at the special dedication ceremonies on September 21, 1987, made a generous donation toward the purchase and renovation of the spacious structure, which is
now the principal building of the Academy's National Research Council's new site. According to Academy President Frank Press, a longtime friend of the Greens, there is to be a nearby Ida Green Memorial
Garden, with a suitable plaque, "where people can gather in a place
of beauty and discuss the great issues of our time." The Greens' impressive support of education in science, medical research, and health
care led to the International Tribute paid as described in the prologue.
Now their building in the nation's capital becomes an enduring monu-
ment of that support. Photo courtesy of the National Academy of Sciences.

Messel knew quite well, when he accepted appointment in 1952, that the Australian government, faced with all the problems of the postwar years, could not be expected to assign to him the large funds he felt he would need to develop his single-chair department into a multidepartmental school. Furthermore, he was convinced that the School of Physics that he envisioned could not develop into a first-rate institution unless somehow he could induce the best young Australian high school students to become interested in the sciences, and eventually in physics.

Faced with the primary problem of expanding his school, despite inadequate funds and few students interested in science—and most of these choosing to study overseas—Messel decided to seek funds outside the university to support basic research in physics in the university and to improve instruction in the sciences in the high schools. The argument for his plan would be that since business, industry, and government all depended on the university for the training of the future professionals that they would employ and the research results that they would use, all should join in financially supporting the university. Support should not be left to the government alone.

By the time of Cecil's visit in 1955, Messel was already a controversial figure known throughout Australia, as well as overseas, not only for his energetic leadership of the University of Sydney's School of Physics but also for his creative ideas about improving instruction in science in the high schools of New South Wales. He had already in 1954 seen the birth of his brainchild in the form of the Nuclear Research Foundation, soon to be referred to simply as the Foundation, and for the primary purpose stated above. It was the first such Foundation in the British Commonwealth and was to be followed by hundreds of others during the next thirty-five years.

To progress further toward his several objectives, Messel needed not only encouragement and administrative and legislative action; most of all he needed increased financial support. He was seeking all of these vigorously in 1955 when he was introduced to Cecil. Cecil recalls that one of Messel's first questions was:

Mr. Green, do you think there is anything wrong in trying to teach mathematics and the sciences, particularly physics, so as to give them a popular appeal? My academic associates regard as almost sacrilegious my stooping to popularize physics!

and that he responded,

You certainly are not wrong; I think you are on exactly the right path. You are doing a great service to physics if you do make high school students, and even business people, aware of the subject and give them some understanding of what is actually going on out there at the University.

Several years then elapsed before the two men saw each other again. Nevertheless, Cecil did not forget the aggressive but sincere young academician, and Messel later told me that Cecil's encouragement really "fired him up."

Some years later, in 1960, when planning a round-the-world trip for himself and Ida that would include another visit to Australia, Cecil was given a list of prominent individuals whom his close friend at UT-Dallas, Lloyd V. Berkner, thought he might contact on the trip; one of these, Cecil immediately recognized, was that energetic young physics professor in the University of Sydney, Harry Messel.

On the return to Sydney, Cecil and Ida had an opportunity to see for themselves the progress that Messel had made in his educational enterprises and to learn of Harry's current problems and hopes. They liked what they saw. A warm friendship quickly developed between the Greens and the Messels as Cecil recognized that Harry was the very kind of imaginative and dynamic educational entrepreneur that he could admire, respect, and financially support. Specifically, he strongly approved the way Messel had developed Sydney's School of Physics, particularly by creating a separate department of computing. He greatly admired Harry's town-and-gown organization, then called the Nuclear Research Foundation, because it was an excellent example of what he himself was promoting at home, and what he would later support at the University of British Columbia and Colorado School of Mines. He was impressed by Harry's success as a fund-raiser, especially in the United States. And he was most enthusiastic about the way that the university academicians under Harry's guidance and urging were blending mathematics and the basic sciences into a single high school text to be titled *Science* that would immediately attract worldwide approbation when it first appeared in a pilot edition in 1963, then as a first edition in 1964. Ida was similarly impressed, and shared with Cecil a hearty desire to help Messel. The opportunity to do so came a few months later when the Greens temporarily stopped in England on their trip around the world. Messel knew beforehand about this stop and arranged one of his many overseas trips to be in London at the same time Cecil and Ida were there. The dramatic story of what actually happened when Messel invited the Greens to have lunch with him at London's Savoy Hotel is narrated on pages 74–75 of D. D. Millar's *The Messel Era*, published in 1987 by Pergamon Press (Australia) Pty. Ltd.:

After a pleasant meal, Messel and the Greens adjourned to Harry's room for drinks and to continue discussion of the most pressing needs of the School of Physics. At that time the greatest need was an expensive new facility for the School's Basser Department of Computer Science—an English Electric KDF9 Computer [figures 38 and 39]. At the appropriate moment, Ida quietly remarked:

"Harry, we are very impressed by what you are trying to do in Australia, and we would like to help you. What would you like?"

"What limit are we talking about?" said Messel.

"No, you tell us what you need most in the School and then we will tell you."

Harry replied: "The thing we need most is a new computer. Our improved SILLIAC is rapidly going out of date. We were one of the leaders in computing and we are rapidly getting left behind."

With Ida's nod of approval, Cecil said "We would like you to have it."

38   The Green KDF9 Computer, the most powerful of the five computers in the Basser Computing Department of the University of Sydney (Australia), was purchased with the help of a major donation to
the university's Science Foundation from Cecil and Ida Green in 1960.
Their gift was the largest single donation that the foundation had received up to that time. The opening of the new computer on April 22,
1964, was honored not only by the presence of top officials of the university but also by the then governor-general of Australia, Viscount de
L'Isle, and by the Greens who had flown to Sydney for the occasion.
The Greens were so pleased with the impressive success that their
good friend Harry Messel had achieved in developing the university's
School of Physics in the previous years that they decided on further
philanthropy. So in 1980, and again in 1983, they endowed a graduate
fellowship in Messel's honor, so that today (1988) the School of Physics has two generously endowed awards—The Professor Harry Messel
Research Fellowships in Physics. Photo courtesy of *The Nucleus*, Science Foundation for Physics, University of Sydney.

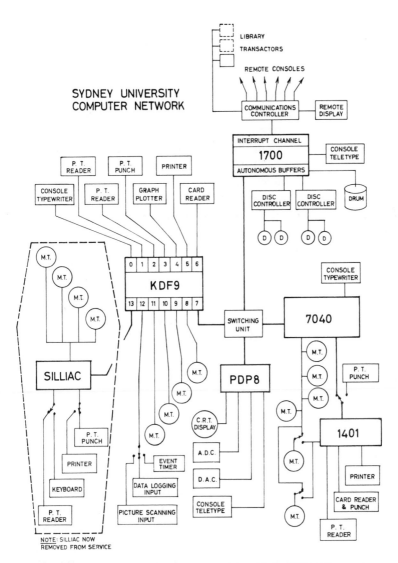

39 Diagram showing the important position of the KDF9 Computer
in the University of Sydney's Computer Network. Graphic courtesy of
The Nucleus, Science Foundation for Physics, University of Sydney.

The machine that the Greens wanted Harry to have was the KDF9 mentioned above, toward the purchase of which they pledged $250,000 (£113,267), more than a third of the ultimate cost of the new facility.

Delighted by the Greens' offer to help meet the cost of the new computer, Messel immediately telephoned Sir Frank Packer, then chairman of the foundation, and Sir Stephen Roberts, vice chancellor of the university, even though it "was in the middle of the night in Australia." Sir Stephen elated in his turn persuaded the university senate at a special meeting to award Cecil the honorary degree of Doctor of Science. The degree was awarded on March 11, 1961 (figure 40), and the next day Cecil was a guest speaker at the seventh annual dinner of the foundation. As was to be expected, he first emphasized how he had guided his own company, GSI, into a collaborative involvement with a dozen educational institutions, and how that collaboration was of mutual benefit to all participants. He then praised Messel for following the same course in creating the Science Foundation for Physics within the University of Sydney, to seek funds from outside sources.

At the eighth annual dinner of the foundation, in 1962, Chairman Packer announced that the Greens' donation would be used, along with other funds, to purchase the KDF9 computer, costing £320,577. Finally, two years later, on April 22, 1964, the Governor General, Lord de L'Isle, officially inaugurated the new facility as the Cecil and Ida Green KDF9 Computer, with the Greens in attendance. The new computer was immediately put to work, and by the end of 1970, *The Nucleus* of January 1971 reported that the Green KDF9 Computer had been serving "as the main computer facility of the univeristy" (see accompanying graphic, figure 39).

During the years following the computer's installation, Harry kept Cecil and Ida fully informed about his activities by sending them regular copies of *The Nucleus*, successive editions of the innovative textbook *Science for High School Students*, starting with the 1963 pilot edition, and copies of other books he himself wrote about physics for both high school and college students.

40 Cecil H. Green received an honorary Doctor of Science degree
from the University of Sydney (Australia) on March 11, 1961. He is
seen here after the award ceremony (*center*) with the acting vice-
chancellor of the University, Sir Edward Ford (*left*), and the chancellor,
Sir Charles Bickerton Blackburn. Photo courtesy of *The Nucleus*, Science Foundation for Physics, University of Sydney.

In addition he never missed the chance to visit the Greens when
he was near Dallas or La Jolla.

With every telephone call, publication, and visit from Messel,
Cecil and Ida became more and more impressed by Messel's
dramatic impact on high school and college education and research in science, not only in Australia, but also in other countries. It was only natural that the Messels were included in the
guests invited to the International Tribute to Cecil and Ida
Green, held in 1978 at the National Academy of Sciences in
Washington, D.C., and that Harry was selected as the speaker
on the Greens' philanthropy in Australia.

Then when plans were being made in Sydney to celebrate the
twenty-fifth anniversary of founder-director Messel's Science

Foundation for Physics, Cecil and Ida decided to honor Harry in a way that would perpetuate his name in the academic world. At the Silver Jubilee Dinner on August 31, 1979, the chancellor of the University of Sydney announced that the Greens had pledged $200,000 to endow the Harry Messel Postdoctoral Fellowship in Physics. Harry was deeply pleased because he knew that Cecil and Ida were expressing their admiration not only for him but also for his wife Patricia ("Pippie"), with whom they had developed a warm friendship. More telephone conversations and intermittent visits from the tireless Harry kept the Greens informed about latest developments in Sydney.

During one of his overseas trips in 1983, when he stopped off in Dallas to spend a day or two with the Greens, Harry experienced the second of the typical surprises that Cecil and Ida enjoy springing on those they most admire. As Messel tells the story, it was nearing midnight in Cecil's study when he was suddenly asked, "Harry, things seem to be going along fine as usual, but do you have any special unfunded need; perhaps another fellowship?" Although stunned for a few seconds, Harry was quick to accept Cecil's offer. So the Greens donated the funds for a second fellowship. In the ensuing transaction, each fellowship was endowed with $250,000, and at the Greens' request, both were titled the Professor Harry Messel Research Fellowship in Physics.

The Green KDF9 Computer together with the two endowed Messel research fellowships stand as a well-deserved tribute to Harry Messel for his extraordinary academic leadership. They will long endure at the University of Sydney as a compelling reminder of how perceptively and generously Cecil and Ida Green aid those who truly excel. The government officials and those of the University of Sydney must have been prescient when they awarded Cecil the honorary degree of Doctor of Science as early as March 11, 1961.

One last comment is needed on Cecil and the Australian connection. On his latest birthday, August 6, 1987, on the eve of departure for an official visit to China, the dauntless eighty-seven-year-old traveler told me that he would be the principal speaker, on December 15, at a celebration in the Great Hall of the University of Sydney, honoring Harry Messel upon his retirement after thirty-five years as professor of physics and founder-director of the University's Science Foundation for Physics. Afterward he planned to spend the Christmas holidays with Patricia and Harry Messel.

The Greens' gifts to the University of Sydney, their first donations to an educational institution outside the United States, were followed by gifts to the University of British Columbia (1967, 1972), and to England's Oxford University (1978).

SUPPORT OF TRADITIONAL EDUCATION IN CANADA—BRITISH COLUMBIA AND ONTARIO

BRITISH COLUMBIA
By the time Cecil was six years old he was starting elementary school in Vancouver, B.C., where his parents settled after emigrating from England in search of a more satisfying life. On completing his education in the Vancouver public schools, Cecil entered the University of British Columbia at eighteen to study applied science. Three years later, when he had determined that UBC did not offer the instruction in electrical engineering that he desired, he transferred to MIT. There he completed study for two degrees in electrical engineering in Course VI-A: an SB in 1923, and an SM in 1924. Cecil was a good student, receptive to the knowledge and experience that he obtained from both undergraduate institutions, and his feeling about them after graduation was one of satisfaction and approval. Like many recent graduates, however, he was much too busy establishing himself in his electrical engineering career to maintain contact with either university. At least twenty-five years passed before he returned to either campus and became involved in the affairs of the institutions.

It is characteristic of Cecil to act promptly once he has decided on an undertaking, and his change of citizenship was no exception. So long as he was uncertain about where to settle permanently, he retained his Canadian citizenship. But when he became GSI's supervisor of field parties in 1936, and realized that his future work lay in the United States, he applied for and was granted American citizenship in the Dallas Federal Court in 1936. As he explained it, "If I intend to follow my career and live permanently in the United States, then I want to be a U.S. citizen." Nevertheless, many years later, when William C. Gibson, M.D., a fellow UBC alumnus, called on him in Dallas one day in 1965 and invited him and Ida to visit Vancouver as guests of the university, Cecil happily accepted the invitation, especially since UBC had awarded him an honorary Doctor of Science degree the previous year (figure 41).

CECIL GREEN PARK
By this time, of course, Cecil and Ida were accustomed to ulterior motives in invitations. However, having already contributed munificently to MIT, Cecil was predisposed to make a donation to his first alma mater, if the proper opportunity arose. That opportunity came when he and Ida learned that the University wished to purchase a nearby stately twenty-four-room mansion for outside functions. Sensing that the imposing building (figure 42) and its surrounding gardens would make an ideal conference and social center for university and community activities, the Greens agreed to provide the funds for the purchase and renovation of the property.

In donating the money to UBC, Cecil expressed the hope that the building would be used not only as a place where alumni could gather but also as a place to foster community relations— the kind of cooperative relationship between university and town that he has so long encouraged. And that is exactly what has happened since its dedication and official opening on October 28, 1967, as Cecil Green Park. As a matter of fact, the property has become such a popular place for alumni weddings that the administration has been obliged to restrict the number so as to leave adequate time for other social affairs as well as for a wide variety of academic activities.

Cecil was particularly pleased to make this donation to UBC, because from the great house he could once again enjoy the magnificent view eastward across the Gulf of Georgia to the

41   Cecil H. Green (*center*) being congratulated by University of British Columbia's Chancellor Phyllis Ross upon receiving an honorary Doctor of Science degree, with UBC President John B. MacDonald looking on, in 1964. Photo courtesy of the University of British Columbia.

290
Chapter 6

42   Cecil Green Park at the University of British Columbia, Vancouver, B.C. Surrounded by three acres of beautiful grounds, the three-
floor mansion, with its twenty-four rooms, serves as an important
center for both university and community activities. When Cecil
learned that the property was for sale, he and Ida immediately provided the funds for its purchase and renovation, at the same time donating the property to the university. Dedicated and opened officially
on October 28, 1967, as Cecil Green Park, it has served for the past
twenty years as an indispensable center for a variety of activities, particularly those involving the UBC Alumni Association, which has its
headquarters and activities rooms in the mansion. The magnificent
building was designed by Samuel McClure, a prominent West Coast
architect, and constructed in 1912 for E. P. Davis, an eminent court-
room lawyer, who called the property on Point Grey peninsula "Kan-
akia" (house on the cliff). Cecil studied at UBC from 1918 to 1921 and
received an honorary Doctor of Science degree in 1964. Ida was
awarded an honorary Doctor of Laws degree in 1979. Photo and data
courtesy of the UBC Alumni Association.

snow-capped peaks of the Coast Range—a view he remembered from his student days.

CECIL H. AND IDA GREEN VISITING LECTURESHIPS

The warm reception accorded the Greens during that initial visit rekindled Cecil's interest in UBC and led him to consider what more he and Ida could do for the university. Why not endow a chair, Cecil thought—one similar to the visiting lectureships that had been so successful at TCU and UCSD? Why not, indeed? Whereupon they endowed the Cecil H. and Ida Green Visiting Lectureships at UBC in 1972. As with the aforementioned lectureships, the Greens expected UBC to use the income from the endowment to bring a variety of distinguished individuals to Vancouver for limited stays, during which they would deliver lectures, conduct seminars, or present performances.

In a December 8, 1986, letter to Keith Griffin (president of Oxford's Magdalen College), Cecil stated that he and Ida had had three objectives in establishing the visiting lectureships:

First, to bring new and inspiring facts and ideas to the [UBC] campus; second, to encourage inter-disciplinary relationships or cross-fertilization between departments; third, to promote all-important "Town and Gown" relations so that the university is not a lone "Ivory Tower" operation in its community.

The program that Cecil envisioned has been extraordinarily successful, so that he and Ida regard it as one of the most productive of the twenty-eight endowed chairs they have funded. During the sixteen years (1972–1987) since the program's inception, 140 distinguished men and women have visited UBC. A quick perusal of the list of visiting lectures (see appendix F) shows satisfying diversity of professional talent and a sincere effort on the part of the program's directors to meet the Greens' expectations by inviting individuals from a wide range of fields.

The list consists of mathematicians; scientists—physicists, chemists, earth scientists, biologists and biochemists, and geographers; engineers—materials, forestry, city planners; medical personnel—clinicians, researchers, health-care experts, psychologists, and psychiatrists; humanists—commentators, economists, historians, judges, politicians, and social sicentists; and musicians—vocalists and instrumentalists. It is especially noteworthy that the list includes eleven Nobel Laureates!

The University of British Columbia awarded Cecil an honorary Doctor of Science degree in 1964 (figure 41) and Ida an honorary Doctor of Laws degree in 1979 (figure 43). Other recognitions of the Greens by UBC are discussed farther on under the heading *Awards and Honors*.

CECIL AND THE UNIVERSITY OF TORONTO

Cecil has long admired and respected Canada's leading geophysicist, Dr. J. Tuzo Wilson, so he promptly accepted the request to chair a 1967 visiting committee for the program in geophysics that Wilson had headed for many years at the University of Toronto. Earlier, in the mid-1950s, when there was a prevailing shortage of university science and engineering graduates, Cecil had arranged to have Toronto's entire junior class in geophysics attend a special seminar in Calgary. He invited Wilson and the head of Toronto's Department of Physics (of which Geophysics was then a division), to accompany the students. His purpose in sponsoring this student-faculty seminar was the same as that of his highly successful GSI Student Cooperative Plan—to acquaint student geophysicists with the potential of a professional career in exploration geophysics. He wanted the students to learn about the real world of the exploration geophysicist, so that they would know how to evaluate the fancy promises they might receive, on graduation, from would-be employers, such as: "Join our company and you can expect to become a vice president in five years, if all goes well."

The long friendship between Cecil and Wilson is well exemplified by an unusual telephone call in the late 1950s.

Cecil likes to relate how he was awakened in the middle of the night by a telephone call. His first thought was that someone had the wrong number—but when he said "Hello," he was startled by a Houston operator asking him if he would accept a collect call from Dr. J. Tuzo Wilson in Little America? In a weak

43   Ida M. Green after receiving two of her four honorary doctoral degrees. *Top:* Ida in cap and gown after being awarded an honorary Doctor of Laws degree from the University of British Columbia (1979). *Bottom:* Ida receives an honorary Doctor of Humanities degree from Chancellor James M. Moudy of Texas Christian University (1977). Photos courtesy of the University of British Columbia and Texas Christian University, respectively.

voice Cecil replied "Yes," as he thought, "this is going to be a terrific telephone bill." He was soon relieved, however. When Wilson came on the line, he explained that the U.S. Navy operator had just completed a regular report call to Washington, D.C., when he announced to the bystanders in his office, "We have inadvertently made contact with a 'ham' radio operator in Houston, Texas; would anyone like to talk to anyone in Texas?" Whereupon Wilson, listening in on the conversation in the Antarctic station, immediately said, 'Please have the ham operator call my friend, Cecil Green, in Dallas.' " So Cecil's expense was merely the cost of the collect call from Houston.

7
THE GREENS' SUPPORT OF MEDICINE

The Greens' support of medicine—from training of practicing physicians, to diagnostic research, to health care involving people of all ages from newborns to the elderly—has been perceptive, innovative, and outstandingly generous. Their donations have provided physical structures—buildings, special rooms, research laboratories, and facilities; endowment of investigatorships, fellowships, and residencies; a unique public TV program on bodily ills; and two international conferences on biological imaging. This part of the Greens' philanthropy has been directed primarily to the following centers of medical training, medical research, and health care:

Scripps Clinic and Research Foundation in La Jolla, California

The University of Texas Health Science Center in Dallas, and Medical Branch in Galveston

The Presbyterian Medical Center in Dallas

The Children's Medical Center in Dallas

Baylor University Hospital in Dallas

The National Jewish Center for Immunology and Respiratory Medicine in Denver, Colorado

Cecil H. Green College at Oxford University in England

THE SCRIPPS CLINIC AND RESEARCH FOUNDATION (SCRF)

In 1959, following brief summer vacations at the historic La Valencia Hotel, the Greens purchased a house and rose garden at 338 Via del Norte in La Jolla with the intent of living there part of the year. It was at this time that they first became acquainted with the staff and facilities of Scripps Clinic and Research Foundation (SCRF).

One Sunday, at a coffee reception following regular services at St. James-by-the-Sea Episcopal Church, the Greens were introduced to Edward J. Hartung, an active lay leader, who was also chairman of the Board of Trustees of Scripps Clinic. Soon afterward, as dinner guests in the Hartung home, Cecil and Ida met Vice Admiral Wilder Baker, later chairman of the Scripps board, Dr. Edmund L. Keeney, president and medical director of the clinic, and Dr. A. Baird Hastings, a distinguished biochemist on the clinic's research staff. When Cecil mentioned his need for a gastroenterologist, Dr. Lee S. Monroe, head of that division at the clinic, was recommended. A consultation with Dr. Monroe proved to be the beginning of a long physician-patient relationship, that developed into a warm personal friendship. Simultaneously, the Greens became close friends of Drs. Keeney and Hastings as they enjoyed the medical services of the clinic and found out more about the scope and facilities of SCRF's overall program.

They also soon learned more about the scientific-medical complex that was emerging in the La Jolla area, of which SCRF was a major part. The University of California had established its San Diego campus on the Torrey Pines Mesa. Jonas Salk had selected an adjacent site on a promontory overlooking the Pacific for the Salk Institute. Dr. Keeney had earlier started Scripps Clinic on the path to distinction in biomedical research by inducing such scientific luminaries as Dr. Hastings, from Harvard and Rockefeller Universities, and Dr. Frank J. Dixon and his group, from the University of Pittsburgh, to form its research staff. And the present Scripps Institution of Oceanography at seaside in La Jolla, founded as the Marine Biological Association of San Diego in 1903, had long since become the world-renowned center of marine research that it remains today.

Impressed by the friendliness of the SCRF staff, by the quality of the care they received as patients at the clinic, and by the bold plans that the institution was developing for future growth, the Greens decided that they wanted to help the clinic in a substantial way. As Cecil put it, "It became easy, and indeed natural, to express appreciation, as well as admiration, for this vitally important service." Then action followed words.

THE CECIL AND IDA GREEN ASSOCIATE CHAIR
In 1961, the Greens became life members of Scripps Clinic's Friends of Research and made the first of their many large do-

nations to SCRF—this for an endowed staff position to be known as the Cecil and Ida Green Associate Chair. Demonstrating that he was keenly aware of the institution's needs, Cecil suggested that the staff member selected to occupy the chair devote half of his time to applied research and teaching, the other half to the care of patients. Vice Admiral Wilder D. Baker, USN (retired), then chairman of the foundation's board of trustees, announced that "This is the first such endowment for Scripps Clinic and Research Foundation."

THE VIA DEL NORTE HOUSE AND ROSE GARDEN AND SUPPORT FOR DIAGNOSTIC RADIOLOGY

Shortly after making the gift to SCRF, the Greens decided to dispose of the Via del Norte house and rose garden, which they were scarcely using as a part-time residence, and move to an apartment. Prudent Ida could see no sense in keeping the property, and paying their caretaker to enjoy his private rose garden, when they were away most of the time. So in 1964, having become much interested in the clinic's research in diagnostic radiology, they conveyed the property to the clinic as a gift, and the proceeds from its sale were used to expand and improve the Department of Radiology.

CECIL H. AND IDA M. GREEN INVESTIGATORSHIP IN MEDICAL RESEARCH

The more the Greens learned about SCRF's ongoing research, the more they involved themselves in the affairs of the foundation. Cecil was elected to the board of trustees in 1969, and remained active until 1983, when the tenure limit required him to step down. Fortunately for Scripps Clinic, however, Ida was willing to accept appointment to the board that same year, thus ensuring the clinic of the continuing representation of the Greens.

By early 1970, Cecil and Ida decided that they wanted to endow a second chair, to be named the Cecil H. and Ida M. Green Investigatorship in Medical Research. In line with their imaginative philanthropy, they intended the income from their donation "to support an eminent investigator or investigators at SCRF in areas of extraordinary scientific importance." They suggested that investigators be appointed for only limited periods so that new appointments could be made as new areas of research requiring investigation emerged. In 1972, Hans J. Muller-Eberhard, M.D., a senior scientist at SCRF, and winner of several major awards for his research, became the first person to occupy the Green Investigatorship.

PURCHASE OF A THERMOSCOPE AND A NEW DEPARTMENT OF RADIOLOGY ON TORREY PINES MESA

Soon after Dr. John L. Smith joined the clinic's staff as head of the Department of Diagnostic Radiology, in the 1960s, Cecil became one of his patients. During subsequent X-ray examinations, he seized the opportunity to question Dr. Smith on the pros and cons of the different kinds of electronic imagery being used in radiological research. These discussions excited Cecil's already strong interest in biological imaging. Consequently in 1971 he and Ida provided the funds for purchase of a thermoscope which Texas Instruments had developed and which Cecil decided, following Dr. Smith's advice, could be useful in the clinic's research.

GREENS' FUNDING OF SCRF'S HEALTH CARE CENTER ON TORREY PINES MESA

By this time both Greens had become fascinated by the whole field of biological imaging, and they urged Dr. Smith to visit a number of medical centers in the East in order to observe the latest developments in the field. Upon his return from the trip, Smith and the Greens were soon enthusiastically planning the new Department of Diagnostic Radiology to be included in the clinic's new Health Care Center on the Torrey Pines Mesa. The big question was, How would the project find funding?

Even though SCRF's fund-raisers could not know, when they launched a capital campaign in late 1971 to raise $10 million for the new Health Care Center project, that Cecil and Ida were waiting in the wings and wanted to provide substantial support, they soon found out. The story of this support is now legend, and is clear proof that the Greens make their donations because they want to, not because they are pressured to do so by some individual or narrowly focused interest group.

The story, as reported by one of the Greens' close friends, James S. Triolo, in an in-house clinic memorandum, goes as follows (excerpted and quoted with permission):

As related in the press at the time, SCRF called a news conference in June 1972 to announce the fact that the Greens had again demonstrated their confidence in the institution and in the future of its "diagnostic imaging" by a very generous pledge to construct the radiology department in the new building. After announcing this and introducing the benefactors, Dr. Keeney went on to explain that "being assured of the funds for the brick and mortar, we could now turn our efforts toward raising an equivalent amount to acquire the necessary equipment." As Dr. Keeney was speaking there was a whispered conversation between Ida and Cecil, who then rose and said,

> "Miss Ida and I have been talking and we decided that since we have given you the barn, we should also give you the hay, so we hereby pledge the additional amount for the radiological equipment."

Pandemonium broke loose. One reporter rushed to the telephone and called Mr. Erik Jonsson, also a founder of Texas Instruments and then Mayor of Dallas, to relate the story and ask if the Greens were "for real!" He was assured they were and that we wouldn't have to worry about the payment of their pledges! And that is how Scripps Clinic's first capital campaign really got underway.

It was characteristic of the Greens to do what has just been described. Having no children and few close relatives, they have long followed the practice of informing themselves of an institution's strengths, weaknesses, and needs, then taking carefully calculated steps to build on strength, remedy weakness, and help fulfill unmet needs. Always their desire is to help—not specific people, but institutions and organizations that render services to large groups of people.

Two years after the Greens had contributed generously to the $10 million campaign, giving it further impetus with their unexpected generosity, it had only reached the half-way mark toward its goal. As a trustee and member of the Resources Committee, Cecil decided that the campaign needed "a shot in the arm," so he and Ida pledged an additional donation to cover the construction of the 4-West Nursing Pavilion, hoping thereby to accelerate the time to ground-breaking on the project. Then when the campaign seemed to be lagging a bit after the ground-breaking in April 1974, the Greens made yet another donation that propelled the campaign beyond its original goal by several million dollars.

Cecil's actions were predictable when Dr. Smith told him about the new technique of computer-assisted tomography (CAT) and pointed out the tremendous improvement in health care that SCRF could achieve if the Clinic had a CAT body scanner. Cecil considered the matter carefully, as was his practice when he became interested in something, and in July 1975 he donated the cost of a scanner.

CECIL H. AND IDA M. GREEN HOSPITAL OF SCRIPPS CLINIC AND MORE FUNDING

Later in 1975 President Keeney recommended to SCRF's trustees that the new Scripps Clinic Hospital be named in honor of the Greens. The trustees agreed enthusiastically, and at a December dinner in 1975, it was formally announced that the new hospital would be named the Cecil H. and Ida M. Green Hospital of Scripps Clinic and Research Foundation (see figure 44).

When it became evident that the $12 million raised in the 1972–75 campaign was going to be insufficient to meet Scripps's projected needs, a supplementary capital campaign was mounted in early 1976. This campaign, called the 50th Anniversary Fund, set a goal of $7 million to meet increased construction costs and growing inflation.

Predictably, because of their longtime interest in and support of SCRF, Cecil and Ida again volunteered financial support. This time, however, they made two major donations instead of one:

44   The Cecil H. and Ida M. Green Hospital of the Scripps Clinic and
Research Foundation. *From left to right:* The first tower is the research
facility; the second, third, and fourth towers, together with the lower
interconnecting buildings, comprise the Green Hospital, which Cecil
and Ida generously funded in the mid-1970s, and which was named
in their honor in December 1976. The four-story hospital includes a
158-bed inpatient facility, the offices of the Scripps's large medical
group (100 + members), a radiology department, also funded by and
named for the Greens, and the extensive facilities of the institution's
ambulatory care program. The hospital also has complete surgical facilities, intensive and coronary care units, and a clinical research center. Afflicted with dreaded leukemia, Ida spent her last days in a
special suite in the Green Hospital. Photo and description courtesy of
Scripps Clinic and Research Foundation.

the first for construction and an equal amount to expand Scripps' endowment resources for research. Few philanthropists would have had the perspicacity to provide additional support for future research when the primary needs were for construction.

With this pledge the Greens became SCRF's leading benefactors. As a token of the institution's gratitude, the renowned Texas portrait artist, J. Anthony Wills, was commissioned to prepare an oil painting of the Greens. This portrait was unveiled at the dedication of the hospital, and it now hangs in the main entrance of the building. A second picture of Cecil and Ida—a color photograph by Antony di Gesu (figure 45)—hangs with pictures of other Scripps benefactors in the hallways of Green Hospital.

And there was even more to come!

THE GREENS' FUNDING OF THE NORTHWEST GARDEN EXPANSION PROGRAM

As the story goes, Cecil and Ida walked into the clinic one day in the fall of 1980, and announced, "We have now paid off our building pledges. We want to do more. What are your priority needs?" They were informed that the institution had outgrown the Green Hospital after only four years, and was already making plans to increase both patient-care and service areas—the Northwest Garden Expansion Program.

The Greens took the plans and proposal home for study. A few days later, Cecil rendered speechless three SCRF men with whom he was having lunch, when he declared that he and Ida were prepared to pledge two-thirds of the amount needed, and then added, "Would that be all right?" Quickly recovering from the surprise, the men agreed that his and Ida's pledge would indeed be all right, whereupon Cecil expressed satisfaction at "being able to help." As they later recalled, the three Scripps men rode back to the institution both elated and amazed at the Greens' tremendous generosity.

The reader might well wonder why the Greens give so gen-

45   This June 1985 photo of Cecil and Ida Green is a copy of the color
photo by Antony di Gesu that hangs in the hallway of Green Hospital
of Scripps Clinic and Research Foundation. Photo by permission and
courtesy of Antony di Gesu and SCRF.

erously, even when they declare, "We give because we want to."
That statement becomes understandable, however, when one
hears Cecil say,

The vital point I wish to make is that the gifts Mrs. Green and I
have made to Scripps and other such organizations are a product of intimate exposure, not selling. If there was any element
of selling—then it was a case of us selling ourselves and we
became partners in the enterprises. I have thus demonstrated to
myself that larger gifts tend to result from knowledge, personal
interest and finally involvement, rather than from routine solicitation.

I myself sometimes answer inquisitive individuals who ask me
how one *sells* Cecil Green on a proposition—one does not sell
the Greens on an idea; they only *buy!*

A THIRD INTERNATIONAL CONFERENCE ON BIOLOGICAL IMAGING

The success of the first International Conference on Biological Imaging held at SCRF in November 1980, and described fully in an earlier section, led to a second similar conference, hosted by the National Academy of Sciences, in Washington, D.C., in 1983. The latter meeting was also a success, with the result that a third conference will again be sponsored by and held at Scripps Clinic in 1988. Drs. Richard Lerner and Peter Wright of the foundation are organizing the meeting this time, but one can be certain that Cecil is following the planning with his usual interest and enthusiasm.

THE GREENS AND SCRF'S NEW RESIDENCY PROGRAM IN INTERNAL MEDICINE

One of the Greens' most recent benefactions (1986) is a major commitment to help underwrite the first year of a new three-year residency program in Internal Medicine at Scripps Clinic and Research Foundation. It is one of their favorite kinds of educational programs because its purpose is *career development* of talented young people who seek professional training in fields that will benefit society—an effort in which both Cecil and Ida have had a longtime interest. They have endowed four such residency programs to date: one at UTHSCD, one at MIT, one at Oxford (England), and the fourth at SCRF.

The SCRF program, known as the Cecil H. and Ida M. Green Residency in Internal Medicine, began on July 1, 1986, and covers the expenses of four recent medical school graduates for a three-year period of residency at SCRF. During this time the residents will receive intensive training in numerous specialties—including cardiovascular and infectious diseases, pulmonary medicine, diabetes, rheumatology and oncology—in preparation for a professional career in internal medicine. Upon completing their training, the residents will be qualified to take the American Board of Internal Medicine certifying examination and to spend a fourth year at SCRF in medical research or in advanced training in medical subspecialties.

The participants in this new program can be expected to add important research in internal medicine to SCRF's already outstanding program in advanced graduate education in clinical medicine and health care.[1]

IDA M. GREEN CANCER RESEARCH CENTER
The Greens' latest donation to SCRF came in 1987 as a major bequest from Ida; it will be used to create the Ida M. Green Cancer Research Center of Scripps Clinic and Research Foundation.

SUMMARY
It is no exaggeration to state that Cecil and Ida Green, acting jointly, have become the most generous of SCRF's benefactors, not only in the timeliness and magnitude of their financial support but also in the remarkable way that they have guided their philanthropy by keeping informed of the latest advances in medical research and practice and then responding to the needs of the Scripps Clinic and Research Foundation.

It is a laudable characteristic of the Greens that however large and whatever the purpose of their donations, they expect no privileges in return. At SCRF, for example, they have steadfastly declined all offers of medical attention on a courtesy basis. Their gentle but firm response to such offers has always been, "We will pay for our medical services like anyone else!"

SCRIPPS MEMORIAL HOSPITAL
It should also be mentioned that in the late 1960s, Ida accepted membership on the board of directors of the Scripps Memorial Hospital (not a part of SCRF) at the invitation from the Greens' longtime friend William Schofield, then board chairman. Furthermore, in 1968, she and Cecil made a modest donation to that medical facility for its general support. By this time, however, they were becoming deeply involved with Scripps Clinic and as a consequence did not make any further donations.

[1] The foregoing description is based in part on a news story in *La Jolla Light* for May 22, 1986.

THE UNIVERSITY OF TEXAS HEALTH SCIENCE CENTER AT DALLAS

When the Greens learned from Charles C. Sprague, who came to Dallas in 1967 to be president of UTHSCD, about his and his associates' ambitious plans for the expansion of the center, they decided to become involved in improving the facility, just as they had done with Scripps Clinic earlier. Their interest was at first concentrated in human reproductive biology and the supportive scientific disciplines, but later they also decided to donate funds for the training of young physicians and clinicians.

CECIL H. AND IDA GREEN SCIENCE BUILDING

Cecil and Ida first pledged a major donation in 1972, to help fund construction of a large building for instruction in the basic sciences; in their honor, the structure was named the Cecil H. and Ida Green Science Building (figure 46). The Greens' old friend and business partner, Eugene McDermott, pledged an equal donation for construction of a second building to be named in his honor, the Eugene McDermott Academic Administration Building. These two structures, together with the Fred F. Florence Bioinformation Center and the Lecture Hall Complex, as well as the landscaping and gardens provided by the J. Erik Jonssons, were dedicated at a colorful ceremony on April 27, 1975.

CECIL H. AND IDA GREEN CHAIR AND CENTER IN REPRODUCTIVE BIOLOGY SCIENCES

In 1973, the Greens endowed a second academic position, the Cecil H. and Ida Green Chair in Reproductive Biology Sciences, hoping to attract well-known medical scientists to the Health Center. The first occupant of the chair was and remains today (1988) the distinguished physician, Paul C. MacDonald, M.D.

When, in the mid-1970s, the Greens learned that space for instruction and research would be available in the Cecil H. and Ida Green Center for Reproductive Biology Sciences, which they had also funded, they pledged another major donation to support an innovative training program enabling medical stu-

46   The two Green buildings of the University of Texas Health Science Center at Dallas. To the left is the five-level, 248,243-square-foot
Cecil H. and Ida Green Science Building, largely funded by a major
donation from the Greens and dedicated in 1975. The structure is
commonly called the "Basic Learning Center" because its principal
function is to provide space and facilities for students during the beginning years of their medical education. To the right is the conjoining
nine-story Cecil H. and Ida Green Biomedical Research Building. The
Greens pledged a major donation toward construction of the building
when they learned that their Center for Reproductive Biology Sciences
could occupy a substantial area in the new structure, which was completed in 1986 and dedicated in 1987. As part of the dedication on
April 21, Cecil unveiled an excellent portrait of himself and Ida. The
building, costing $18 million, now houses the centers for research in
cancer, diabetes, molecular genetics, human nutrition, growth and
development, arthritis, and chemistry, in addition to the Green's Center for Reproductive Biology Sciences. Photo courtesy of the University of Texas Health Science Center at Dallas.

dents, nurses, physicians seeking a new specialty, and a variety of other students and professionals interested in medicine to participate in the Green Center's overall program.

CECIL H. AND IDA GREEN INTERNATIONAL SCHOLARS PROGRAM

An international training program, formally designated the Cecil H. and Ida Green International Scholars Program, was started in 1975, and is conducted in cooperation with the Departments of Obstetrics and Gynecology, Internal Medicine, Pediatrics, Pharmacology, and Physiology. It monitors pregnant mothers to detect early development of defective fetuses. It also provides generous funds for bringing talented and promising predoctoral graduate students (Predoctoral Trainees), postdoctoral investigators (Postdoctoral Fellows), and foreign medical experts (International Scholars) to UTHSCD's Southwestern Medical School for a year or more of training in the Reproductive Biology Sciences.

As of 1985, ten years after its inception, more than two hundred people from the United States and twenty-five foreign countries have participated in the training program.[2] Every year brings a new contingent of trainees. Among the twenty-three participants in 1985–86, for example, were five predoctoral trainees, six postdoctoral M.D. fellows, eight postdoctoral Ph.D. fellows and four international scholars. Since 1978 fourteen people have been appointed trainees in the Greens' Health Profession Career Opportunities Training Program—a special subdivision of the International Scholars Program.

[2] In a November 11, 1979, letter to me, Ida mentioned that she and Cecil had recently attended a dinner at the Southwestern Medical School to honor the first twenty Green fellows registered in the Reproductive Biology Program. She reported that the scholars were from Japan, New Zealand, England, Holland, and a number of other countries, and that this large international representation was for her "most heartwarming". She was glad that the program was off to such a fine start. "'Better babies make better people,' or so we hope," clearly expressed the special interest that induced her and Cecil to donate the funding for the Reproductive Biology Program.

So once again, in making possible yet another innovative training program, the Greens have demonstrated their strong belief in the worthiness of support for the education of talented people on the way to professional careers.

CECIL H. AND IDA GREEN BIOMEDICAL RESEARCH BUILDING

The second major building on the UTHSCD campus funded in part by the Greens, and completed in 1986, has been named the Cecil H. and Ida Green Biomedical Research Building by the University of Texas Board of Regents to honor the Greens for the donations that have played such an important role in supporting UTHSCD. This latest addition to the Health Science Center campus is a nine-story structure (figure 46) housing specialized centers of research in areas such as nutrition, cancer, heart disease, and molecular biology. The sixth floor is occupied by the greatly expanded Cecil H. and Ida Green Center for Reproductive Biology Sciences, thereby tripling the space devoted to one of the Greens' predominant medical interests.

The far-reaching importance that a new center like the Green Center in the Biomedical Research Building can have is shown by the attention it has received in the medical world. No sooner was the building completed than its occupants agreed to be the host for the 1987 meeting of the directors of the Prenatal Emphasis Research Centers of the National Institute for Child Health and Human Development. And late in 1986, Paul C. MacDonald, director of the center, was asked if the center would be interested "in serving as one of the official sites for the celebration of the Centennial of the founding of the NIH."

THE UNIVERSITY OF TEXAS MEDICAL BRANCH AT GALVESTON

Cecil's interest in biological imaging, dating from his involvement with the Scripps Clinic and Research Foundation, led him in 1983 to provide funds for an NMR installation at the University of Texas Medical Branch at Galveston (UTMBG). This first donation, covering start-up and initial operating costs, made possible the NMR research center that had been promoted by his good friend, Dr. Andrew D. Suttle, Jr., professor of nuclear physics and radiochemistry and special assistant to the president of UTMBG. This challenge gift and its results are discussed in a preceding section. Dr. Suttle had been of particular assistance to Cecil in arranging the two international conferences on biological imaging—the first hosted by Scripps Clinic in 1980 and the second by the National Academy of Sciences in Washington, D.C., in 1983.

THE PRESBYTERIAN MEDICAL CENTER IN DALLAS
In 1978 and 1979 the Greens made substantial donations toward the Dallas Presbyterian Hospital's North Village Retirement Home, the reception area of which is named in their honor. They followed this in 1982 with a major donation toward a new Center for Diagnostic Medicine, which is now offering full services and has been named for them in recognition of their generosity.

THE NATIONAL JEWISH CENTER FOR IMMUNOLOGY AND RESPIRATORY MEDICINE IN DENVER
The Diamond Jubilee Award of the National Jewish Hospital and Research Center (now the National Jewish Center for Immunology and Respiratory Medicine) in Denver was presented to the Greens on April 18, 1974, in appreciation of their substantial support of medical research and health care. The award—a bronze plaque—bears the following statement:

To

Cecil H. and Ida M. Green

Who in Defense of Man's God-Given
Birthright to Breathe, with Gentleness
and Compassion, Have Given Their Help
for the Healing of the Sick and the
Conquest of Disease

The next year, when the center held a banquet in Dallas, Cecil and Ida were included among the honored guests. Special attention was directed to them when Dr. Jerome B. Wiesner, then president of MIT, and guest speaker of the evening, handed Cecil a reproduction of his 1924 MIT Master's thesis in electrical engineering. Cecil was touched by Wiesner's thoughtful gesture, and soon afterward he and Ida visited the Hospital and Research Center in Denver. In Cecil's words,

this [visit] resulted in our deciding that this [the Hospital and Research Center] was a very worthy project—also that a large number of people from all around the Mid-Continent, particularly including the Dallas–Ft. Worth area, were utilizing this medical center. In some cases, if the individual couldn't afford to pay for treatment, it was rendered free. So we decided then to set up a Research and Investigator Chair.

That is how the Greens came to endow the Cecil H. and Ida Green Research and Investigator Chair at the Center in Denver.

CHILDREN'S MEDICAL CENTER OF DALLAS

Soon after the Greens established a permanent residence in Dallas in the early 1940s, and Ida became an active member of a number of women's organizations, she began to hear about the problems resulting from the inadequate care that was available in the Dallas area for many young mothers and their infants. This prompted her to ask herself: How can Cecil and I help with this urgent social problem?

Their first major contributions were made to the Children's Medical Center of Dallas in 1968 through 1971, the period during which they were expanding their philanthropy in Dallas to include donations for health care. Their first donation was used to establish the Maggie Green Recovery Room (honoring Cecil's mother) to provide "an intensive care area for young mothers and their new-born babies." Three later donations went toward construction of a laboratory, and additional funds made possible the purchase of a lung machine and a kidney machine. Ida would serve for a number of years on the center's board of directors, as vice president of its three hospitals, and as life member of the Women's Auxiliary, demonstrating her continued interest in the center's special program by such service as well as by a generous legacy.

As recently as the spring of 1987, Cecil acted as honorary chairman of the Dallas exhibit Dr. Seuss from Then to Now, which raised $108,000 for the Children's Medical Center. Cecil suggested that these proceeds be designated for reconstruction of the hematology-oncology clinic, scheduled to be moved to a new six-floor building that will begin construction in 1988. Thus the funds from the exhibition, added to Ida's bequest, will be used to partially fund the new clinic unit, which is to be named for Ida, at Cecil's request.

BAYLOR UNIVERSITY HOSPITAL IN DALLAS

In addition to major donations in support of medicine and health care, the Greens have made modest or nominal gifts to a number of other Dallas organizations involved in medical care. For example, after receiving special care as a patient in the Baylor University Hospital in Dallas, Cecil gave the hospital a modest sum which was used to create the Cecil and Ida Green Suite in the wing originally funded by J. Erik and Margaret Jonsson.

DANIEL FOSTER, M.D.—A PUBLIC TV PROGRAM ON HEALTH CARE

The Greens' longtime interest in the medical programs of the Scripps Clinic and Research Foundation and of The University of Texas Health Science Center in Dallas prompted them to support an innovative educational program on health care, to be prepared at the Dallas center and broadcast on public television.

The program, initially titled "Daniel Foster, M.D.," but quickly redubbed "M.D.," was arranged by Daniel W. Foster, a medical internist on the staff of The University of Texas Health Science Center, working closely with KERA-TV Channel 13 in Dallas–Fort Worth.

The program attempted to enlighten people in an interesting way about common health problems that can be diagnosed and perhaps treated at home. In thirty-minute conversations with selected specialists, Dr. Foster would discuss the body organs and the ailments related to each organ. In all, the TV program covered twenty-six subjects; among these were heart trouble, diabetes, thyroid problems, nervous disorders, obesity, high blood pressure, arthritis, cardiovascular complications, cancer, and problems of the liver, prostate, kidneys, and lungs. The program was syndicated in 1977 and then was telecast by public broadcast stations in numerous other states. The Greens' contribution consisted of eleven approximately equal donations for the programs, between 1974 and 1978. The Greens also contributed generously toward construction of KERA-TV's new Ralph Rogers Administration Building. Recently, Cecil and Ida gave a complete set of the "M.D." videotapes to Texas Instruments Inc. for employees' departmental viewing sessions. They intend to give their personal set to the Southwestern Medical School in Dallas.

CECIL H. GREEN COLLEGE OF OXFORD UNIVERSITY, ENGLAND

Cecil's support to medical institutions outside the United States has been confined to Oxford University, to whose needs he was introduced by a UBC alumnus, William C. Gibson. After receiving his M.D. degree from McGill University, Gibson had been an intern at the University of Texas Medical Branch in Galveston. On a later visit to Texas he decided to stop off in Dallas to meet Cecil and, he hoped, to rekindle his interest in their alma mater, the University of British Columbia. Since more than twenty-five years had passed since Cecil had last been on the UBC campus, he and Ida were happy to accept Gibson's invitation to visit the university and meet the president and some of the other senior administrators. It was during this visit that he and Ida donated the funds for the purchase of what became Cecil Green Park.

Gibson had completed his medical training at Oxford University. Ever the catalyzer, he soon acquainted Cecil with the pressing needs of that institution for housing and other aid not provided for medical students during their final years as clinicians. The prospect of helping to support Oxford's program of medical education appealed to Cecil for several reasons. It would give him the pleasure of making an enduring benefaction to the general welfare of the country of his birth, in which he has always had a friendly interest. Such aid would assist highly trained students at a critical career juncture and in an area of special interest to Cecil and Ida—medical education. What could be more appropriate than support of the Greens' two principal areas of concern: training of the mind and care of the body? Inasmuch as his own company, Texas Instruments, had two plants in England, Cecil also liked the idea that TI would be associated with a major contribution to such a widely respected educational institution as Oxford.

Although Gibson, the enthusiastic and effective fund-raiser, had excited Cecil's interest in Oxford's medical needs, the Greens decided to defer any commitment until they had seen for themselves the nature and extent of those needs—the customary procedure that they followed before making a major pledge. Accordingly, at Gibson's suggestion, they made their first visit to Oxford in 1977 and there met Sir Richard Doll, M.D., then Regius Professor of Medicine, and his wife Joan. They were shown the plans for two new buildings, along with those for renovating the university's famed Radcliffe Observatory. The visit further strengthened Cecil's interest. Soon after the Greens returned to Dallas they were visited by the Dolls, and shortly afterward they informed the appropriate Oxford officials of their willingness to fund the first phase of the proposed building program. They would pay the first installment on January 1, 1978.

Construction of the first building began immediately; by May 1980 the building was completed and occupied. A second generous donation in late 1978 financed the refurbishment of Radcliffe Observatory. Renovation of the observatory would produce a first-floor library, preserve the remarkable spiral stairway leading to the topmost room which had once housed the astronomic instruments, and transform that room into a commons where students could gather for tea.

Quite unexpectedly, only a few years after its renovation, Radcliffe Observatory was chosen as a center for training promising musicians in the baroque technique from all over Europe—

a decision leading Oxford's director of music to predict, "It will make Oxford the baroque centre of the world."

In the interim, the Greens had agreed to fund the second building with another major donation, and the foundation of this building was being laid when my wife and I visited Oxford in May 1980 and met Sir Richard, who had become the first warden of Green College. Cecil had informed us that he and Lady Doll had played an active role in the architectural planning and decorating of the buildings and also in designing the beautiful gardens between the buildings. The cluster of buildings funded by the Greens was dedicated on June 13, 1981, as the Cecil H. Green College (figure 47), with former British Prime Minister Harold Macmillan as the principal speaker.

It is one of Cecil's endearing characteristics that he delights in surprising a recipient and the assembled audience by announcing a completely unexpected major donation or gift "out of the blue." So it was that during the celebration of Sir Richard Doll's seventieth birthday and retirement as warden of Green College—on June 13, 1983—Cecil made the following announcement:

Ida and I have strong feelings that the memory of this historic event should be preserved by submitting to you a little surprise idea, which we hope will also be a symbol, or an expression, of our admiration and appreciation for two companion careers which have contributed so importantly, not only to the development of Green College here at Oxford, but also the welfare of mankind all around this planet—Earth.

Accordingly, Ida and I are very gladly committing ourselves to the creation of a "Joan and Richard Doll Research Fellowship"— to be financed by a related endowment fund . . . You will be interested to know that candidates are to be nominated by any science department in the University (including inorganic chemistry, nuclear physics and engineering) subject only to the particular research interest being related to medicine.

Since the creation of the Doll Fellowship in 1983, three Green College students—two of whom were Doll Fellows—have won prestigious prizes in medicine.

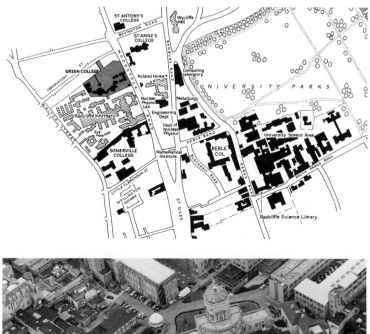

47 Cecil H. Green College located at Oxford University (England).
Top: Map showing location of Green College in upper left quadrant of
the campus. *Bottom:* Aerial photo of Green College looking southeas-
terly at the famous Radcliffe Observatory, centerpiece of the college.
Entrance to the college grounds is through the L-shaped building facing Woodstock Road adjacent to the tennis courts (*lower left*). Map and
photo permission and courtesy of Oxford University and the Oxford
University Press.

In late 1985 Cecil was notified that the Hebdomadal Council
of Oxford University had recommended to the congregation
that he be given an honorary degree of Doctor of Science, at the
1986 Encaenia ceremony on June 25, 1986. One can understand
why Cecil ranks this prestigious honor ahead of all his other
honorary degrees, because it comes from the older of the two
most venerated universities of his native England (see figures
48 and 49).

Universitas Oxoniensis

Ego Universitatis Oxoniensis Registrarius

per praesentes testor

Doctorem Cecil Howard Green

die xxv mensis Junii   A.S.MCMLXXXVI

Encaeniis nostris admissum fuisse

ad gradum

DOCTORIS IN SCIENTIA

honoris causa

Datum Oxoniae
die XXV mensis Junii
A.S. MCMLXXXVI

Registrarius

48   Green College, Oxford University. *Left:* Cecil's honorary Doctor of
Science Letter Patent of Oxford University. *Right:* Green College's coat
of arms. *Following page, top:* View looking through the entrance arch
along Woodstock Road toward Radcliffe Observatory in the background. *Following page, bottom:* View from inside the college looking
toward the entrance along Woodstock Road. All items permission and
courtesy of Green College, Oxford.

See preceding page for caption.

49   Cecil at Cecil Green College, Oxford University. *Top:* Cecil after receiving an honorary Doctor of Science degree from Oxford University. *Bottom:* Cecil stands between two close scientist friends after receiving his honorary degree—Frank Press, president of the U.S. National Academy of Sciences (*left*) and Raymond Hide, head of the Geophysical Fluid Dynamics Labortory of the U.K. Meteorological Office in Bracknell and formerly professor of geophysics at MIT, Fellow of Jesus College, Oxford, and F.R.S. Top photo courtesy of Green College, Oxford (Ivor Fields); bottom photo courtesy of Ann Hide.

8
THE GREENS' SUPPORT OF CIVIC ORGANIZATIONS

As might be expected from the nature and extent of the Greens' philanthropy, they have not only made large gifts to schools and medical institutions but have also been longtime, generous donors to many civic organizations that regularly solicit funds from the general public to support their benevolent activities. The Greens have committed time and effort, as well as money, to the organizations that they have chosen to support—namely, those that offer the greatest good to the greatest number. Among these have been the American Red Cross, the Salvation Army, Goodwill Industries, the United Way, the YMCA and YWCA, the Dallas Library, the Dallas Art Museum, the Dallas Symphony Orchestra, the Dallas County Heritage Project, Thanksgiving Square, and the Dallas Arboretum and Botanical Gardens. Their desire to help maintain and improve such organizations in Dallas was one of the reasons they were given that city's prestigious Linz Award in 1973—the first husband and wife to be jointly honored.

THE SALVATION ARMY AND THE GOODWILL INDUSTRIES OF DALLAS

As longtime supporters of the benevolent activities of the local chapter of the Salvation Army, the Greens considerably increased their donations after settling permanently in Dallas. They made substantial donations to the Salvation Army's building fund in 1973 and 1974, and a major commitment toward construction of its new Dallas headquarters in 1984. To the Goodwill Industries of Dallas they first gave clothing and furniture of nominal value. Then in the late 1960s they provided much needed funds to help Goodwill get established in their new building on North Hampton Road.

THE Y'S IN DALLAS

The Greens' concern for the recreation of underprivileged youngsters has led them to support generously the programs of the YMCA and YWCA in Dallas. They have been major supporters of the Dallas YMCA since 1970, when they initiated a five-year commitment of major annual donations. Ten years later, in 1980, they again made a large contribution, this one over two years, toward construction of a new building for the Dallas Metropolitan YMCA. The conference room in this new building carries the name Cecil H. and Ida M. Green Multipurpose Room. The Greens have also been generous supporters of Dallas's YWCA. In 1972, Cecil served as chairman of a committee to raise funds for a new building for the Dallas Community YWCA. The Greens' major donation to this fund drive was divided into two parts: one for the new building, the other for a swimming pool—later named the Ida Green Natatorium.

THE DALLAS SYMPHONY ORCHESTRA

The Greens have both been consistent and generous patrons of the Dallas Symphony Orchestra because they enjoy good music. As a youth, Cecil played the violin in Vancouver's Symphony Orchestra. In his diaries of 1928 and 1929 he describes how he built his own radio, and how he and Ida frequently entertained evening guests by listening to broadcasts of musical programs of the time.

The Greens' earliest donation in support of music in the Dallas area was made in 1961 when Southern Methodist University was soliciting funds to renovate its McFarlin Memorial Auditorium so that the Dallas Symphony Orchestra could perform there. They joined their Texas Instruments business partners, the J. Erik Jonssons and Eugene McDermotts, in pledging funds for the renovation. More recently, in 1975, they made a generous donation to the university in support of its program in music therapy.

While Cecil served as president of the Men of the Symphony during the 1965 season, he and Ida arranged benefit affairs and participated otherwise in fund-raising efforts. In other years he served the Symphony Association as a trustee and chairman. In 1983, when a campaign was started to raise funds for construction of the new Dallas concert hall, the Greens made one of the early major donations. And today, when talking to students,

Cecil often remarks, "I still experience a tingling sensation when I hear a specially melodious piece of music."

BESSES O' TH' BARN BRASS BAND (ENGLAND)

When in 1980 Cecil visited his birthplace in Whitefield, a northern suburb of Manchester, he inquired about the musical group that his mother and father had often mentioned—Besses o' th' Barn Brass Band. Did it still exist, and if so, where was it located? He discovered that the band was still as active as ever, and still located in its quarters near Manchester. Arrangements were promptly made for Cecil and Ida to meet the members of the band, who graciously presented a special program for Cecil in honor of his eightieth birthday, which fell on the very day of their meeting. Following his usual practice in such situations, Cecil left a generous donation.

THE DALLAS MUSEUM OF ART

Cecil and Ida have been loyal patrons of the Dallas Museum of Art since 1960, no doubt in part because the museum is a prime interest of their close friend and business partner Margaret Milam McDermott (Mrs. Eugene McDermott). During the past twenty-five years the Greens have given the museum major art objects from the classical period: a Greek marble kouros (man) statue from the fifth century, B.C., a marble Roman copy of a Greek kore (woman), from the first century A.D., and a collection of ancient Greek pottery, brass artifacts, earrings, and a tiny statuette.

In 1980 they made a major challenge gift toward construction of the new museum building; then early in 1984 they pledged another substantial donation to help meet the cost of refurbishing a special area in the new museum for housing and exhibiting a unique collection of French furniture and art, the gift of a donor in Paris. Finally, in late 1986 the Greens made a major donation for the purchase of eighteenth-century American furniture. In addition to their material benefactions to the museum, Cecil has also served the Dallas Art Association as a trustee.

THE DALLAS COUNTY HERITAGE PROJECT

The Greens' interest in history and antiquities led them to pledge sizable donations for the Dallas County Heritage Project initiated in the late 1960s. The purpose of the project, for which funds were to be raised by public solicitation, was to acquire Dallas' old City Park and to restore the historic buildings that once stood there. During the mid-1970s Cecil served as vice-chairman of the committee to raise funds for the project, and of course it was expected that he and Ida would contribute significantly, which indeed they have done.

THE DALLAS PUBLIC LIBRARY SYSTEM

Cecil and Ida have been strong supporters of the Dallas public library system ever since they settled in the city more than forty-seven years ago. They have contributed regularly to the library's annual operating expenses, and in the late 1970s pledged a substantial gift to the new central Dallas public library, dedicated on April 18, 1982. This latest donation is commemorated by a plaque designating a small conference room adjoining the director's office as the Cecil H. Green Board Room.

When, after viewing the room on dedication day, Ida and I were alone for a few moments, she remarked in an aside that she was not disturbed that her name was left off the plaque, even though their donation was a joint one, because:

It is enough that the gift has helped to produce something that will make the world better in some respect. It is a kind of religious thing for me—just to feel inwardly that we have done something good for our fellow humans. I don't need the accolade of the public nor the inscription on a plaque.

Cecil recently advised me that this new library building has been named in honor of J. Erik Jonsson in recognition of his outstanding achievements for Dallas as former mayor of the city.

THE DALLAS ARBORETUM AND BOTANICAL GARDENS

The Greens' most recent contribution to enhancing the urban environment of Dallas was a major 1984 challenge donation to develop the Dallas Arboretum and Botanical Gardens; the first garden at the main entrance is named for them.

In connection with their 1984 pledge of $300,000 to the Capital Fund Campaign for the development of the new arboretum and botanical gardens in Dallas, Cecil wrote as follows to Ralph B. Rogers, chairman of the Dallas Arboretum and Botanical Society, Inc.:

. . . our decision was based on two important factors—first, we are now convinced that this new Arboretum and Botanical Gardens in their dramatic location, overlooking White Rock Lake, will indeed bring added pleasure and educational benefits to the Greater Dallas Community—secondly and equally important, we are greatly impressed by the dedicated and capable leadership of you and our other good friends [Stanley Marcus, Virginia Nick and Margaret McDermott] . . .

THE GREENS' SUPPORT OF RELIGIOUS ORGANIZATIONS

Cecil and Ida have seldom attended church services for any length of time, no doubt in part because they moved about so much in their earlier married years and because, more recently, they spent much time traveling. Nor have they chosen to support a particular denomination to the exclusion of others. Quite to the contrary, they have made substantial donations to a wide range of religiously affiliated organizations, in every case with the hope that their donation would benefit the greatest number of people. These organizations, most of them in the Dallas area, include:

Cistercian Preparatory School (general support and construction of Science Center)

Episcopal School

Jarvis Christian College

Jesuit College Preparatory School

National Jewish Center for Immunology and Respiratory Medicine (Denver)

Presbyterian Medical Center and Retirement Home

St. Christopher's Church

St. James-by-the-Sea Episcopal Church (La Jolla)

St. Luke's Church

St. Michael's Episcopal Church

Thanksgiving Square in downtown Dallas (a chapel and surrounding space for quiet meditation)

Ursuline Academy

A SUMMATION OF THE GREENS' PHILANTHROPY

The model that the Greens have developed in their philanthropy consists of six actions, not necessarily taken in the order
listed:

1. Recognizing excellence and the potential for excellence.

2. Recognizing critical and basic needs, which they often discover by themselves.

3. Funding buildings, laboratories, and special facilities, where
teachers, researchers, and students can work together.

4. Endowing chairs, visiting lectureships, investigatorships,
and the like, to bring distinguished professors, researchers, and
performers to leading educational and medical institutions for
variable periods of time.

5. Establishing endowed scholarships, fellowships, internships, and residencies to attract talented students to professional careers in critical fields.

6. Donating generous funds to encourage and support research
and interchange of ideas devoted to the overall improvement
and advancement of education, medical research, and health
care.

7. Providing both funds and personal effort to enhance the programs and activities of civic organizations functioning for the
public good.

There remains one final component of their philanthropy that
is probably the secret of why they are so widely respected and
admired. They have given as generously and unselfishly of their
time, thought, and personal effort as of their wealth. And in
their giving, they have actively sought the advice, cooperation,

and collaboration of others, involving them as partners in their many innovative enterprises. Finally, that they have so often voiced their pleasure in giving shows that they regard their philanthropy as a joy and not just a responsibility.

The generous and altruistic sharing of their wealth has brought them both national and international acclaim, as is described in detail in the following section. Yet despite all the admiration and esteem that have been accorded Cecil and Ida Green, both separately and jointly, it could well be that their greatest reward of all will be the knowledge that their benefactions have left the world a better place than they found it.

9
AWARDS AND HONORS

The many honors accorded the Greens during the past thirty-five years bear witness to the magnitude and variety of their philathropy. These honors recognize the Greens' services to numerous public and private organizations, and express gratitude for their support of educational and medical institutions throughout the English-speaking world. Many of the honors have been given to both Greens; others have been granted to Cecil or Ida individually. The Greens' honors are organized chronologically in the following sections.

JOINT AWARDS AND HONORS

A CIVIC TRIBUTE—THE LINZ AWARD, DALLAS (1973)

The Linz Award, sponsored by the *Dallas Times-Herald* and Linz Jewelers, and regarded as the community's highest honor, has been given annually for more than forty-five years to people whose benefactions and humanitarian acts have helped to improve the Dallas community in an unusually significant way. An independent committee of twenty-four members, evaluating almost a hundred nominees, selected Cecil and Ida Green for the 1973 award. They are the first husband and wife to have been honored jointly in the forty-five-year history of the award.

The formal citation, read at the presentation of a hand-engraved silver plaque, begins with the following summation of the Greens' contributions:

. . . they have directed their personal energies and their material resources as well as their knowledge and integrity of purpose in numerous selfless ways to help alleviate important human needs. To view their contributions in the most appropriate perspective . . . it is necessary to bear in mind that a great part of improving the quality of life for all Dallas citizens can only be attained through long-term efforts—to raise the quality of education at all levels; to support the creation of new knowledge in the physical and social sciences; to support efforts to improve the delivery of health services; to broaden the opportunities for Dallas citizens to increase their awareness of our heritage and culture.

The citation, presented by Albert Linz Hirsch at the Fairmont Hotel in Dallas on February 19, 1974, then proceeds to list the many good deeds of the Greens that have benefited the citizens of Dallas.

It is worth noting here that Cecil's close business associates, Eugene McDermott and J. Erik Jonsson, co-owners of GSI and co-founders of TI, had also received the Linz Award for their respective community activities—evidencing a continuing tradition of commitment by TI to the welfare of the Dallas community.

HONORARY ADMIRALS OF THE TEXAS NAVY (1973)
On the occasion of the christening of the R/V *IDA GREEN*, on May 31, 1973, at which Ida did the usual honors, the state of Texas named Cecil an Honorary Admiral of the Texas Navy and Ida an Honorary Vice Admiral, recognition accorded the Greens in appreciation of the substantial donation that made construction of the ship possible (see figure 54).

DIAMOND JUBILEE AWARD OF THE NATIONAL JEWISH HOSPITAL AND RESEARCH CENTER (1974)
On April 18, 1974, the Greens jointly received the Diamond (seventy-fifth) Jubilee Award of the then National Jewish Hospital and Research Center (now the National Jewish Center for Immunology and Respiratory Medicine) in Denver, Colorado, in recognition of their financial support of medical research and health care. The bronze plaque presented to them at the celebration bears the tribute reported earlier, in the discussion of the Greens' support of medicine.

AN INTERNATIONAL TRIBUTE (1978)
Unprecedented recognition was accorded the Greens at a gala social event held at the National Academy of Sciences in Washington, D.C., on the evening of November 9, 1978, as described briefly in my prologue. Distinguished representatives from leading educational and medical institutions in Australia, Can-

ada, England, and the United States gathered to pay Cecil and Ida "An International Tribute" for their imaginative, creative, and exceedingly generous philanthropy (figure 1).

CITATION OF THE NORTH DALLAS CHAMBER OF COMMERCE (1978)

At a special luncheon of the North Dallas Chamber of Commerce in 1978, Cecil and Ida were honored guests, and each received a handsome plaque with the following citation:

> 1978
> Award for Excellence
> Presented to
> Cecil H. Green
> Ida M. Green
> In recognition of Significant and
> Outstanding Achievement in the
> Field of Humanities
> North Dallas Chamber of Commerce

A NATIONAL TRIBUTE—THE NAS PUBLIC WELFARE MEDAL (1979)

A year after the Washington gathering, the Council of the National Academy of Sciences selected the Greens to receive its prestigious Public Welfare Medal, established to honor people who have contributed significantly to advancement of the public welfare. Thomas F. Malone, foreign secretary of the Academy, concluded his citation by stating:

The distinguishing features in the Greens' involvement with scientific endeavors are the degree to which they have personally immersed themselves in the institutions they serve, the range of scientific fields they have sought to foster, and the number of lives they have personally affected. They have been discriminating donors, seeking and fulfilling those opportunities where their support could make qualitative difference—to people and to institutions—and thereby, in the highest sense, contribute to the public welfare.

The prestige of the Academy's Public Welfare Medal can be judged by the record of past recipients: George W. Goethals and

William C. Gorgas (1914), Herbert Hoover (1920), John D. Rockefeller, Jr. (1943), James R. Killian, Jr. (1956), Warren Weaver (1957), and Detlev W. Bronk (1964), to name but seven of the thirty-six persons elected between 1914 and 1978. The Greens are the first couple to be honored jointly, and Ida is only the second woman, after Leona Baumgartner, to have received the medal. This award of the NAS Public Welfare Medal also made the Greens honorary members of the Academy, once again the first husband and wife to be jointly elected.

THE PHILANTHROPY AWARD OF THE DALLAS
HISTORICAL SOCIETY (1982)
Cecil and Ida, individually but simultaneously, received the 1982 Dallas Historical Society Award for Excellence in Community Service, in recognition of their philanthropic activities in Texas and elsewhere in North America, as well as abroad. This award was presented to the Greens in special ceremonies at the Hall of State located in Fair Park, Dallas, on September 22, 1982. The society emphasized that the Greens are known worldwide for their contributions to the geophysical sciences, and their donations to hospitals, museums, and elementary and secondary schools, colleges, and universities.

HONORARY MEMBERSHIP IN THE AMERICAN
INSTITUTE OF ARCHITECTS (1983)
Cecil and Ida were jointly named honorary members of the American Institute of Architects in New Orleans on May 22, 1983. As the first husband and wife to be elected honorary members of the AIA since 1891, the Greens were cited " . . . for consistently stressing the importance of architectural excellence" in the dozen major academic and medical buildings that they have made possible through their judicious philanthropy. These architecturally distinctive buildings include the towering Center for the Earth Sciences at MIT, designed by I. M. Pei; the Graduate and Professional Center at Colorado School of Mines, designed by Donald L. Prezzler; the extension of the expanded Stanford University Library, designed by Gyo Obata—all

named in honor of Cecil and Ida; and finally the Library Study Center, Mathematics and Science Quadrangle, and other buildings at the St. Mark's School of Texas, Dallas, designed by O'Neil Ford.

THE GREAT TREKKER AWARD OF THE UNIVERSITY OF BRITISH COLUMBIA (1984)

In the fall of 1984, the Greens were notified that they had been selected by the Students' Council of UBC, Cecil's first undergraduate institution, as the recipients of that year's Great Trekker Award. This award was established to commemorate the anniversary of UBC's move to its present campus from a cluster of four buildings in the Fairview district of Vancouver, which the university had occupied as temporary quarters from 1915 to 1925. The award has been made periodically to UBC alumni who have achieved preeminence in their careers and who have also supported the university. The award, given to both Cecil and Ida, rather than to Cecil alone, is the first one ever to be given jointly. It was the special desire of the Students' Council to emphasize that much of the Greens' philanthropy has been, in fact, the result of joint effort, even if Ida was often the silent partner. The Greens' relationship with the University of British Columbia was discussed in detail in the section entitled "The Greens and UBC."

FIRST CHARTER MEMBERS OF FRIENDS OF THE AAUW FOUNDATION (1984)

The 1984 Annual Report of the American Association of University Women announced that the names of Cecil and Ida had been placed at the head of the list of the First Charter Members of the AAUW Educational Foundation, in recognition of their long-standing support of and commitment to the goals of AAUW and its Educational Foundation. That support, consisting of the creation and endowment of the Ida M. Green Fellowship for Women, successive additions to the original endowment, and other donations, was discussed more fully above in the section headed "The Greens and the AAUW." Thus far, between 1969 and 1988, eighteen outstanding Texas women have received the AAUW's Ida M. Green Fellowship. Their names are listed in appendix G.

HONORARY LAY AWARD FOR DISTINGUISHED SERVICE FROM THE AMERICAN MEDICAL ASSOCIATION (1984)

In late December 1984 Cecil and Ida were told that the Board of Trustees of the American Medical Association had nominated them for the AMA's Honorary Lay Award for Distinguished Service.

This citation, established by the House of Delegates in 1948, is the highest award the AMA can bestow on non-physicians. It is awarded in recognition of unusual contributions by lay persons to the advancement of the ideals of American medicine in the fields of medical science, medical education, or medical care.

The citation, read at the opening session of the AMA House of Delegates at its meeting in Chicago, June 16, 1985, adds that

Cecil and Ida Green have touched the lives of countless human beings around the globe. Their generous investments in their fellow citizens through strengthening the teaching and scientific research capabilities of many institutions is [sic] reminiscent of the strategic role private individuals played in fostering the Age of Exploration centuries ago. They have made it possible for thousands of young people to pursue their own educational goals and develop their talents.

HONORARY FELLOWS OF OXFORD'S GREEN COLLEGE, ENGLAND

In 1980 both Cecil and Ida were elected Honorary Fellows of Oxford University's Cecil H. Green College in recognition of their major donations in 1978–79 that made possible the establishment of the new medical college named in Cecil's honor.

CECIL AND IDA GREEN TOWER, TULSA (1986)

In midsummer of 1986, Marvin R. Hewitt, president of the Society of Exploration Geophysicists, wrote the Greens

. . . that the SEG Executive Committee has voted unanimously to name the SEG international headquarters building the "Cecil and Ida Green Tower" [see figure 50].

With this action, the Society is thus establishing its lasting tribute to and formal recognition of each of you as dedicated, concerned individuals who chose to get involved and remain involved. SEG, exploration geophysics and science in general have benefitted [sic] greatly from this involvement and the multitude of positive results of your involvement will live on forever fostering continuing progress in the scientific and educational communities.

DISTINGUISHED SERVICE MEDAL OF THE CHANCELLOR'S ASSOCIATES OF UCSD (1986)

The Chancellor's Associates of the University of California-San Diego recently awarded Cecil and Ida their Distinguished Service Medal in recognition of their contributions "in time, creative effort and monetary gifts to UCSD." In choosing the Greens as one of the two couples honored at their fourteenth annual dinner at Revelle College on the campus, on Saturday, June 28, 1986, the associates took notice of the contributions already discussed in the section titled "Support of Traditional Education in California."

These contributions include the endowment of the Green Foundation for Earth Sciences, which supports the Green Visiting Scholar Program by paying the expenses of visiting scholars to the IGPP. They also include funds for equipment purchases at the institute; the securing of the land for the Piñon Flat Observatory; the purchase of Donal Hord's sculpture, *Spring Stirring*, for Walter Munk's office area at IGPP; and most recently a major contribution toward the cost of a faculty-community club building at UCSD, which will be named after Cecil and Ida.

NSFRE OUTSTANDING PHILANTHROPISTS OF THE YEAR AWARD (1987)

On November 17, 1987, the Dallas chapter of the National Society of Fund Raising Executives (NSFRE), at their Seventh Annual Luncheon and Awards Ceremony, celebrating National

50   Cecil and the Society of Exploration Geophysicists. *Top:* The SEG
named its Tulsa international headquarters building (the Geophysical
Resource Center) the Cecil and Ida Green Tower on October 15, 1986,
in recognition of their continuing support for scientific and educational communities. Included in that support was a major donation to
SEG's Education Foundation. *Bottom:* Cecil has been a longtime and
active SEG member, and in the photo stands with other members of
the 1948–49 Executive Committee (*from left to right*): E. V. McCollum,
secretary-treasurer; Cecil H. Green, past president (1947–48); L. L.
Nettleton, president; M. King Hubbert, editor; and Andrew Gilmour,
vice president. Photos courtesy of Bob McCormak and SEG.

Philanthropy Day, honored Ida M. and Cecil H. Green as Outstanding Philanthropists of the Year. Nominated for the award by the Children's Medical Foundation of Texas, the Greens were recognized jointly for their many contributions to local schools and hospitals.

AHE CECIL AND IDA GREEN AWARD ANNOUNCED AT TAGER TWENTIETH ANNIVERSARY CELEBRATION (1987)
AHE President Fred Baus recently announced the establishment of the Cecil and Ida Green Award for Significant Contributions to Higher Education in North Texas. This award, announced at the twentieth Anniversary gala dinner of AHE in November 1987, "will be given periodically to those individuals who have demonstrated leadership, commitment, and vision in support of higher education" (*AHE News* 2/2, Winter 1988). The award honors Cecil and Ida jointly for their leading role in establishing and developing TAGER in the late 1960s.

INDIVIDUAL AWARDS AND HONORS TO
CECIL H. GREEN
In addition to the foregoing awards and honors conferred upon the Greens jointly, each also received numerous awards and honors individually. Descriptions of the more important of Cecil's individual honors follow; Ida's honors are discussed farther on.

IEEE HONORARY MEMBER (1983)
Cecil joined the American Institute of Electrical Engineers after completing his training in electrical engineering at MIT and General Electric in 1926. A lifetime later, in 1983, the Institute of Electrical and Electronic Engineers, successor of the original AIEE, elected him an honorary member, in recognition of his achievements in applying his professional training to geophysical exploration—as party chief, regional supervisor, corporate executive, and educator of fledgling explorationists—and in cofounding Texas Instruments Inc.

SEG MEMBER (1936); SECRETARY-TREASURER (1945–46)
TO PAST PRESIDENT (1948–49); HONORARY LIFE
MEMBER (1954); FIRST RECIPIENT OF VIRGIL
KAUFFMAN MEDAL (1966); FIRST RECIPIENT OF
MAURICE EWING GOLD MEDAL (1978)

Cecil has long felt that scientists and engineers should strongly support their respective societies and associations both locally and nationally. He himself has done exactly that. He applied for membership in the two leading professional organizations primarily concerned with the petroleum industry—the Society of Exploration Geophysicists (SEG) and the American Association of Petroleum Geologists (AAPG). In 1936 he became a member of both organizations.

In SEG he soon became an enthusiastic participant in the activities of the society by serving on special committees and urging GSI's younger employees to join SEG. He himself was successively elected secretary-treasurer (1945–46), vice president (1946–47), president (1947–48), and past president (1948–49), as well as serving on the executive committee during his four years in office (1945–49) (see figure 50).

In 1954 SEG made Cecil an honorary member in recognition of his distinguished contributions " . . . to the advancement of the profession of exploration geophysics through service to the Society." In 1966 the Awards Committee of the society selected him to be the first recipient of the Virgil Kauffman Gold Medal in recognition of " . . . his outstanding contribution to the advancement of the science of geophysical exploration as manifested during the previous five years [1961–66]." He received the medal at the society's thirty-sixth Annual Meeting in Houston on November 7, 1966. In the citation presented by CSM's John C. Hollister, Cecil was honored specifically

. . . for his active and persistent support in the development of geophysical data processing methods based on communication and information theory whose full impact upon the profession has become manifest in the past year.

Cecil replied modestly to Hollister's comments:

My own contribution [to the advancement of geophysical exploration] is indirect and lies mainly in a long-time interest in two things: (1) science and engineering education, and (2) the development of all-important working relationships between industry and government agencies, with our professional society and the processes of education.

Cecil's most recent award from the SEG came in 1978 when the society selected him as the first recipient of its highest honor, the newly established Maurice Ewing Gold Medal, which also confers honorary membership upon the recipient. The medal was presented to Cecil at the SEG's Forty-eighth Annual International Meeting held in San Francisco on October 30, 1978.

The recommendation of the SEG's Honors and Awards Committee included the following statement:

He has done more to advance the profession of geophysical exploration than any other living person.

And Robert C. Dunlap, Jr., in his lengthy citation printed in the luncheon program, touched on one of Cecil's outstanding personal traits when he wrote:

Cecil's love of people results in a very personal relationship which many have experienced but none can adequately describe. One of our great scientists recently said that without question Cecil is the most beloved person in our profession.

Cecil was deeply pleased to receive this prestigious medal because it honors and perpetuates the memory of his personal friend and highly esteemed geophysical colleague, W. Maurice Ewing, one of the foremost geophysicists of the twentieth century. The extent of that esteem can be judged by the fact that the Greens, in 1973, endowed a distinguished chair—the Cecil H. and Ida Green Professorship in Marine Sciences—at the University of Texas Medical Branch in Galveston, with Ewing as its first occupant (see figure 29).

THE AAPG'S HUMAN NEEDS AWARD (1974)

Cecil has been a faithful member of the American Association of Petroleum Geologists since 1936, though not as active in the

affairs of the association as he has been in the SEG. Nevertheless, in 1974 he received one of the association's highest honors,
the Human Needs Award, which Cecil received at the AAPG-
SEPM Annual Meeting in San Antonio, Texas, on April 1, 1974,
in recognition of "his beneficial services to mankind through
significant contributions directed toward pressing human
needs." In my citation, I wrote,

Cecil long has been concerned about three of the more basic of
human needs: (1) the need to know and understand, in short,
to be educated; (2) the need to communicate and cooperate, that
is, to collaborate; and (3) the need to innovate. What he has
done about these makes a most impressive story. . . .

Citation: For extraordinary generosity of time, effort, counsel
and means in advancing education and technology, and in improving communication and cooperation between industry,
education, and government. (*AAPG Bull.* 58/9, pp. 1881–1882,
September 1974)

CSM'S CECIL H. GREEN GEOPHYSICAL OBSERVATORY
(1961)

Early in his long association with the Colorado School of Mines,
that institution recognized Cecil's achievements as an exploration geophysicist by naming its forty-acre geophysical research
facility the Cecil H. Green Geophysical Observatory.

The earthquake-recording observatory is in an abandoned
horizontal mining tunnel driven into the mountain at Bergen
Park, near Evergreen, a few miles west of Golden. One of five
similar geophysical observatories developed with federal funds,
it is equipped with modern instruments for measuring various
physical properties of the earth's crust. Data from the instruments are telemetered to the observatory building near the entrance to the tunnel (see figure 25).

A GOLD MEDAL FROM POPE PAUL VI (1965)

A papal diploma and a gold medal were awarded to Cecil by
Pope Paul VI in 1965 in recognition of outstanding work done

for Christianity. The decoration (*Benemerenti*) was presented to Cecil by the University of Dallas, which was authorized to do so by His Holiness, on April 28, 1965.

GSI HONORS CECIL BY NAMING TWO OF ITS SHIPS AFTER HIM

Geophysical Service Inc., which Cecil served from 1930 to 1955, when he retired as president and became board chairman, named two of its fleet of marine vessels after him.

The first ship, M/V *CECIL H. GREEN*, was christened on July 11, 1965, at the Mangone Shipbuilding Company in Houston, Texas. The ship was designed and equipped specifically for seismic exploration at sea and started her seventeen-year career in September 1965. By the time she was deactivated in 1982, to be succeeded by the M/V *CECIL H. GREEN II*, her crews had accumulated a proud record of work at sea, thereby fulfilling the charge given her by GSI's fourth president, Robert C. Dunlap, Jr., who remarked at the Houston launching:

In a few minutes you will touch the water for the first time and be christened with the name you will carry the rest of your life. You will be proud of your name, but when you begin your life's work, you must also assume the responsibilities imposed by your name. These are simple to state but difficult to discharge. To honor your name you must excel in being a safe ship, a happy ship, a hard working ship, a productive ship, and a creative ship. . . . When you have accomplished these things, you truly will honor the name that you bear.

The M/V *CECIL H. GREEN II*, the new flagship of the GSI fleet, was dedicated on February 15, 1982, at the St. Andrew Marina in Panama City, Florida. At her dedication, GSI President Dolan McDaniel stated:

This vessel is an excellent one, equipped with a full complement of TI-built digital field systems, data logging, streamer tracking, computer-controlled airgun arrays, and integrated navigation and control systems. She is capable of collecting the most sophisticated and advanced seismic data available to the industry. We are truly proud of her, as we are of her namesake.

He went on to declare that M/V *GREEN II* was "the finest vessel in GSI's fleet," an evaluation that was confirmed in August 1984 when Cecil and the *GREEN II*'s crew received notice that their ship had been designated the "Winning Vessel of the Month for July," because she broke all production records during that month, in competition with the twenty or more vessels of the entire GSI fleet! (figure 51).

Dot Adler's account of the dedication in GSI's *Shotpoints* includes the following description of one of Cecil's most persistent characteristics—the desire to know whereof he speaks:

The first thing Cecil wanted to do when he arrived in Panama City was tour the ship, which he happily explored from bow to backdeck and from bridge to engine room. Inspecting the crew's cabins, he tried out a bunk for size and hinted he would like to ride the *Green II* on her first voyage through the Panama Canal after starting work in the Gulf of Mexico.

Later, in commenting on how he felt about having two GSI ships as his namesakes, Cecil remarked:

I cannot even imagine any greater honor than to have this latest GSI exploration vessel named for me. In fact, I am all the more overwhelmed with emotional feelings of gratitude as I realize this is actually the second vessel in the GSI fleet to carry my name. I feel just a little dizzy as I address you from cloud 57, somewhere up in the stratosphere.

Then thinking of others, as was his wont, he continued,

I am going to be extremely proud of this latest GSI ship and so, naturally, hope that each and every member of its crew will be very happy in guiding its destiny through a lifetime of operations in waters all around the free world.

Robert C. Dunlap, Jr., summarized the feelings of older GSIers by emphasizing that everything the Greens did was a *joint* effort, and that "The important thing is that this old world is a much better place because of Cecil and Ida Green."

More information about the specifications and performance of M/V *CECIL H. GREEN II* can be found in the 1982 *GSI Grapevine*.

51   The M/V *CECIL H GREEN II*, the current flagship of the GSI fleet, was christened on February 15, 1982, following deactivation of the M/V *CECIL H. GREEN (I)*. The 185-foot, 247-ton ship is equipped with the latest digital systems, data logging, streamer tracking, computer-controlled airgun arrays, and integrated navigation and control systems. She is capable of operating anywhere on the world ocean and of collecting the most sophisticated and advanced seismic data available to the industry. Cecil is even prouder of this second ship named for him, and greatly enjoyed the special tour of the vessel before it was launched. Photo and description courtesy of GSI and the *GSI Grapevine*.

[CECIL] GREEN KNOLL [SEAMOUNT], [IDA] GREEN CANYON, AND GREEN CANYON LEASE BLOCK IN THE GULF OF MEXICO

Numerous seabottom topographic features along the southeastern edge of the Sigsbee Scarp, in the north-central part of the Gulf of Mexico, were discovered, mapped, and informally named in the 1970s by J. L. Worzel and members of his crew, operating from the University of Texas Marine Biomedical Institute at Galveston.[1] Two of these features, [Cecil] Green Knoll, a seamount, and [Ida] Green Canyon, an adjacent submarine em-

---

[1]Worzel, J. L., and Watkins, J. S., Exploration of the Sigsbee Scarp (abs.): *Geophysics*, 38, 1973, p. 1229; Shih, T.-C., Worzel, J. L., and Watkins, J. S., Northeastern Extension of Sigsbee Scarp, Gulf of Mexico, *AAPG Bull.*, 61, 1977, p. 1962–1978.

bayment, were informally named for the Greens by Worzel and members of his crew. These designations are now in common use by oceanographers and petroleum geologists interested in the seabottom topography of the north-central part of the Gulf of Mexico. Furthermore a large rectangular lease area containing the two features mentioned above is now designated the Green Canyon block (see figure 30, pages 258–259).

GREEN KNOLL (SEAMOUNT)

The seabottom prominence with the kidney-shaped outline that is located in the very southwest corner of the accompanying chart (figure 30, top), and designated Green Knoll, is a salt diapir formed by near vertical intrusion of salt from an underlying source (Shih, Worzel, and Watkins 1977).

In the lower left panel of figure 30, titled Cecil Green Seamount, is an enlarged sketch of the prominence and of the immediately surrounding bathymetry. The lower right panel of the figure shows an east–west section through the seamount and adjacent bottom sediments. The two foregoing figures were taken from a special sketch prepared for Cecil by Worzel and his co-investigators at UT's Marine Biomedical Institute at Galveston. Obviously, they intended to honor Cecil in naming the knoll for him.

As to the naming of Green Canyon, Worzel has informed me by telephone and letter as follows (October 1985):

The Green Canyon is an embayment in the Sigsbee Scarp to the west of the northern edge of the Green Knoll. Sediment is rapidly filling in the mouth of the canyon against the NW corner of the Green Knoll.

Concerning the name for the embayment, Worzel added,

I have a vague recollection of discussing with Joel [Watkins] and Shih [Tai-chang] using the name [Green Canyon] in a published paper without going through any of the formalities of naming it and by that means the name became a fait accompli.

Further information came to my attention when Green Canyon was mentioned in GSI's *Shotpoints* for July–August 1985, by its editor-in-chief, Dot Adler, who kindly wrote me as follows:

Skinnie Holbert of GSI contacted Don Clark of the U.S. Department of the Interior Minerals Management Service . . . who looked into the origin of the name "Green Canyon." Mr. Clark reported to Skinnie that Green Canyon is named in honor of Ida Green and was named by the people at the Marine Biomedical Institute of the University of Texas at Galveston who operate the research vessel *Ida Green*.

From the foregoing information, it now seems evident that Cecil Green Seamount is the complete name for the knoll, Ida Green Canyon is the complete name for the embayment, and the name Green Canyon for the lease block is a shortened version of the canyon name that honors Ida. It seems especially appropriate that the Greens' names have been given to the seamount and canyon, inasmuch as the R/V *IDA GREEN* and crew conducted extensive exploration in the Gulf of Mexico.

SANTA RITA AWARD OF THE UNIVERSITY OF TEXAS (1974)

The Santa Rita Award, the highest honor that can be bestowed on an individual by the University of Texas System, is given periodically by UT's board of regents to honor those who have made valuable contributions over an extended period to the university system in its developmental efforts. Among other things, the recipient is judged "to have demonstrated concern for the principles of higher education, as well as a deep commitment to furtherance of the purposes and objectives of The University of Texas System specifically."

Cecil and Eugene McDermott were selected to receive 1973 Santa Rita Awards at the UTD commencement on May 18, 1974. McDemott's award was made posthumously, however (he died on August 24, 1973). Both Green and McDermott, together with their wives, were cited for their founding and funding of GRCSW-SCAS, the predecessor of UT-Dallas; for their longtime service as trustees of UT's Southwestern Medical Foundation; for their generous funding of buildings, facilities, and endowed chairs in the Southwestern Medical Center; and for numerous other services and donations.

SMU CITES CECIL H. GREEN TWICE FOR SPECIAL ACCOMPLISHMENTS

Twice in recent years Southern Methodist University has publicly cited Cecil H. Green for special accomplishments. The first citation was made in September 1979 when the Institute for the Study of Earth and Man held its second symposium on "Unconventional Methods in Exploration for Petroleum and Natural Gas" in Dallas. The symposial banquet featured the institute's first award given for outstanding accomplishment in devising and applying new approaches to the search for petroleum and natural gas. That chosen honoree was Cecil, who received a specially bound copy of *Websters Third New International Dictionary* together with the following citation:

Cecil H. Green, exploration geophysicist, pioneer in the application of modern instrumentation in the search for hydrocarbons, enlightened philanthropist and inspiring colleague: The Institute for the Study of Earth and Man is pleased and proud to acknowledge these notable accomplishments with this citation (ISE&M *Newsletter*, December 1979).

The second citation by the university was made during its first School of Engineering and Applied Science open house for graduate students in the fall of 1981. Cited as the "father of the TAGER network," Cecil was credited with creating and funding the linkage between area colleges and universities and high-technology industries, which together constitute the TAGER system.

A MEMORABLE LUNCHEON AT THE DALLAS PETROLEUM CLUB (as reported in the *GSI Grapevine* 32/3, 1976)

On May 7, 1976, forty GSIers gathered for a luncheon at the Dallas Petroleum Club to take note of the change in Cecil's company status from "active retirement" to "semi-active retirement," as the *GSI Grapevine* reported. Actually, the real purpose of the gathering was to give Cecil's GSI friends, old-timers, and recent comers alike, an opportunity to honor him for his services to Geophysical Service Inc. and TI.

A special collage prepared for the occasion included a record of his forty-five years of service at GSI and TI as follows:

1930–31/1932–1936	Party Chief
1936–1941	Supervisor
1941–1950	Vice President, GSI
1950–1955	President, GSI
1955–1959	Chairman of the Board, GSI
1959–1975	Honorary Chairman of the Board, GSI
1951–1965	Vice President, TI
1951–1976	Founder of TI, and Member of the Board
1976–	TI Founder, and Honorary Director

Friendly banter, fraught with special meaning regarding Cecil's personal traits and accomplishments, was the tone of the gathering as one close friend after another recalled Cecil's past. J. Erik Jonsson remarked that in his own beginning with GSI he had no field experience and Cecil had no office experience "so the combination worked out to everyone's advantage." H. B. Peacock described Cecil as "an old smoothie," then praised his effective use of his many talents in getting a job done. J. C. Karcher characterized Cecil as "a real operator," and, like Jonsson, "emphasized the way he made good things happen." And R. C. Dunlap, Jr., commented that he always "managed to come up with some real smart ideas after Cecil had paid him and his field party a visit." Friendly comments from others clearly showed that Grant Dove's title for Cecil—"Mr. GSI"—was fully deserved. Everyone emphasized that Cecil is "a man that considers human relations to be the most important factor in any venture, either personal or business."

In addition to the esteem and respect which his fellow GSIers showed him that May day in 1976, Cecil also received several keepsakes. Bill Blakeley, representing GSI's Marine Operations, presented him with a specially prepared book about the M/V *CECIL H. GREEN* and her many accomplishments as the first GSI ship expressly designed and equipped for seismic work. Karcher gave him a copy of a special treatise on the development of the seismic reflection method and its great importance to the petroleum industry. And old friend and fellow partner Jonsson gave Cecil a unique GSI emblem mounted on a plaque, which Cecil called a "gem of a thing" because of the many differently colored pieces of stone used in its composition.

As on so many such occasions, when it came time at the end for Cecil to make a few remarks, he accepted the accolades and gifts with sincere gratitude and humility, then followed with his own tribute to Ida and the wives of his fellow GSIers:

It is their understanding and willingness to endure, in some cases the hardships of moving about, that has kept good men on the job in some very adverse places.

AMERICAN ACADEMY OF ACHIEVEMENT'S
GOLDEN PLATE AWARD (1985)
On June 29, 1985 Cecil received the Golden Plate Award of the American Academy of Achievement (AAA) at that organization's "Salute to Excellence" meeting in Denver. He was one of forty new guests of honor selected from the "Captains of Achievement" from America's "Great Walks of Life," because he is a distinguished "representative of the many who excel" in their chosen profession. Cecil's name was added to the latter group as a distinguished exploration geophysicist and generous benefactor to educational institutions worldwide.

FREEMAN OF THE CITY OF VANCOUVER
On November 9, 1987, the mayor of Vancouver, Gordon Campbell, informed Cecil by letter that the Vancouver City Council had unanimously voted to bestow on him its "highest honour: *Freeman* of the City," with the following comments:

Your lifelong record of achievements, innovations and contributions in business and education are exemplary. City Council believes that it is important for its citizens, particularly the younger generations, to be aware of the impact residents have had in the course of their careers.

You are our first citizen who has met the challenges of the high tech industry on a world scale, and we are all very proud of you. By awarding you this honour we pay tribute to your accomplishments and hold them up as an example for our next generations to work towards.

Cecil was named "*Freeman* of the City of Vancouver" at the *Science World* dinner in March 1988.

HONORARY DOCTORAL DEGREES

It is the tradition of many colleges and universities throughout the world to honor certain men and women for their good deeds and outstanding achievements by awarding them an honorary doctoral degree in some broad field of knowledge.

Thus far, by 1987, the Greens together have received a total of sixteen such honorary doctorates—twelve to Cecil and four to Ida.

In view of Cecil's varied accomplishments—as engineer and explorationist, as corporate founder and executive, as educator, as concerned humanitarian, and as innovative benefactor to schools, medical institutions, and civic organizations—it is not surprising that he has received the twelve honorary doctorates listed here. The four doctorates awarded to Ida, and their accompanying citations, are listed farther on.

DOCTOR OF ENGINEERING (1953)

COLORADO SCHOOL OF MINES   CSM was the first academic institution to confer an honorary doctorate on Cecil. In so doing, the school recognized the leadership that he had shown in directing the growth of GSI and the support that he had given to the training of student engineers.

DOCTOR OF SCIENCE (1961)

THE UNIVERSITY OF SYDNEY (AUSTRALIA)   This degree was awarded to Cecil "in recognition of his many accomplishments and continued activities directed toward assisting education in the sciences."

DOCTOR OF SCIENCE (1961)

THE UNIVERSITY OF TULSA   This degree was awarded to Cecil as "recognition for his services to education, particularly in science, both at the collegiate and secondary level, for his contributions to his chosen profession, geophysics, to the electronics industry, international relations and the American way of life."

DOCTOR OF SCIENCE (1964)

UNIVERSITY OF BRITISH COLUMBIA   Cecil's first alma mater awarded him its honorary degree in recognition of his achievements as an applied geophysicist and his and Ida's many benefactions to educational institutions, medical training and research organizations, hospitals, and health care centers.

DOCTOR OF LAWS (1966)

AUSTIN COLLEGE (SHERMAN, TEXAS)   This doctorate recognized Cecil's strong support of the visual media and the performing arts, an area in which Ida also had a deep interest.

DOCTOR OF SCIENCE (1967)

SOUTHERN METHODIST UNIVERSITY   Both Cecil and Ida have had a long affiliation with SMU, starting in 1941 when they established a permanent home in Dallas not far from the university's campus. Since then Cecil has been continuously involved in the affairs of the institution, especially with science and engineering matters, TAGER, and UTD. It is this important and productive service directed to the advancement of learning that SMU recognized in awarding Cecil the honorary Doctor of Science degree.

DOCTOR OF SCIENCE (1974)

UNIVERSITY OF MASSACHUSETTS (AMHERST)   This degree recognized Cecil's accomplishments as an exploration geophysicist. It should be noted that at the time it was awarded, Randolph W. Bromery, well known for his own geophysical work on aeromagnetics, was chancellor of the University of Massachusetts at Amherst.

DOCTOR OF SCIENCE (1974)

TEXAS CHRISTIAN UNIVERSITY   Cecil's citation reads: "he has combined the risk-taking of the bold entrepreneur, a mental curiosity that continually surprises his peers, and a multi-form generosity; so that . . . the Cecil Greens may be counted among the world's philanthropists in the fullest and healthiest meaning of that word."

DOCTOR OF CIVIL JURISPRUDENCE (1976)

UNIVERSITY OF DALLAS   This honorary degree followed the Papal Award of 1965, which had been presented to Cecil by the University of Dallas, in recognition of his support of Christianity.

DOCTOR OF COMMERCIAL SCIENCE (1978)

SUFFOLK UNIVERSITY (BOSTON)   The citation read in part: " . . . you have been a longtime proponent of collaborative thinking and effort between education and industry. Indeed your untiring efforts have brought about a unique cooperation and a wider understanding in these vital areas."

DOCTOR OF SCIENCE (1986)

OXFORD UNIVERSITY, ENGLAND   The degree was awarded at the prestigious ceremony of Encaenia at Oxford on

June 25, 1986, in recognition of his founding of Oxford's Green College.

DOCTOR OF PHILANTHROPY (1987)

HAWTHORNE COLLEGE (ANTRIM, NEW HAMPSHIRE)
This unique degree, the first ever in philanthropy, was awarded jointly to Cecil H. and Ida M. Green (posthumously), on May 24, 1987, by Hawthorne President Vincent A. Fulmer, whose citation included the following tribute:

Cecil and Ida Green are exceptional for many reasons. First, they have always acted together. Before we lost Ida last December, they had been married almost sixty-one years and had roamed the world together doing good works and inspiring others to greater efforts. Second, they have not stopped with two or three institutions in their home town communities but they chose instead three dozen colleges, universities, hospitals, academies, schools and museums in four countries as their field of focus. And third, they made a science of the art of philanthropic networking by devoting prodigious energies to their task and personally participating in the governance of those institutions.

SPECIAL ACADEMIC APPOINTMENTS
In addition to the foregoing honorary doctorates awarded to Cecil, we should mention three special academic honors that he has received: an honorary lectureship at MIT (1973), a consulting professorship at Stanford (1983), and an honorary lectureship at the University of British Columbia (1984).

AN HONORARY LECTURESHIP AT MIT (1973)
On March 1, 1973, a small group of MIT administrative officers and professors gathered in the office of Vincent A. Fulmer, then secretary of the MIT Corporation, to witness the appointment of Cecil H. Green and Eugene McDermott as Honorary Lecturers in the Department of Geology and Geophysics. Since MIT does not award honorary degrees, appointment as honorary lecturer is regarded by the Institute as its most prestigious

honor. Until 1973 only one person had been so recognized by the Institute, Britain's wartime Prime Minister Sir Winston Churchill, who came to MIT in April 1949 to be the principal speaker of the Institute's Mid-Century Convocation held in the Boston Garden.

In the case of Green's and McDermott's appointment, MIT wished to recognize their eminence as entrepreneurs in applied geophysics and the benefactions and services that they had contributed to the Institute. But sadly for Cecil, he was unable to join the group in Fulmer's office. Early on the previous day, February 28, he had received an urgent telephone call from Ida in La Jolla saying that his mother's death was imminent. He flew to San Diego the same day but arrived there too late to see his mother alive. Maggie Howard Green (figure 52) had died during his flight, the last day before her ninety-fifth birthday. Ida was with Cecil's mother at the last, just as she had been with his father earlier, in both cases because Cecil was away from home on business. Cecil's last act for his mother, who had been steadfastly solicitous of her only child from the day he was born, was to arrange for her burial beside his father in San Diego's Greenwood Memorial Park.

CONSULTING PROFESSOR OF GEOPHYSICS AT STANFORD (1982)

Stanford University, like MIT, has a long tradition of not awarding honorary degrees; instead, the university gives special importance to the title Consulting Professor.

In mid-1982 the Stanford Board of Trustees appointed Cecil Consulting Professor of Geophysics, thereby recognizing that his participation in the university's activities was much more extensive and pervasive than that of the traditional professor. Vice President and Provost Albert H. Hastorf assured Cecil of the importance of the honor by letter as follows:

In the absence of honorary awards, we attached particular importance to the appointment of Consulting Professors. These are working appointments that help us maintain our all-important ties with distinguished scientists, engineers, and

52 Cecil's mother, Maggie Howard Green (March 1, 1878-February 28, 1973), in front of her apartment at 4091 45th Street, East San Diego, California, ca. 1960. Photo from Cecil.

scholars from industry and government who give lectures from time to time on campus and otherwise participate in our intellectual life.

HONORARY LECTURER IN GEOPHYSICS AT UBC (1984)
In 1984, after being chosen by the Students' Council of UBC to receive the university's Great Trekker Award (jointly with Ida), Cecil accepted an appointment as Honorary Lecturer in Geophysics. He had already delivered his well-known lecture, with appropriate slides, on his 1939–40 visit to GSI field parties in Egypt, Saudi Arabia, and northwestern India (now Pakistan), so the UBC students and faculty in the Geophysics Department knew what they could get from one of his slide shows. No doubt he will be giving another of his travelogues the next time he visits UBC, for he enjoys nothing more than telling anecdotes about the life of the old-time doodlebugger.

MIT'S RECOGNITION OF CECIL'S MANY DONATIONS AND SERVICES

MIT has honored Cecil H. Green in many ways for his donations and personal services to the Institute. Recognition first came in 1958 when he was elected a member of the Corporation. Life membership followed in 1961, and he became Life Member Emeritus in 1975. Since 1950, the Institute has repeatedly requested Cecil's services on committees, especially as chairman of the Corporation's visiting committees to the Departments of Geology and Geophysics, and of Physics. Cecil served as president of the MIT Alumni Association during SY 1968–69, and as chief marshal led the traditional parade of participants from the main building to Rockwell Cage for the 1969 commencement exercises (figure 53). When the Corporation recently asked Cecil to serve as honorary chairman of its $550 million five-year *Campaign for the Future* announced in October 1987, he accepted and pledged a major donation that will fund MIT's new Center for Physics.

The examples cited here impressively demonstrate the loyalty and support that Cecil has given his second alma mater and why he has been a repeatedly honored alumnus, class of 1923.

INDIVIDUAL AWARDS AND HONORS TO IDA M. GREEN

The high esteem in which Ida has been held by a host of friends and admirers, for the quiet but active role that she has played in the Greens' life-style and philanthropy, is manifest from the many awards and honors that she has received.

Those awards received jointly with Cecil are particularly significant, because in a number of cases it was the first time that the honor was awarded jointly to a husband and wife. The Greens are so widely considered inseparable in thought and action that it was only natural to have them share an honor equally, even if such a joint award broke with long-standing tradition. Ida has also received a number of honors on her own, the more important of which are described briefly here.

53   Cecil as an active MIT alumnus forty-five years after graduation.
Cecil was president of the MIT Alumni Association during school-year
1968–69, and served as chief marshal of the 1969 MIT commencement
exercises on June 13, 1969. He wore the marshal's formal academic re-
galia and carried the Institute's mace, which he recalls he "found convenient to rest on one shoulder." In the above photo he is flanked
(*from left to right*) by Francis W. Sargent, governor of Massachusetts;
Howard W. Johnson, president of MIT; Vannevar Bush, honorary
chairman of the MIT Corporation; and James R. Killian, Jr., corporation chairman. Courtesy of The MIT Museum.

THE IDA GREEN ROOM, IN THE GREEN BUILDING, AND THE IDA FLANSBURGH GREEN HALL, BOTH AT MIT

Twice in recent years the MIT Corporation has formally recognized Ida's important role in the Greens' benefactions to the Institute: first, in December 1974, when the Corporation named the ninth-floor lounge of the Cecil and Ida Green Earth Sciences Building, the Ida Green Room, in her honor; and second, in June 1983, when they authorized dedication of the venerable Albert H. Tuttle Residence as the Ida Flansburgh Green Hall, MIT's first residence hall for women graduate students (figures 22 and 23).

THE IDA GREEN NATATORIUM IN THE DALLAS YWCA

Among the Greens' numerous benefactions to civic organizations in Dallas was a donation to the YWCA's building fund that was sufficiently large that the new building could have a swimming pool. This much-desired facility was named the Ida Green Natatorium in Ida's honor.

THE IDA M. GREEN COMMUNICATION CENTER AT AUSTIN COLLEGE, SHERMAN (1972)

When, in 1972, Cecil donated the funds to Austin College for a new building to be "dedicated to the Advancement of Human Understanding and Appreciation through the Arts and the Science of Communication," the college's Board of Trustees named the building the Ida Green Communication Center. In addition they designated a large space in the Center as the Ida Green Theater, and a smaller room as the Cecil Green Conference Suite (figure 27).

THE R/V *IDA GREEN* (1973)

When a new ship for oceanographic work was planned for the UT Medical Branch at Galveston, the Greens agreed to donate one-half of the vessel's cost of construction. The completed ship was christened the R/V *IDA GREEN* in Ida's honor on May 31, 1973. Since then the ship has played an active role in the ocean-

ographic program of the Earth and Planetary Sciences Division
of the UTMB in Galveston (figure 54).

THE IDA M. GREEN PROFESSORSHIP IN GEOSCIENCES AT UT-DALLAS (1981)

This chair at the Dallas campus of the University of Texas System was established by the Board of Trustees in 1981 to honor Ida for her personal interest in UTD and for her many joint donations with Cecil in advancing higher education. In announcing establishment of the chair, Bryce Jordan, then president of UTD, clearly indicated the high esteem in which Ida was held in the Dallas community by the following comment, which bears repetition here:

We want very much to do this not only because of your great personal interest in UT-D, but also in recognition of your share in the brilliant partnership in the philanthropy that you and Cecil have carried on over the years. That partnership, practiced with such modesty and thoughtfulness, has enriched higher education, and thus the minds of young people, the world over.

THE *IDA M. GREEN* RACING SHELL (1987)

Friends of Ida, and of the St. Mark's School of Texas, wishing to honor her posthumously, contributed the funds to purchase a four-oarsman racing shell for the school. Cecil christened the *Ida M. Green* at nearby Bachman Lake on Saturday, October 17, 1987 (see figure 55).

THE IDA M. GREEN HEMATOLOGY/ONCOLOGY CLINIC IN THE CHILDREN'S MEDICAL CENTER IN DALLAS (1987)

Early in 1987 (April 4 through June 21), Cecil served as Honorary Chairman of the Dr. Seuss From Then to Now exhibition that raised some $108,000 for the Children's Medical Center in Dallas. This money will be added to Ida's 1987 bequest to help fund preparation of new space for the Center's hematology/oncology clinic in the six-floor building now starting construction. At Cecil's suggestion, the new space will be named the Ida M. Green Hematology/Oncology Clinic.

54   The R/V *IDA GREEN* and Greens as "admirals" of the "Texas
Navy." *Top:* The R/V *IDA GREEN* was christened by Ida at the Galveston (Texas) Yacht Basin on May 31, 1973. The 116-foot, 278-ton ship
was designed and equipped for geophysical research in the open
ocean by scientists in the Earth and Planetary Sciences Division of the
University of Texas Medical Branch at Galveston. The Greens donated
half of the estimated cost of construction as a challenge gift, and the
new vessel was named for Ida. The R/V *IDA GREEN* and crew, operating out of Galveston as a base, followed courses totaling more than
20,000 miles of multichanneling seismic and gravimetric observations,
mostly investigating continental margins. The ship took measurements in the western Caribbean, in the Gulf of Mexico, and off the
west coast of Central America and southern Mexico as far north as
Acapulco. The ship was deactivated several years ago. *Bottom:* At the
christening celebration, Truman G. Blocher, president of UTMBG
(center) named Cecil "Admiral of the Texas Navy," and Ida "Vice Admiral." Photos courtesy of the University of Texas Medical Branch,
Galveston.

55   Cecil at St. Mark's School of Texas. *Top:* Cecil poses with a group
of lively lower school pupils (grades 1–4) and two of their teachers.
The woman on his right is Mrs. Linda Williams, first grade teacher; on
his left is Sister Rachel Auringer O.S.U., head of the lower school.
Bottom: Cecil helps carry St. Mark's *Ida M. Green,* a four-oarsman rac-
ing shell purchased by friends of Ida's, which he christened at nearby
Bachman Lake on October 17, 1987. Photos courtesy of St. Mark's
School of Texas.

HONORARY DOCTORAL DEGREES

DOCTOR OF HUMANITIES (1977)

AUSTIN COLLEGE (SHERMAN)   This degree was awarded to Ida in recognition of her special interest in the theater and the performing arts and of her part in funding the Communication Center named for her.

DOCTOR OF HUMANITIES (1977)

TEXAS CHRISTIAN UNIVERSITY   The citation reads as follows:

To this devoted wife and partner in a lengthening history of philanthropic gifts, to this gentle and unpretentious leader and follower in manifold community and national enterprises, to this patroness of education and the arts, Texas Christian University is pleased to add its honor, joining the many that have cited her for vision and generosity, for unquenchable faith in human possibilities.

DOCTOR OF LAWS (1979)

UNIVERSITY OF BRITISH COLUMBIA   Cecil's first alma mater, UBC, conferred an honorary LL.D. degree on Ida in recognition of her role in a husband-wife partnership that has bestowed an impressive number of benefactions on educational, medical, and civic institutions in Australia, Canada, Great Britain, and the United States.

DOCTOR OF PHILANTHROPY (1987)

HAWTHORNE COLLEGE (ANTRIM, NEW HAMPSHIRE) As discussed earlier, in a precedent-setting action Ida was posthumously awarded a first-ever honorary doctoral degree in philanthropy jointly with Cecil, on May 24, 1987, and the next day a sturdy oak sapling was planted on the Hawthorne campus in their honor.

Because of the unique nature of this joint degree, I ask the reader's indulgence in allowing me to repeat President Vincent A. Fulmer's citation, which included the following tribute:

Cecil and Ida Green are exceptional for many reasons. First, they have always acted together. Before we lost Ida last December, they had been married almost sixty-one years and had roamed the world together doing good works and inspiring others to greater efforts. Second, they have not stopped with two or three institutions in their home town communities but they chose instead three dozen colleges, universities, hospitals, academies, schools and museums in four countries as their field of focus. And third, they made a science of the art of philanthropic networking by devoting prodigious energies to their task and personally participating in the governance of those institutions.

HONORS FROM SOUTHERN METHODIST UNIVERSITY

Ida has been honored repeatedly by Southern Methodist University, even though she was unable to complete requirements for a degree because of her ever-increasing responsibilities as a corporate wife.

Upon request, she served as a trustee and as a member of several important policy committees, and the university bestowed upon her its three highest honors: Mortar Board, Distinguished Alumna, and Life Member of the SMU Alumni Association.

OTHER ACADEMIC RECOGNITIONS

Ida's affiliations with academic institutions other than her alma mater, SMU, have brought her, along with Cecil, recognition at the following:

MIT: Corporation Member and Life Member emerita; Honorary Alumna

UBC: Honorary Life Member of Women's Faculty Club; Great Trekker Award

TCU: Honorary Alumna

UCSD: Project IDA (International Deployment of Accelerometers)—a global geophysical network for earthquake recording

AAUW: Honorary recognition; Member and past Vice President of Dallas chapter

SUMMATION OF AWARDS AND HONORS TO CECIL AND IDA

As might be expected from the number and diversity of their benefactions, Cecil and Ida have received considerable recognition in the media; honors and awards from academic institutions, professional and civic organizations, and the business world; and unusual tributes of different kinds.

Special anniversary publications, textbooks, and pamphlets have been dedicated to them in honor of their support of education and scientific research. Finally, numerous professional, civic, and social organizations of which they have been members, have elected them to leadership roles as officers, chairpersons of fund-raising and advisory committees, and other similar assignments. Although these assignments were often regarded by the uninitiated as "honors"—and that they were—nevertheless, they commonly demanded considerable time and effort on the part of both Cecil and Ida. It is widely known and deeply appreciated that they have given as generously and unselfishly of themselves as of their means—a fact emphasized in one citation after another. Indeed, it can be said that Cecil and Ida Green, in the way that they have shared themselves and their means, have left a magnificent legacy that will enrich the lives of generations to come.

EPILOGUE

It is with deep sadness that I commence this epilogue, because Ida died the day after Christmas 1986 as I was nearing completion of this account of her and Cecil's sixty years of happy married life. On December 30, 1986, more than 240 friends from around the world gathered at the St. James-by-the-Sea Episcopal Church in La Jolla at 2:00 p.m. to bid their silent farewell to an exceptional woman whose compassion and generosity had touched so many people so deeply. They came from Australia, England, and Canada, and from coast to coast in the United States. Following the conclusion of the memorial service, the earthly remains of Cecil's "dear Ida" were placed in a mausoleum in San Diego's Greenwood Memorial Park where Cecil, too, will be laid away (in a crypt).

On May 1, 1987, a memorial service was held in MIT's Ida Green Room, in the Green Center for the Earth Sciences, to give members of the Institute an opportunity to express their appreciation of the support that both Ida and Cecil had provided MIT during the past thirty-five years. Following are excerpts from the eulogies delivered (and electronically recorded) by the six speakers and Cecil's response to their tributes.

David S. Saxon, Chairman of the Corporation:

We are drawn together here this afternoon, here in this special room that bears her name, to speak of our regard and affection for our friend and colleague, Ida Mabelle Green; to express our sense of loss and our sympathy to her husband Cecil; and to give thanks for a life well and fully lived. Here in this gathering are members of the Corporation's Executive Committee, present and past Corporation members, our presidents emeriti, Ida Green Fellows and Green Professors, faculty members from the Department of Earth, Atmospheric and Planetary Sciences, from the Department of Physics, from other departments as well. Drawn from so many different elements of MIT, those present bear impressive witness to the many diverse and important ways in which Ida and Cecil have helped to build and strengthen the basic structure of this institution. When I reported Ida's death last December to her colleagues on the Corporation, I noted that she held a very special place in the MIT family—one that is uniquely hers. At the dedication in 1983 of the Ida Flansburgh Green Hall, a residence for women graduate students, Ida was aptly described as "a grand MIT woman," who has played a major role in MIT's development for the past twenty-five years. Together, as partners, Cecil and Ida became in Jim Killian's words "two masters of what John D. Rockefeller called 'the difficult art of giving.'" She and Cecil were the only husband and wife ever to serve on the MIT Corporation at the same time. This building, the Cecil and Ida Green Building, was the first at MIT to bear the names of a husband and wife . . . In this service we can touch on only a few of the facets of Ida's life and call on only a few of the many who came to know and admire her . . .

Robert R. Shrock, Professor Emeritus of Earth, Atmospheric and Planetary Sciences:

Ida Mabelle Flansburgh, of Danish and Dutch ancestry, grew up near Schenectady and was working in the General Electric plant there in the early 1920s, when a Course VI-A Canadian student, born in England, one Mr. Cecil Howard Green, appeared on the scene. When Ida first saw him striding down a GE corridor, she suddenly said to herself, "There's the man I'm going to marry." Marry they did, in 1926, and during the next sixty years they combined their values, standards, attitudes, and actions in such harmony that they blended into a single personality and became Ida and Cecil or Cecil and Ida. It was love at first sight, and that love for each other persisted through thick and thin until the inevitable parting a few months ago. On their eighth anniversary, Ida wrote in her diary—on Mustang Island, offshore from Corpus Christi, Texas—"Cecil and I wandered around the sand dunes and along the beach. Had a nice talk. I feel that happiness is so elusive to most that once you have it hang on to it for dear life, because happiness between a man and a wife can be a beautiful thing." Ida always wanted a piece of the action, as Van Bush would say it, and wrote in her diary, "I just couldn't stand to be left behind. There was always the challenge, and I was determined to meet it. If Cecil decided he was going, then I was going too." . . . Like the diamond that reflects its inner strength and beauty by its natural facets, Ida reflected her many laudable traits of character from her multi-faceted personality. She had an inherent generosity that led her to declare, when only thirty-eight years old, that if she and Cecil became rich . . . she wanted most to be a philanthropist. She had an unaffected compassion for the unfortunate and deprived, particularly for young mothers and their babies; and a quiet determination to help women students get a college education, urging them to get the best possible training they could get and then try to make the world a little better place to live in . . . Ida had many, many other attributes . . . Independent, she had a mind of her own. When she spoke it was always with few words but to the point, and she held her own wherever she went—unaffected, modest . . . Ida left us an imperishable legacy and we have compelling reason to be forever grateful to her and to Cecil, for both made MIT a better place.

Margaret L. A. MacVicar, Dean for Undergraduate Education:

Almost ten years ago, I found myself in Dallas. It was my first time, and the reason was an invitation by Cecil and Ida to visit St. Mark's School, TAGER, and Texas Instruments. The trip was exhausting and exhilarating and it was finally winding down when I came to the Greens' house . . . As I approached the front door of the Greens' home, I mentally prepared for being peppered with his [Cecil's] questions about what I had seen on my visits earlier that day, and I knew that Cecil would give me no quarter about what I had seen and heard. So the door opened and Cecil welcomed me in, and he led me into the living room, already questioning me about TAGER. As we entered the room, a very firm and quiet voice came from the chair and said, "Cecil, be still now for a few moments," and lo, Cecil was still. Ida turned to me, her eyes were twinkling, and there was the hint of a smile beginning to engulf her face, and we began an afternoon of quiet unusual conversation and for me a new friendship. She asked me about young people, she asked me about their hopes and aspirations, she wanted to know about student housing, she wanted to know about elementary school education, she asked me about secondary school science. As the conversation went on, we turned to some of the earlier things in her own life, and she told me about some of the towns that she and Cecil had lived in. Some of the loneliness and temporariness of those moves and stops, she said, had had very erosive effects on people's spirits; but it was the base of love and friendship between herself and Cecil that had seen her through those moments . . . Then Cecil perked up mischievously and he said, "Now may I speak?" And the three of us laughed at his question. You could always feel the warmth of the special code in the teasing and in the laughter, but then suddenly Cecil was very serious and he said, "It was Ida who got us through, who got us all through those days." Now over the years since, I have been privileged to see Ida in many different circumstances, not the least of which is here on our own campus. And I note that inside the edges of a great modesty and with a tad of shyness, her concern for the quality of life and well-being of our students, and especially our women students, has imprinted on the history of MIT.

Frank Press, President, National Academy of Sciences:

It is difficult to talk about Ida without talking about Cecil, and vice versa, and to separate their good works and attribute one project to Cecil and one to Ida. They were truly lifelong partners, but, as you have heard, it was Ida who led the Greens to a life of philanthropy which was to become their great passion and a great source of satisfaction to them. In the program you see the record of their philanthropy at MIT which is quite extraordinary. In other places you can read about their world record of philanthropy, their good works on three continents—universities, hospitals, laboratories, cultural centers, chairs, scholarships, and fellowships. Their philanthropy is in good taste. It represents investment in people. It emphasizes science, education, health, and culture. Their projects are well researched, but not intrusive. The Greens have received recognition in many forms—Ida in her own right as well—but you'll excuse me if I mention one that came from our own organization, when they both received, for the first time jointly offered, the Public Welfare Medal of the National Academy of Sciences, which entitled them to honorary membership in the Academy . . . Ida had a rich, unique personality in her own right. She accepted diversity in people; she was well educated, well read, she had strong views but she was willing to listen, and change her mind. She was a feminist ahead of her time, but nevertheless accommodating to the very demanding career of her husband, a career that is unique to his profession, exploration geophysics—constantly uprooted, long separations. In viewing their projects, one can definitely recognize a theme throughout all of them that represents Ida. She was concerned about people and their development from their earliest beginnings . . . The Greens' work in reproductive biology, in the problems of premature parturition, shows that the first starts of a human can affect the rest of their subsequent development . . . The Greens led very simple lives, unpretentious lives . . . Ida did her own housework. They never had a driver. They were products of a very simple past. Occasionally they splurged, on a cruise to an exotic place, but their big splurge was always philanthropy, and doing good for other people . . . In the nation's capital in September we will be dedicating a Cecil and Ida Green Building which will house the National Research Council. It is on a lovely four-acre plot in Georgetown in Washington just at the edge of a national park. On a small corner of that campus we are setting aside a small place of beauty—benches, shrubs, trees—so that in the nation's capital there will be a special place as a memorial to Ida Green where people can gather in a place of beauty and discuss the great issues of our time.

Howard W. Johnson, Honorary Chairman:

As you all know, Ida Green was a rare individual, an unforgettable person, and throughout her wonderful life, a staunch friend of education. She was characterized by a strong sense of personal integrity, a straightforwardness in her approach to life and in her relationships with her friends, an uncompromising sense of high values, a strong sense of duty, a belief in equal partnership, a sense of strong commitment to wise philanthropy; and overall, a sense of seeking quality in everything with which she dealt . . . Ida and Cecil were well recognized as, perhaps, . . . the premier philanthropists of their time, for a wide variety of institutions of higher education . . . I remember the discussions with her and with Cecil that led to their funding in 1974 of the Ida Green Fellowships . . . for women graduate students at the Institute. Years later when the program had already supported dozens of scholars, I sat next to Ida at a dinner in Dallas . . . and told her that I knew how much the individuals involved had appreciated that program, and she said . . . "I hope so . . . but I appreciate as much how much thay have given to me." Ida has done many things in the world of education and human development, especially here at MIT . . . We've seen many great women at this institution—Ellen Swallow Richards, Margaret Compton, Katharine McCormick, to name just a few. There will be many more, but added to the list now, for all time into the future, is the name of Ida Green.

Paul E. Gray, President:

We've heard this afternoon moving testimony to the many ways in which Ida Green made this world a better place. In thinking over what I might say, it occurred to me that underlying the caring, the generosity, the style, and the courage that we have

all come to associate with Ida Green, there is a quality that per-
meated her work and her life in every domain. And that quality,
I submit, was leadership—leadership in the deepest sense.
When I think of leadership, I think of those human qualities
that enable one person to motivate others to do their very best,
to give them confidence in themselves and in the future, to provide the vision and the means for unleashing creativity and energy to serve the common good. This university and numerous
other academic, medical, and cultural organizations around the
country, and indeed around the world, have benefited enor-
mously from Ida Green's leadership. Together with Cecil, Ida
was instrumental in building a business and creating a philanthropic career in stunning significance and influence. In so
doing, she made a world of difference in a world of people that
she cared deeply about . . . While focusing their interests on
education, on communications, on medicine and health, Ida
and Cecil did not impose their visions or their wills on the
people and the institutions that have had the good fortune to
become partners with the Greens. They picked their partners
carefully; but once done, they exercised the kind of leadership
that Ida personified. Rather than invest us with the means to
pursue their dreams, they invested us with their confidence and
with the means to pursue our shared dreams and our common
goals. And so we have on this campus a major academic building for the earth, atmospheric and planetary sciences, where
teachers and students come together to learn and to teach, to
stretch their horizons and ours. We have nine professorships
endowed by the Greens, chairs in earth sciences, in electrical
engineering, in physics and education—chairs which give their
holders the support and the flexibility and, in a fundamental
way, the confidence to pursue their visions of the future. And,
because of Ida's special interest in expanding opportunities for
women in science, engineering, and related fields, we have the
Ida Green Fellowships—fellowships for women who are just
beginning their graduate study at MIT. Since 1974, when Cecil
and Ida endowed this program, there have been nearly one
hundred Ida Green Fellows. Some are here today. Most are well
launched into their careers—careers in physics, chemistry,
mathematics, applied biological science, aeronautics, electrical
engineering, mechanical engineering, civil engineering, architecture, political science, economics, and management. I submit
that through them, through those one hundred Ida Green
scholars, and hundreds more to come in the generations ahead,
the vision, the confidence, the leadership of Ida Green, will be

celebrated for decades to come. Someone has said that the teacher affects eternity; no one knows where her influence ends. I submit that the creators, Cecil and Ida, of the Green Fellows program in the same way affect eternity. And now, Cecil, in this gathering of old colleagues and friends, if you would care to address us, we would as always consider it our privilege.

Cecil H. Green:

Well, dear friends all, you are looking at someone here who feels . . . like crying, because I am almost overcome with sentimental feelings of gratitude by this series of heart-warming memorial tributes to dear Ida, my wonderful wife and constant companion for nearly sixty-one years. But, this event is typical of good old MIT, which I rate highest, not only in education proficiency but also by being the very highest of all universities with which I am so familiar, in thoughtfulness and in sensitive feelings of concern and compassion for its many associates and friends. So it comes easy and indeed natural for me to express the warmest feelings of gratitude, especially to David Saxon and Paul Gray, for arranging this sentimental reunion as a memorial to dear Ida. . . . Ida instructed me in no uncertain terms, before she departed our happy life, to carry on, and so it is very important to mention that this wonderful get-together certainly encourages and indeed helps me to do just that—carry on. My remarks would certainly not be complete without reminding all of you that MIT gets full credit for bringing Ida and me together in the first place, and especially as I remind you that we did not grow up together, but were widely separated in our early days by the entire width of America, Ida growing up in Schenectady, New York, and I in faraway Vancouver, British Columbia. So I feel special thanks are due MIT's Course VI-A—Cooperative Electrical Engineering—and . . . also its General Electric option, because of being so directly responsible for bringing us together at just the right time in our young lives. And now I say again that I am almost overcome with the warmest feelings of gratitude to each of you dear friends for coming together and especially in this Ida Green Room with its wonderful portrait back there of my longtime companion, dear Ida . . .

Ida's last wish to Cecil—that he "carry on" as they had so long done together—is being followed faithfully by Cecil as witness his most recent philanthropic actions: the 1987 commitment that will make possible the new MIT Green Center for Physics; the March 1988 major donation that will fund the cost of the improved Physical Sciences Center at St. Mark's School of Texas; funding of one of the twenty proposed distinguished chairs for the University of Texas Southwest Medical Center in Dallas; a major donation to make possible a proposed Cecil and Ida Green Research and Training Institute for the Regional Center for Infants and Young Children (RCIC) of Washington, Maryland, and Virginia—a special interest of Billie and Frank Press; and a generous commitment to the fund being raised to endow a Jerrold R. Zacharias Chair in Physics at MIT.

In addition to his philanthropic activities at home, Cecil has made three long trips abroad to deliver invited addresses. August 1987 found him in China delivering the keynote address at the joint meeting of China's Society of Petroleum Geophysicists and the United States' Society of Exploration Geophysicists held in Zhou Zhou City, Hebei Province, People's Republic of China. Returning to America, he had only a few months to rest before flying to Australia in mid-November to give a laudatory address at the retirement ceremony of his greatly admired friend, Professor Harry Messel, founder of the University of Sydney's famous Science Foundation for Physics. After spending the Christmas holidays with the Messels, he returned to La Jolla to prepare for his next trips. In early March 1988 he flew to London and then continued to Oxford, where he laid the foundation stone for additional quarters for Green College. Then on to the Kingdom of Saudi Arabia where, as Honorary Chairman of the Board of GSI, he was the guest of honor, and of course one of the principal speakers, at a banquet celebrating GSI's fiftieth anniversary of operations in that country. After a brief stop in London, en route home to La Jolla, he spent the rest of March on the West Coast, as well as the first week in April, then flew to Stanford to attend a meeting of a visiting committee in geophysics. While there he learned that the Stanford Associates had voted to award their prestigious Degree of the Uncommon Man to him jointly with Ida, which will require his presence at special ceremonies in Dallas and San Diego late in 1988 and early in 1989. He then had to hustle back to La Jolla to be present at the dedication of UCSD's Cecil H. and Ida M. Green Faculty Building on April 14. Ten days later he was on the phone to British Columbia where he was scheduled to receive a special honor, Freeman of the City, from the Mayor of Vancouver at an evening ceremony on April 29, 1988.

The remaining months of 1988 kept Cecil traveling back and forth between Dallas and La Jolla, as he passed his eighty-eighth birthday, and 1989 looks as if it will be little different from past years regarding speeches, honors, and travel.

The year 1989 started auspiciously for Cecil when he received word from the People's Republic of China, through the SEG,

. . . that the Standing Committee of Chinese Geophysical Society had passed a resolution to honor Cecil H. Green with the first Foreign Honorary Member of Chinese Geophysical Society.

Conditions permitting, Cecil has every intention of accepting the invitation of the Chinese to go to Beijing to receive this internationally prestigious honor, yet another "first," but that long trip will have to be delayed until late in 1989 because of Cecil's previous commitments.

After the reawarding of the Degree of the Uncommon Man in San Diego, Cecil will go to Vancouver to participate in the kick-off celebration on March 20 for the UBC fund-raising campaign of which he is the Honorary Chairman. April calls for trips to Washington, D.C., and Boston; June, for a return trip to Vancouver and a week at Green College, in each case to participate in dedication ceremonies.

Judging from his present good health and unabated interest in the results of his and Ida's many benefactions, one can expect Cecil to continue a program of philanthropic activities, invited lectures, and occasional visits to Vancouver and the Boston area, thereby demonstrating that he is, indeed, carrying on as Ida instructed (figure 56).

56   "Until death do us part." *Top:* Cecil and Ida (*center*) enjoying their sixtieth anniversary party, February 6, 1986, at the A. W. Harris Faculty Club in Dallas, hosted by Paul C. MacDonald (*left*), director of the Green Center for Reproductive Biology Sciences at the University of Texas Health Science Center, and Charles C. Sprague, president of the Green Center. *Bottom:* A saddened Cecil in their Dallas apartment, a year after Ida's death on December 26, 1986. Top photo courtesy of Ann Harrell, *Center Times*, Office of Medical Information (UTHSCD); bottom photo courtesy of the *Dallas Morning News* (Ken Geiger).

APPENDIX A
ACRONYMS AND ABBREVIATIONS

AAA	American Academy of Achievement
AAAS	American Academy of Arts and Sciences, or American Association for the Advancement of Science
AAPG	American Association of Petroleum Geologists
AAUW	American Association of University Women
AHE	Association for Higher Education of North Texas
AIA	American Institute of Architects
AIEE	American Institute of Electrical Engineers
AMA	American Medical Association
AT&T	American Telephone and Telegraph Company
BuAer	Bureau of Aeronautics (U.S. Navy)
CALTEX	Standard Oil Company of California and Texaco (= Texas Oil Company)
Caltech, CIT	California Institute of Technology
CASOC	California Arabian Standard Oil Company
CAT	Computer Assisted Tomography
CMCD	Children's Medical Center in Dallas
CSM	Colorado School of Mines
EMI-CT	Electro-magnetic Imaging—CAT
GAG	Geophysical Analysis Group (at MIT)
GE	General Electric Company

GRC	Geophysical Research Corporation
GRCSW	Graduate Research Center of the Southwest
GSI	Geophysical Service Incorporated
IBM	International Business Machines Incorporated
IDA	International Deployment of Accelerometers
IEEE	Institute of Electrical and Electronic Engineers
IGPP	Institute of Geophysics and Planetary Physics (UCSD)
IOC	Indian Oil Company, Limited
IT&T	International Telephone & Telegraph Corporation
KERA-TV	Channel 13 (Dallas–Fort Worth Station)
LJFES	La Jolla Foundation for Earth Sciences
MAD	Magnetic Aerial Detector
MB	Medical Branch of University of Texas at Galveston
MBI	Marine Biomedical Institute (University of Texas)
MIT	Massachusetts Institute of Technology
M/V *CECIL GREEN*	Marine Vessel *Cecil Green*
NAS	National Academy of Sciences
NAS-NRC	National Academy of Sciences-National Research Council

NSFRE	National Society of Fund Raising Executives
NIH	National Institutes of Health
NJCIRM	National Jewish Center for Immunology and Respiratory Medicine in Denver
NMR	Nuclear Magnetic Resonance
NRA	National Recovery Act
NRC	National Research Council (of NAS)
NSF	National Science Foundation
NSFRE	National Society of Fund Raising Executives
PAW	Petroleum Administration for War
PET	Proton Emission Tomography
R/V *IDA GREEN*	Research Vessel *Ida Green*
SCAS	Southwestern Center for Advanced Studies
SCRF	Scripps Clinic and Research Foundation
SEG	Society of Exploration Geophysicists
SIO	Scripps Institution of Oceanography
SMU	Southern Methodist University
SMS	Southwestern Medical School (of UTHSCD)
SOCAL	Standard Oil Company of California
SY	School Year

TAGER	The Association for Graduate Education and Research of North Texas
TCU	Texas Christian University
TI	Texas Instruments Inc.
UBC	University of British Columbia
UCSD	University of California at San Diego
UTD	University of Texas at Dallas
UTHSC UTHSCD	University of Texas Health Sciences Center (Dallas)
UTMB UTMBG	University of Texas Medical Branch (Galveston)
UTMBI	University of Texas Marine Biomedical Institute (Galveston)
UTSMS	University of Texas Southwestern Medical School (Dallas)
WHOI	Woods Hole Oceanographic Institution
WSAC	Wireless Specialty Apparatus Company

APPENDIX B
1930 LETTER TO CECIL GREEN FROM ROLAND BEERS

The following letter was responsible for Cecil Green's decision to leave Federal Telegraph in Palo Alto, California, and accept the offer of Geophysical Service Inc. of Dallas, Texas, to become chief of their Party 310 in 1930.

Geophysical Service Incorporated
Republic Bank Building
Dallas, Texas

> July 1930
> Apartment 107
> 501 North First Street
> Seminole, Oklahoma

Dear Cecil,

It has been a long time since I received your letter, and I suppose you have long ago given up any idea of ever hearing from me again. I have not forgotten you; the truth is that whenever I undertake to write a letter, I am so intent on telling everything that has happened since last seeing you, it seems like a long task, and I never can find time to do a really satisfactory job. As a result, I rarely write letters to anyone, even now to my own people, who have been shamefully neglected for the past three months!

However, there is now a real occasion for writing you. I have long been thinking of you and your work, and of Ida's remarks that you were not entirely satisfied with your position. When I was with the Geophysical Research Corporation, I often looked for a good opportunity for you, as I was sure you would love this work as much as I do, and I have felt that you are well qualified for it. But G.R.C. was long on the decline, and it was not until this company became well established in the field that any further opportunities developed. We are now in the lead in this work, and I have great hopes for a tremendous success by the whole organization. Naturally, as I am one of the first in the new company, I think of my friends in this new opportunity, and particular of you, for whom my friendship has been long-standing, and now in need of repair for want of intimate associations. I will briefly tell you what I can of this company, and of the work, and hope that you will find it attractive.

The old organization, G.R.C., began a rapid decline in this

field of geophysical prospecting for oil, about a year ago. They had always been a subsidiary of the Amerada Petroleum Corporation, and part of the geophysical parties were employed by the Amerada for the purpose of locating oil for the parent company. The remainder of these parties were leased out on contracts to other oil companies for the same purpose. From a technical standpoint, the work has always been successful, excepting those cases where there were failures of individuals to meet the responsibilities of their positions. But even though some of the G.R.C. parties did good work, there was always an unpleasant feeling that perhaps some of the information regarding oil locations might leak back to Amerada, and by that means the contracting oil company would lose part of its legitimate information. In addition to this situation, the Amerada restricted G.R.C. activities to Texas and New Mexico, not permitting them to operate for other oil companies in Oklahoma or Kansas.

This situation gradually went from bad to worse until Dr. Karcher, President of G.R.C., and Eugene McDermott, my former field supervisor, left G.R.C. and formed their own company, the Geophysical Service, Inc. (familiarly known as G.S.I., now). They must have had tremendous confidence in their abilities, for they financed the organization almost out of their own pockets, to the extent of $200,000 capital stock, no par value. They developed an entirely new and radical design of instruments, which have since been proved to be far superior to the old designs, and with these instruments only in the process of construction, they went out and sold contracts for seven parties, at a monthly price of $9500 per party! That performance alone was short of miraculous, but it has been equalled by the actual performance of these parties, now operating in the field after three to four months of successful operation. All the oil companies are now awake to the fact that we have something unusual that cannot be passed aside, and our immediate future looks very brilliant. Although there is a bad depression in the oil industry, we are able to get contracts for new parties as fast as we can build the instruments and get the men to operate the parties. So far we have been fortunate in getting good men, but from now on we will be in a critical period.

The parties consist of a party chief, chief observer, mathematician, surveyor, labor foreman, and about eight or ten roustabouts or laborers. The party chief has full charge of everything, and his authority is supreme. It is a very responsible position, because the chief must be able to stand on his own feet in any situation. There will be no one for him to fall back on in case of emergencies, and he will have to bear the result of errors in judgment or of what seems worse to me, a lack of decision. Further, he is personally responsible for the results of the geophysical survey, their interpretation, and for the continuance of friendly and satisfactory relations with the geologist of the company to whom his party is contracted. In spite of all the factors, there are many attractive features in connection with the position that compensate for the burden of the responsibility. You are your own boss, having to take orders from no one except Dr. Karcher, President, and our field supervisor, now Mr. McDermott.

In return for this unusual rendering of service, the company is willing to pay an unusual compensation. They give you a monthly salary of from $450 to $500, a Buick eight, with all car expenses paid, and an excellent opportunity with a promising organization. Personally, I am more than pleased with my situation, as I am sure that five years of this work will make it possible for me to occupy a strictly executive position, at a good income, from which we might be able to retire at an early age.

Well, Cecil, old dear, I have tried to tell you the facts of the position as clearly as possible, without making it either too bright or too gloomy. I would like to know if you would be interested in accepting an offer as party chief under these conditions. I realize that it will be difficult for you to make a decision without further information, so won't you write me, and ask any questions that might occur, so that I can inform you fully.

There are certain things connected with this work that are purely personal, and I must tell you those as you will be greatly influenced by them in case you should come with us. First, you will have to give up any idea you may have of having a permanent residence. In this work, we move about frequently, often as much as every two weeks, but in the present case, we have been in Oklahoma City for three and a half months, in a good furnished house. We are now in Seminole, and expect to remain here until March at least. We may be here a year, and then again, we might tear up and move in a month or two. There is always a great uncertainty to that phase of it, which is very disconcerting to Helen, but which has not affected me at all. You must consider this phase of it seriously, however, as it is an ever-present factor. In most towns, where you expect to remain for a month or more, you can rent nice little furnished houses or apartments at not too extravagant rates, but occasionally you will have to live in a hotel. You being a party chief, could live at some distance from your work, say 25 to 40 miles, and drive back and forth every day, but I have never found this arrangement satisfactory, as it brings me home so late at night. This moving about business naturally runs up your living expenses a little above normal, but in view of the liberal salary, you can afford $50 a month extra or so without hardship.

The living conditions sometimes get pretty difficult, but not always for long, so you rarely have to put up with hardships for any length of time. If you should work in Oklahoma, as I think you might, you would find living in these little towns not bad at all. For your own part, you will soon become so interested in the work, that you will be willing to endure anything. The real hardship would be with Ida, but I'm sure she is a good sport and would enjoy seeing the country, as we have the past two years. Your work keeps you in the office most of the time, but when you get accustomed to the routine of it, you will find plenty of spare time to do as you please. At times the party chief has to go to the field, to supervise procedure in cases of difficulty or in new prospects. You report all your work directly to a company geologist, and you receive supervision from the field supervisor only when necessary, such as during your first few weeks. I would suggest to him that you come to my party for your early training and instruction, and if you did, we surely would have wonderful few weeks together. The technical part of the work isn't at all difficult. It is mainly a matter of good judgment and common sense, all guided by the general principles of electrical engineering that you are so well trained in.

Well, Cecil, this is indeed quite an accomplishment for me. I hope you will receive this favorably, and expect to hear from you pronto. Please wire me collect if you are at all interested, following up with a letter, so that I can keep our office posted. In the meantime, take good care of yourself and of Ida, and accept our kindest regards.

Most sincerely yours,

[Signed Roland]

APPENDIX C
THE EARLIEST DAYS OF REFLECTION SEISMIC PROSPECTING FOR PETROLEUM

Cecil Green became chief of the tenth field party of Geophysical Service, Inc. (GSI) in October 1930, only six months after that company, under the leadership of John C. Karcher, initiated reflection seismic exploration. So it can be said that Cecil was there at the beginning. The following brief commentary, prepared by him at my request, is of special interest because it gives an authentic picture of the beginning of what was to become the principal method of seismic prospecting for petroleum for the next fifty years.

Geophysical Service, Inc. (GSI) was launched in April 1930 as a strictly service company for the petroleum industry, and I started out as chief of its tenth party in October 1930 after driving from Palo Alto [California] to Maud, Oklahoma (population about 265).

In those initial days each field party consisted of about a dozen men, whose duties were as follows:

PARTY CHIEF   Directed each day's field operations, also interpreted the seismic records in a temporary office rented in a small town nearest to the particular area of interest to the client oil company. He maintained constant contact with his computer and the client's geologist, for he constructed a subsurface contour map as each day's field data were computed and analyzed. He, of course, kept in close contact with GSI headquarters in Dallas with regard to progress of his assigned program, as well as . . . necessary changes in instruments and motor equipment. The client geologist aided the daily field work by obtaining entry permits from affected landowners.

COMPUTER   The party chief shared his office with his computer, the man who calculated the raw data on the seismic records to determine the depth to the reflecting layer. This information then passed from computer to party chief for interpretation.

OBSERVER   The senior man next to the party chief and in direct charge of the field operations. His principal duty was the operation of the instruments in the recording truck. His assistant, called "first helper," developed each paper strip record as it was turned out by the seismic camera. The observer would then inspect each record so as to be able to instruct the shooter of the size of the new dynamite charge in order to obtain optimum amplitude of the desired reflections. The observer would also vary the length of the Spread—that is, the distance from shot point to nearest seismometer—in order to get a good separation between erratic "ground roll" and the desired reflections. The observer and first helper had one or two "jug hustlers" to help them move and connect the seismometers from location to location.

A typical recording set-up in those earliest days comprised five seismometers, or "jugs"—cylindrical in shape and each weighing about 30 pounds. They were placed about 100 feet apart, all in a straight line with the shot point—generally 1,000 to 2,000 feet from the nearest seismometer—and each buried in a shallow hand auger hole to avoid wind disturbance.

The recording truck was connected by telephone line to the shot point so that the observer could instruct the shooter when to fire the dynamite charge with his hand-operated blaster, while at the same time he started hand cranking the strip of recording paper through the camera and on into the developer box.

The moment of discharge was also recorded via the telephone line from an electric cap detonated at the same time as the charge. Additional charges of varying size were detonated in different holes until the observer obtained an optimum reflection amplitude.

The depth and time to traverse the surface weathered layer under the seismometers and then obtained by a shallow refraction profile from small dynamite charges in shallow holes located in line with the seismometer array and 100 to 200 feet from the nearest seismometer.

Finally, the surveyor's measured distance from shot point to each jug in the seismometer line was checked by the shooter detonating a very small dynamite charge in the air above the shot point, which was also recorded by the observer.

SURVEYOR   Received instructions from the party chief regarding the specific locations to be covered each day by shot points on land for which permits had been obtained by the client's geologist. After establishing each shot point as near as possible to a creek bottom, he and his rodman, with the use of a steel tape, then indicated the location of each of the five seismometers all in a straight line with respect to the shot point—100 feet apart, with the nearest seismometer being at a distance of 1,000 to 2,000 feet.

The elevations of each shot point and related seismometers and their location in relation to area landmarks were then obtained with plane table and alidade.

PROCUREMENT OF SHOT HOLES   In the earliest days, and before the advent of truck-mounted drill rigs (first spudders and then rotary units), shot holes (three to five per shot point) were dug to a depth of about 20 feet using hand augers. There were usually two teams of three men so that two of the men could turn the auger while the third sat on the handle to provide sufficient downward weight.

THE SHOOTER   Had his own truck, and with his helper, prepared a charge of dynamite of size prescribed by the observer via the telephone line, then pushed the charge to the bottom of a particular hole with the help of tamp rods. In order to get the maximum amount of downward energy from the subsequent detonation, the charge was then covered with water to the top of the hole from a tank on a special truck.

PROCUREMENT OF WATER   Water for tamping of charges was provided from a large tank on a special truck, the driver refilling the tank from any nearby creek or pond while the recording crew was moving from one set-up to another.

The party chief directed the daily field operations with instructions to the observer and surveyor. At the same time he was responsible for processing and interpreting each day's reflection results with the help of his computer. Thus the party chief identified the usable reflections on each day's records while the computer measured their arrival times to the nearest 1/1,000 second. Then with the aid of a time-depth chart, appropriate to the particular area, the depth to each corresponding subsurface reflection horizon was computed; the results were then placed on a map of the area by the party chief, who was then able to construct progressively a subsurface contour map for the prime reflection horizon.

This interpretation and computation of field data on a daily basis enabled the party chief, with the help of the client's resident geologist, to decide on the optimum location for the next day's field recording.

The party chief and client geologist were quite aware of the regional downward slope for the particular region, as is well confirmed by the accumulative seismic results, and so the prime objective was to discover any reversal in direction of normal downward slope as a possible forerunner indication of a sub-surface anticline on which petroleum might be "trapped" in a porous reservoir.

I [CHG] am also including a diagram to show the straight line arrangement of shot point and seismometers (jugs)—this being referred to as a "profile." [See p. xxx.]

Generally, the program for the area, selected by the client, of course, was divided into two parts: first, a reconnaissance phase with profiles a mile or so apart; then, if such reconnaissance information showed any reverse from normal subsurface dip in the principal horizon, the party chief would undertake the second phase wherein he would "fill in" with more closely spaced profiles to see if the indicated reverse dip might be related to a subsurface anticline in which oil or gas might be "trapped" in a reservoir.

Incidentally, "scouts" from competing oil companies were always on the lookout for any such indication of detail work so that their company could then compete in leasing the particular area. So, you will realize, part of our job was to try to lead such scouts astray by various maneuvers.

Cecil H. Green

May 4, 1986

APPENDIX D
IDA GREEN'S COMMENTS ON HER VOYAGE FROM SAN FRANCISCO TO NEW YORK CITY, ON THE SS *VIRGINIA*, SEPTEMBER 12–28, 1931

My trip via San Francisco to New York. Left there and arrived here. Had lunch on the boat and then made reservations for deck chairs. Next day was spent unpacking things and reviewing gifts received which were two corsages of flowers, letter, hankie, box of nuts. Mrs. McClellan from Phila[delphia] was my ship mate. Days were spent resting and eve up on deck. The shore along Southern Calif[ornia] rough and wild looking. The Caribbean sea equal to blueness and beauty of the Mediter[ranean], so the room steward said . . . Reached Panama, anchored and went ashore to shop in Balboa. Bought table cloth, 2 lunch cloths, 3 slips, 3 panties, 2 hats, kimono, hankies, and perfume. Visited the points of interest. Had a few drinks and then took a carriage out to old Panama. There was beautiful foliage, varied colored flowers and colorful people. Had lunch in town then went back to boat. The canal was very nice and in the early morn with vivid green grass, blue water and sky, birds, and monkeys along the shore all in all made it beautiful. Entering Havana, Cuba, was interesting. We docked out in the Harbor and had to take a launch aboard for shore. Little black boys hung around the boat diving for pennies. Stopped for refreshments, then saw Capitol buildings. Most beautiful cemetery in world next to one in Italy, two cathedrals, one where Columbus's bones rested many years, Morro Castle, along boulevard. Beer garden had all the drink we wanted free of charge, thru the Creole section, as slums, out into the country, more refreshments and then cigar factory. Grand rush for boat and homeward bound. Bought Pop [her father] cigars. In few days reached N.Y.C. where my Cecil met me and was sure glad to see him. Spent two weeks with Mrs. Whelan, a lovely home, and then our apartment. A grand glorious time.

(From Ida's diary entry for January 3, 1932.)

APPENDIX E
A COOPERATIVE PLAN IN
GEOPHYSICAL EDUCATION

The following discussion of the GSI Student Cooperative Plan, conceived, organized and initiated by Cecil H. Green, is excerpted from my *A Cooperative Plan in Geophysical Education* published in 1966 by Geophysical Service Inc. of Dallas, Texas.

Appendix I

How Did the GSI Student Cooperative Plan Originate and What Has It Accomplished?

A Bit of History, Describing How Some Existing Needs Were Combined With Some Ideas to Make an Actual Program of Action

A BIT OF HISTORY

In the fall of 1950, Cecil H. Green, President of Geophysical Service Inc., of Dallas, Texas, realized that in the near future his Company was going to need many new employees with knowledge of the latest advances in science and engineering. Where could they be found and how could they be identified?

At the same time, Robert R. Shrock, Head of the Department of Geology and Geophysics at the Massachusetts Institute of Technology, was becoming aware of the need for some kind of practical field training for the dozen or so undergraduates majoring in geophysics. Inasmuch as the students would have to forego regular employment if they were to get such training in the summer, which was the only logical time, how could they get the desired training without too much loss of income, and what employer might be interested in hiring them for such work?

Green decided to visit his alma mater, M.I.T., to see if he could interest some of the electrical engineering students in working for GSI. He, himself, had received two degrees from the Department's well-known Cooperative Plan, Course VI-A, in 1923, and felt that his Company could offer an engineering graduate exciting opportunities in geophysical exploration. But when he arrived on the campus in early November, he found little interest among either students or staff in his inquiries, for this was a time when young engineers were in short supply and every graduate was being offered several excellent employment opportunities.

Green then decided to call at the Headquarters of the Department of Geology and Geophysics to see if some of the geophysics students might be interested in employment. So, on the 13th of November, 1950, he walked into the Department's front office, told the secretary he was Cecil Green from GSI, and asked to talk with the Head of the Department. He was soon ushered into Shrock's office, and it was not long before both men found that they had a number of common interests and needs. The actual conversation of that afternoon belongs in another history, but the results were immediate and took the form of a concrete program of summer training. Within less than three months the program was organized, approved by both GSI and M.I.T., and named The MIT-GSI COOPERATIVE PLAN. Before the end of February, more than a dozen M.I.T. students had agreed to participate in the program. Late in March, Shrock went to Dallas to help work out the final arrangements for the summer, and on the 15th of June, 1951, the first group of students arrived in Dallas. Thus began what is known today, in its 16th summer, as *The GSI Student Cooperative Plan*.

ORGANIZATION OF PLAN

The original aim of the program was to provide a means by which college students interested in geophysics and related sciences could learn about exploration geophysics firsthand by actual field experience during the summer months. To accomplish this aim students would have to be selected and brought to the attention of a company that could provide the opportunities for the desired field work. It was with this general aim in mind that GSI and M.I.T. first joined forces to organize a cooperative program. M.I.T. would provide the first students; GSI would provide the summer employment. As soon as this cooperative plan was agreed on, certain important problems had to be solved at the very start. What students would be chosen, and by whom? How would the selected students be made aware of the professional opportunities in geophysical exploration, and, more specifically, how should their temporary employer, GSI, demonstrate these opportunities by an actual program of exposure, instruction and practice? What should be expected of the student participating in the program — merely experience as an observer, or should he become a fully active and productive member of a field crew or an office or laboratory team? What pay and other financial rewards should he receive? To what extent should he prepare independent reports? In short, just how should he be fitted into the personnel structure of the party or team, in view of the fact that he was to be working for only the summer months and as a temporary employee.

These, and many others, were the problems faced by Green, Shrock, and their respective associates, in organizing and carrying out the first summer's experiment. Selecting the first group of students was simple. Inasmuch as the whole junior class in geophysics at M.I.T. in 1951 qualified for the program, it was agreed that they would be hired as an experimental group. It was next agreed that they would need to know something about their employer, Geophysical Service Inc., and the petroleum industry it served, with special reference to company policies, field procedures, contractor-client relationships, etc. As a consequence of these needs, a three-day indoctrination or orientation program was organized and scheduled to precede work assignments in the field. Finally, on 15 June 1951, fourteen (14) M.I.T. geophysics students, a mathematics major from Williams College, and a geophysics student from Harvard assembled in Dallas, together with a dozen staff members of GSI, representatives of several petroleum companies (DeGolyer and Mac-Naughton, Magnolia Petroleum Company, and Standard Oil Company of Texas), and professors from Southern Methodist University, University of Minnesota, and the University of Toronto.

RECORD OF FIRST FIFTEEN SUMMERS

A total of 308 carefully selected students, from 78 different colleges and universities, have participated in the program through its 15th summer. Of these, 40 spent the summer doing special work in the Dallas laboratories of GSI and Texas Instruments Incorporated, while 268 were assigned to geophysical field parties in the western half of the United States. The 308 students, at the time they participated in the program, were distributed among the academic fields as follows:

Academic Discipline	Regular Co-op Field	Summer Dev. Prog. Dallas
Geology	36%	—
Geophysics	34%	40%
Physics	12%	12%
Mathematics	8%	20%
Electrical Engineering, etc.	8%	27%
Miscellaneous	2%	1%
	100%	100%

From a questionnaire sent to the 268 students who participated in the Field Program, the distribution of the participants in January 1966, by type of employment or activity, seemed to be about as follows:

Continuing academic work (largely graduate)	33.5%
Employed by 15 different petroleum companies	14.0
Employed by 4 different geophysical companies	11.5
Academic Teaching and Research	11.0
Employed by Federal Agencies, etc.	6.0
In Armed Services	3.5
Employed by 34 different firms, mostly as engineers and applied scientists	20.5
	100.0%

If the former participants who are still in college distribute themselves after graduation as the questionnaire returns indicate they probably will, then the ultimate distribution will probably be somewhat as follows:

40% in geophysical exploration and the petroleum industry

40% in federal agencies and a wide range of service, producing and manufacturing organizations

20% in teaching and academic research

A few additional statistics that may be of interest show how many skilled persons have been involved in training the 308 students:

1. A total of 100 professors from 41 different academic institutions attended at least one Orientation Conference. Many of them gave papers on their research activities.

2. A total of 185 representatives from 35 different petroleum companies attended at least one Orientation Conference. Most of them read papers on company activities, policies, or research.

3. A total of 23 representatives from 11 different Federal Agencies, Bureaus, and Armed Services attended at least one Orientation Conference. Most of them reported on current federal research projects.

4. At least 77 technical employees of Texas Instruments Incorporated and Geophysical Service Inc. from the Dallas and Houston offices, and another 117 GSI field party chiefs, have participated in one way or another in the program.

EVALUATION – HAS THE PROGRAM ACHIEVED ITS AIMS?

The student participants, who returned more than 70% of the questionnaires sent out, seem to feel that the program was well worth while. Ninety percent (90%) of those reporting said they profited from the summer's work, even though many did not go into geophysics ultimately, and most of these said they hoped the program would be continued. The professionals who have participated in the Orientation Conference are unanimous in their enthusiasm for and approval of the Plan as a whole and have often expressed the hope that it will be continued.

In summary, the available information conclusively shows that the GSI Student Cooperative Plan has consistently achieved its principal aim of exposing college students to the real world of geophysical exploration. The same evidence indicates that it should be continued because it is unique in bringing together the technical leaders of industry, government, and the academic world, in a cooperative effort that is certain to produce the kinds of scientific and technological leaders that the future will require.

Summary and Conclusions

Evaluation of any program of activity must be based on what that activity accomplishes and what ultimate effects the accomplishments have. How well have the primary objectives been achieved? Following are a number of accomplishments of the Program —

1. 308 well-trained scientifically oriented students were carefully selected from several thousand applicants and recommended by their professors.

2. These 308 students came from 78 different colleges, universities and technical institutes, distributed widely throughout the United States and Canada, and including many of the leading educational institutions of North America.

3. Each year, for a period of 15 years, 18 of these students, on the average, attended a 3-4 day orientation session in Dallas, after which they spent 10 weeks in practical work on a geophysical field party.

4. In the orientation session they heard an average of 42 half-hour presentations by nationally and internationally known scientists, engineers, and industrial leaders, whose names read like an honor roll of the earth sciences profession. This series of lectures was easily the equivalent of a 2-credit course at any major university.

5. The speakers, numbering some three hundred in all, represented 33 different petroleum companies and GSI-

TI, 41 different educational institutions and 11 different federal agencies, bureaus, and armed services. Each year, on the average, the 40 or more speakers were distributed as follows: 16 (40%) from GSI-TI; 16 (40%) from petroleum companies; and 8 (20%) from educational institutions and federal organizations.

6. The students, besides hearing this galaxy of distinguished speakers, also had on the average 10 opportunities to mingle with them and ask questions (coffee breaks, luncheons, dinners, etc.). Many students mentioned on the questionnaire that they regarded meeting the senior people at the session as the high point of the summer's program.

7. Each summer, on the average, 18 students each spent 10 weeks in the field as a regular member of a geophysical party. In this assignment, they came into close personal contact with the party chief (117 chiefs in all) and other members of the party — an experience that all professional earth scientists recognize as invaluable in the training of a scientist or engineer and the first real test of success. Serving in as many different roles as the party's work schedule permitted — rodman, surveyor, jug hustler, driller's helper, water boy, recorder, or office assistant — the student had ample opportunity to learn about geophysical field work firsthand, and thus to have a sound basis for deciding whether or not he would like geophysics and/or the petroleum industry as a career.

8. An additional 40 students, in groups of 5 or 6, worked on special research projects in the Dallas laboratories of GSI-TI, but did not in general get into the field, hence are not counted in the group whose experiences are being recapitulated here.

9. All of the 268 regular Co-op students were required to prepare a report at the end of the ten-week period. This report was to contain a critique of the party's operations

Summary and Conclusions

and a separate section on some technical problem in which the student had developed a special interest.

10. As a wrap-up of the summer program, the Party Chief evaluated the student's summer work and sent his report together with the student's own report to the Dallas office, where a final evaluation of the student and his work was made.

11. This evaluation was ultimately printed and sent to each student's academic institution where it could become a part of his permanent educational record.

Thus GSI and TI have successfully brought together the geophysical exploration industry, the petroleum industry, the educational institutions, and the federal government into a co-operative enterprise focused on the selection and training of first-rate college students for ultimate careers in one or more of the earth sciences. The pre-Sputnik recognition of the new trend in science and technology in the United States by the founders of The GSI Student Cooperative Program, and the response to that trend in the form of the 15-year Program being evaluated, are known and respected around the world in earth science circles.

A COOPERATIVE PLAN IN GEOPHYSICAL EDUCATION

EXHIBIT 11.

Current Activity of Former Student Participants

33.5%	59	Continuing academic work
14.0%	25	Employed by 15 different petroleum companies
11.5%	20	Employed by 4 different geophysical companies; 14 employed by GSI and TI
11.0%	19	Involved in academic work; 14 teaching and 5 doing full-time academic research
5.0%	9	Employed by federal agencies and bureaus (NASA, USGS, USCGS, USBM and USNO)
3.5%	6	In Armed Services
1.0%	2	1 on a State Geological Survey and 1 for a private research laboratory
20.5%	36	Employed by 34 different firms, mostly as engineers and applied scientists
100.0%	176	

If those participants who are continuing academic work and those in the Armed Services be excluded, then 111 of the 176 who returned questionnaires are now employed as follows —

40% in geophysical exploration and the petroleum industry

40% in federal agencies and a wide range of operating companies, a large percentage of whom could be clients of GSI

20% in teaching and research that could be of ultimate value to GSI

More than 300 outstanding students can be regarded as "alumni" of the GSI Co-op Program. Among the 60 participants from MIT, Cecil's second alma mater, were the following:

Milo M. Backus   John R. McGinley
Howard W. Briscoe   Norman Ness
Maurice J. Davidson   Robert L. Sax
Edward A. Flinn   Laurence Strickland (deceased)
Parker Gay, S. Jr.   Lynn R. Sykes
Freeman J. Gilbert   Newell J. Trask, Jr.
Philip G. Hallof   Sven Treitel
Kenneth Larner   Carl Wunsch

APPENDIX F
ENDOWED PROFESSORSHIPS, LECTURESHIPS, INVESTIGATORSHIPS, RESIDENCIES, TRAINING PROGRAMS, AND FUNDS FOR ENHANCING EDUCATION IN SCIENCE AND ENGINEERING BY CECIL H. AND IDA M. GREEN (1961–1988)

Note: From the following discussion it can be recorded that Cecil and Ida Green have thus far (1988)

1. Initiated and completely endowed 25 professorships and other teaching and research positions, and made donations to three additional academic positions, at a total of 15 different educational institutions.

2. Initiated and completely endowed special funds for residencies in two medical institutions, and made a substantial donation to a third.

3. Initiated and endowed a special fund in three different universities to be used for improving education in the sciences and engineering.

To the extent possible, the names and special fields of the occupants of the chairs and residencies are listed. This information is clear evidence of the high distinction of those who occupied the chairs and the great diversity of special fields they represented. Finally, an abundance of letters from both institutional administrators and occupants of the endowed positions assure the Greens that their philanthropy in creating such thoughtful and timely opportunities for exchange of ideas has been deeply appreciated by and helpful to both the recipient institutions and the individual visiting scholars. And, of course, for the Greens this particular component of their overall program of philanthropy represents "investments in altrustic pleasure" that have yielded "dividends" of lasting endurance.

Following are the endowed positions that the Greens have created, listed chronologically under the respective institutions.

	NAME OF POSITION
SCHOOL	(AND OF OCCUPANTS)
SCRF	Cecil and Ida Green Associate Chair in Medical
1961	Teaching and Research provided funds to sup-

	port research and other expenses, the funds to be allocated at the discretion of the director of the Foundation.	
SCRF 1970	Cecil H. and Ida M. Green Investigatorship in Medical Research at SCRF, the Foundation's first endowed chair, was established in 1970 and the first occupant appointed in February 1972: Hans J. Muller-Eberhard 1972–1986 Peter E. Wright 1986–	
SMU 1969	Cecil H. Green Professorship in Engineering: Thomas L. Martin, Jr. 1969–1974 F. Karl Willenbrock 1976–1986 (currently unoccupied)	
SMU 1979	Cecil H. and Ida Green Professorship in Electrical Engineering: W. Milton Gosey 1986–	
TCU 1969	Cecil H. and Ida Green Honors Chair for Visiting Scholars. During the 17 years since the Chair was established, 100 Visiting Green Scholars have been members of the TCU faculty—at least 14 for a complete semester, and 86 others—writers, scientists, educators, and other notables—have come for week-long periods. Many of the scholars delivered lectures open to the general public, thereby enhancing town-and-gown relations. Following is the list of occupants for two recent years, kindly provided by Paul W. Hartman, TCU's vice-chancellor for University Relations and Development.	
TCU 1985	*Economics* Dr. W. H. Locke Anderson Department of Economics, U. Michigan	Fall 1985
	Physical Education Dr. Robert P. Pangrazi, Chair Department of Health and Physical Education, Arizona State	November 4–8

	Political Science Prof. Barbara Sinclair Department of Political Science, Cal.-Riverside	November 17–19
	Dr. Wilson C. McWilliams Livingston College, Rutgers	February 3–5
	Mathematics Dr. Phillip J. Davis Applied Mathematics Division, Brown	February 17–21
	Nursing Dr. Peggy Lois Chinn Professor of Nursing, SUNY-Buffalo	February 23–28
	Sociology Dr. Jack Gibbs Department of Sociology, Vanderbilt	February 24–26
	Radio-TV-Film Mr. George Dessart Vice President of CBS Television, New York	March 3–7
	Sociology Dr. Marvin Harris Department of Anthropology, U. Florida	March 12–13
TCU 1986	*Chemistry* Dr. Harry Grey Caltech, Pasadena	Spring 1986 (1 week)
	Honors Program Appointment pending	April 1986

UBC
1969

The Cecil and Ida Green Visiting Professorships (Lectureships), at Cecil's first alma mater, bring distinguished visiting professionals to Vancouver for a week or two on the average. The program has long since become one of the best public forums in North America. The following information, kindly provided by Rosanne Rumley, program chairwoman, of the committee that selects the lecturers, clearly indicates the variety of fields covered and the distinction of the visiting participants. At least one typical lecture is included for most of the participants, but some gave as many as five or six, in addition to informal seminars.

UBC
1972

Gerhard Herzberg (Nobel Prize winner in Physics from McGill University) lectured on molecular physics

Donald O. Hebb (chancellor of McGill University, and one of Canada's best-known experimental psychologists) discussed "The Nature of University Education"

J. Tuzo Wilson (one of the world's leading geophysicists, University of Toronto) lectured on "The Mechanics of Plate Tectonics," and "Earthquakes and Earth Sciences in China"

Lawrence A. Cremin (professor of education, Columbia University) gave four lectures on "The History of Education"

Richard Schechner and The Performance Group, under the direction of Dr. Schechner, gave a number of presentations of his avant-garde play called *Cummune*

C. B. Macpherson (Canada's best-known political scientist and a teacher at the University of Toronto) delivered four lectures on different aspects of "Democracy"

Harold O. Seigel (an internationally famous Canadian geophysicist and president of Scintrex Limited) gave several lectures on exploration geophysics

UBC
1973

Morton Fried (a leading U.S. political anthropologist and former chairman of the Department of Anthropology at Columbia University) presented "*Noch Einmal!* The Problem of Tribe"

Rodolfo Stavenhagen (professor of sociology,

National University of Mexico City) lectured on "Revolutionary Potential in Latin America"

Allen B. Rosenstein (professor of engineering at UCLA and an expert in the field of engineering curricula) discussed "Applied Humanities"

Norman MacKenzie (director of the Centre for Educational Technology at the University of Sussex, England) talked about H. G. Wells and Education in England

Michael Scriven (professor of philosophy at UC-Berkeley, and a widely recognized authority on the evaluation of educational programs) discussed "The Exact Role of Value Judgments in Science"

Enrique Luis Revol (professor of modern literature at University of Cordoba, Argentina, and an authority on Spanish-American literature) spent four months in the Department of Hispanic and Italian Studies, and gave a lecture on "The Poet as Thinker in Our Times"

Frederick M. Combellack (professor of Greek literature, University of Oregon, a North American authority on the Greek poet, Homer) lectured on "Homer and the Fashionable Critics"

Sir John Eccles (Nobel Prize winner and one of the greatest brain scientists; professor of physiology, Australian National University at Canberra) gave several lectures on aspects of the brain, for example, "Understanding the Human Brain"

B. F. Skinner (professor of psychology at Harvard University, considered to be among the most influential of living American psychologists) discussed "Obstacles to a Science of Behavior"

UBC
1974

Donald Keene (professor of Japanese literature at Columbia University and one of the foremost

Western authorities in the field) lectured on "The Beginnings of the New Japanese Literature"

Paul T. K. Lin (chairman of the Department of East Asian Languages and Literature, McGill University, an authority on the People's Republic of China) discussed "China, Canada, and Future World Order"

Richard Evans Schultes (director of the Botanical Museum at Harvard University and a leading ethnobotanist specializing in the study of drugs and poisons of primitive peoples) lectured on "Hallucinogenic Plants of the New World"

Michael Riffaterre (chairman of the Department of French and Romance Philology at Columbia University and specialist in stylistics and textual analysis) lectured on "The Structuralists' Approach to Literature"

Ian McHarg (chairman of the Department of Landscape Architecture and Regional Planning, University of Pennsylvania, and a leading authority on human logical planning) discussed "A Case Study in Logical Planning"

Theodore W. Rall (professor of pharmacology at Case Western Reserve University, and one of the top authorities on cyclic AMP (adenosine monophosphate) discussed "Cyclic AMP, An Intra-Cellular Messenger in Neuro-hormone Action"

John Chadwick (Fellow of Downing College, Cambridge University, and a specialist on the history of the Greek language) discussed "The Late Bronze Age in the Aegean"

Igor Kipnis (world famous harpsichordist in New York and one of the leading authorities on music of the baroque era) gave several recitals, and lectured on "The Harpsichord Today"

Lord Denning of Whitchurch (head of the British Court of Civil Appeal and one of the world's

most distinguished jurists) spoke on "Let Jus-
tice Be Done," and "The Failure of Leadership"

A. S. V. Burgen (director, National Institute for
Medical Research, London, England, one of
the leading British authorities on the effects of
chemicals and drugs on body cells) discussed
"Drug Specificity—Chemicals as Magic Bullets
or Blunderbusses"

Zhores A. Medvedev (Russian gerontologist,
now at the National Institute for Medical Research in London, England) gave a lecture on
"Molecular and Genetical Aspects of Ageing"

J. J. Harwood (director, Physical Sciences Laboratory, Ford Motor Company, Dearborn,
Michigan, and expert metallurgist) gave several
lectures on the automobile, including "Physics,
Materials and Automotive Power Plants"

Lord Wolfenden (former director of the British
Museum) lectured on "The Glory That Was
Greece"

Victor Weisskopf (Institute Professor of Physics, MIT, and former director-general at the International Laboratory [CERN] in Geneva,
Switzerland) lectured on "Modern Insights into
the Structure of Matter" and "On Atoms,
Mountains and Stars—A Study in Quantitative
Physics"

The preceding 30 visiting lecturers came to Vancouver during
the first three years (1972–1974); during the ensuing eleven
years (1975–1985), 90 additional distinguished men and women
have been named Green Visiting Lecturers. And it should be
mentioned that eleven of these 120 visitors were Nobel laureates! Furthermore, during the past two years (1986, 1987), at
least 20 more visiting lecturers have increased to 140 the total
number of Green Visiting Lecturers who have shared their
knowledge and skills with citizens of Vancouver, thanks to the
thoughtful generosity of Cecil and Ida Green.

The impressive impact of the Green Visiting Lecturers on the

Vancouver community is clearly evident in the following excerpt
from a recent letter to the Greens:

The Vancouver Institute
c/o Centre for Continuing Education
The University of British Columbia
Vancouver, B.C. V6T 2A4

December 4, 1984

Dear Drs. Cecil and Ida Green:

We have been delighted with our speakers and the really tremendous response we have had. Our main hall seating 500
people is always full and closed circuit television carries the
picture and sound to other rooms in the building as well as the
lobby. As many as 7000 to 10,000 people have turned out to
hear the Dalai Lama and celebrities like Tennessee Williams.
Unfortunately the Fire Marhsall won't permit more than 1600
to 1800 people in the building for more crowded evenings and
the others have to be turned away.

The speakers you have sponsored have been tremendously
helpful to us. As you can imagine, it is a rather formidable task
for us volunteers to secure at least *twenty* [RRS underline] high
calibre speakers a year.

. . . Our visitors have called our public forum the best in
North America. It is largely due to your help that this has been
made possible.

Yours sincerely,
[Signed] Bel N. Nemetz
Program Chairman

MIT 1970	Cecil H. Green Professorship in Electrical Engineering:	
	Peter Elias	1970–1972
	Samuel J. Mason	1972–1974
	Richard B. Adler	1974–1976
	Alan V. Oppenheim	1976–1978

	Fernando J. Corbato	1978–1980
	Thomas H. Lee	1980–1981
	Alvin W. Drake	1981–1982
	Albert R. Mayer	1982–1983
	Frederic R. Morganthaler	1984–1986
	David Staelin	1987–1990

MIT
1970

Robert R. Shrock Professorship in Earth and Planetary Sciences:

Frank Press	1970–1980
Keiti Aki	1982–1984
Thomas H. Jordan	1984–

MIT
1974

Cecil and Ida Green Professorship in Education:

Seymour A. Papert	1974–1980
Margaret L. A. MacVicar	1980–

MIT
1976

Cecil and Ida Green Professorship in Earth Sciences (I):

William F. Brace	1976–

MIT
1976

Cecil and Ida Green Professorship in Earth Sciences (II):

Carl I. Wunsch	1976–

MIT
1976

Cecil and Ida Green Professorship in Physics (I):

Herman Feshbach	1976–1983
Jeffrey Goldstone	1983–

MIT
1979

Cecil and Ida Green Professorship in Physics (II):

Robert J. Birgeneau	1982–

MIT
1979

Cecil and Ida Green Career Development Professorships

Cecil's longtime interest in and support of career development, particularly for younger scholars and professionals, as demonstrated by his efforts in the establishment of GRCSW (UTD), TAGER, and The GSI Student Cooperative Plan, culminated in 1979 in the endowment of two 1- to 3-year career development professorships. The following people have held these positions thus far:

Professorship (I)
Frank Morgan   1985–1986
Sarah Deutsch   1987–1990

Professorship (II)
Robert Ledoux   1985–1987

SCRF
1970

Cecil H. and Ida M. Green Investigatorship in Medical Research at SCRF, the Foundation's first endowed chair, was established by the Greens in 1970, and in February 1972 the first occupant of the chair was appointed:
Dr. Hans J. Muller-Eberhard 1972–1986
Dr. Peter E. Wright   1986–

Austin
College
1971

Ida M. and Cecil H. Green Chair of Creative Educational Leadership, established in 1971:
DeWitt C. Redick 1976–1978 (a senior trustee)
John D. Moseley 1978–81 (chancellor, Austin College)
Harry E. Smith 1981–(current president, Austin College)

UCSD-IGPP
1972

Cecil H. and Ida M. Green Foundation Scholars—visiting scientists, appointed for variable periods of time, who come to La Jolla for exchange of ideas with staff members of IGPP on earth science problems.

In 1969, as a part-time resident of La Jolla, Cecil advised his longtime friend, geophysicist Walter Munk, that he and Ida would like to do something to help IGPP at UCSD's Scripps Institution of Oceanography at La Jolla. Walter was then director of IGPP. Did he have any suggestions? His response was immediate and enthusiastic: the ideal thing would be to establish a fund independent of the university to bring both distinguished and promising young scholars to IGPP for a few months to exchange ideas with members of the Institute's staff, and to "shake them up," as Walter put it.

Accordingly, in January 1972, after several years of negotiations with state and university officials, Cecil and Ida initiated and generously endowed the independent La Jolla Foundation for Earth Sciences, with the proviso that

The purpose of this gift is to provide financial assistance to the Institute for the creation and support of one or more lectureships. Each recipient of a lectureship shall be a visiting scholar in the Earth Sciences on short appointment, who engages in one or more of the following activities: research, teaching, conferences and lectures. (Excerpted from Cecil's letter of January 19, 1972, to Mr. P. H. Johnston, president of LJFES)

The purpose of the Foundation was promptly put into action in 1972 by inviting, as the first Green Scholar, the distinguished geophysicist Xavier Le Pichon, then chairman of the Department of Earth Sciences of The Oceanology Center of Brest, France, renowned for his work on plate tectonics.

During the 15 years since the establishment of the Foundation, which was renamed the Cecil H. and Ida M. Green Foundation for Earth Sciences in 1983, some 30 or more Green Scholars have spent time at IGPP in La Jolla. These scholars have ranged from scientists of world renown to yong promising postdoctorates. They have been largely from the United States, but there have been others from England, Scotland, Canada, France, and Italy. Munk and his IGPP colleagues have found the innovative program of visiting scholars "an unqualified success," and Cecil regards it as one of his and Ida's best "investments in pleasure."

UTMBG
1972–
1974

Cecil H. and Ida Green Professorship of Marine Sciences:
This honors chair was established specifically for occupancy by world-renowned geophysi-

cist W. Maurice Ewing, who moved to the University of Texas Medical Branch in Galveston, in 1972, to become founding chief of the Earth and Planetary Sciences Division of the Marine Biomedical Institute, as well as the occupant of the Green Chair. Soon after his untimely death on May 4, 1974, the authorities decided that henceforth the income from the Green endowment would be used to bring distinguished scientists and engineers to UTMB for short visits during which they would deliver lectures in their special fields, if such an arrangement was satisfctory to the Greens. Having received the Greens' enthusiastic approval of their proposal, the Marine Branch proceeded to invite leading academicians and other professionals to be Visiting Green Scholars, as shown on the accompanying list, kindly provided by Dr. A. D. Suttle, Jr., of UTMG.

In his much appreciated letter, Suttle made the following comment regarding the valuable effects of the Green Visiting Scholars:

We at the University of Texas Medical Branch at Galveston count ourselves fortunate to have received such largesse from the Greens and have found it to contribute materially to stimulating new ideas and programs and to exchanging information between our faculty and others who are interested in a particular discipline. Interactions between the physical sciences and engineering on the one hand, and the biological disciplines and the healing arts on the other, have been among the most beneficial.

UTMBG 1978	Cooper, Leon N., Ph.D. Biophysics Brown University	January
	Jones, Edward G., M.D., Ph.D. Neuroanatomy Washington University	June
	Young, John Zackary, F.R.S. Neuroscientist Cambridge University	November

UTMBG 1979	Drost-Hansen, Walter, Ph.D. Biophysicist University of Miami	October
	Eccles, Sir John, M.D., D. Phil. Neurophysiologist SUNY-Buffalo	November
	Seaborg, Glenn T., Ph.D., Sc.D. Nuclear chemist University of California System	November
UTMBG 1980	Eccles, Sir John, M.D., D. Phil. Neurophysiologist SUNY-Buffalo	May
	Davis, Ruth M., Ph.D. Computer scientist Pymatuning Group, Inc.	January– February
	Schmitt, Francis O., Ph.D. Neurologist MIT	February
	Baker, William O., Ph.D. Physical chemist Bell Telephone Labs	November
UTMBG 1981	Doll, Sir Richard, M.D., F.R.S. Epidemiologist Oxford University	April
	Bullock, Theodore H., Ph.D. Neuroscientist University of California, San Diego	May
UTMBG 1982	Abelson, Philip, Ph.D. Chemist American Association for the Advancement of Science	January
	Richards, Sir Rex, Ph.D., Sc.D. Physical chemist Oxford University	September– October
	Seitz, Frederick, Ph.D. Solid state physicist Rockefeller University	October
UTMBG 1983	Press, Frank, Ph.D. Geophysicist National Academy of Sciences	March

	Baltimore, David H., Ph.D. Bioengineer and molecular engineer MIT	April
	Hall, Laurence, Ph.D. Chemist University of British Columbia	May
	Hinde, Robert A., F.R.S. Ethnologist Cambridge University, England	June
	Bassett, C. Andrew L., M.D. Orthopaedic surgeon Columbia Presbyterian Hospital Columbia University	
	Robbins, Frederick, M.D. Endocrinologist National Institute of Medicine at National Academy of Sciences	October
UTMBG 1984	Seaborg, Glenn T., Ph.D., Sc.D. Nuclear chemist University of California	March
	Rosenblith, Walter A., Prof. Electrical engineer and Neuroscientist MIT	April
	Walton, Sir John, M.D., Sc.D. Neurologist Oxford University	June
	Hofstadter, Robert F., Ph.D. Physicist Stanford University	
	McGeer, Patrick, M.D., Ph.D. Neuroscientist Province of British Columbia, Canada	November
	Buncke, Harry J., Jr., M.D. Plastic and reconstructive surgeon	December

University of California, San Francisco

Stanford
1973

Cecil H. and Ida M. Green Professorship in Geophysics:

Allan V. Cox	1973–1987
Jon F. Claerbout	1987–

UTD
1974

Cecil H. and Ida M. Green Honors Chair in Natural Sciences:

Francis S. Johnson   1974–
Cecil H. and Ida M. Green Distinguished Lectures Series:
15 Lecturers during 1980–1986

UTHSC-
SMS
1973

Cecil H. and Ida Green Chair in Reproductive Biology Sciences:
At the Southwestern Medical School in Dallas, the Greens established an honors Chair in the Reproductive Biology Sciences, a special interest of Ida's, which has been occupied since its inception in 1973 by the distinguished physician Paul C. MacDonald, M.D., now Professor of Obstetrics and Gynecology and Biochemistry, and Director of the Cecil H. and Ida Green Center for Reproductive Biology Sciences, in the UT Southwestern Medical School in Dallas.

CSM
1975

Charles Henry Green Professorship in Exploration Geophysics:
At the Colorado School of Mines the Charles Henry Green Professorship, established by Cecil in honor of his father, has had as its first occupant:

J. Edward White	1976–1986 (retired in 1986)
Kenneth L. Larner	1988–

NJCIRM
1975

Cecil H. and Ida Green Research and Investigator Chair:
In 1975 Cecil and Ida endowed the Research and Investigator Chair upon becoming impressed by the work of the National Jewish

Hospital and Research Center in Denver, now called the National Jewish Center for Immunology and Respiratory Medicine.

Howard V. Rickenberg, Ph.D., is the first and only occupant of the Cecil and Ida Green Chair in Developmental Biochemistry. His title at the Center is "Chairman of the Division of Immunology and Cellular Biology."

St. Mark's
School
of Texas
1977–1984

Cecil H. and Ida Green Faculty Chair (for Master Teacher):
In 1977 Cecil and Ida established the first of their two chairs to be occupied by outstanding Lower School (elementary grades) teachers. The first occupant of the first chair was Mrs. Evelyn Boone, honored for her teaching skills. The second Master Teacher Chair, endowed in the mid-1980s, is not yet occupied. The Greens' purpose in creating the two innovative chairs was to encourage recognition of outstanding elementary grade teachers.

The Greens also made small donations to the following three academic positions:

1971 To complete the funding of a professorship at UCSD named for Harold C. Urey, Nobel Prize chemist

1978 A professorship at Caltech named for geologist Robert B. Sharp

1980 A lectureship at Caltech named for geophysicist J. Hewitt Dix

---

Generously endowed Cecil H. and Ida M. Green Residency programs at UTHSC in Dallas, at Oxford's Green College in England, and at SCRF in La Jolla, have made it possible for many medical and health care personnel—ranging from beginning interns to certain practicing physicians—to participate in training and research activities during an important transition period early in their respective careers. These programs can be expected to continue into the indefinite future, so long as the endowments produce the income to support them.

In addition to the three foregoing residencies, Cecil and Ida have endowed three important funds created to provide income to support research on the learning process and overall improvement in teaching:

1971 The Excellence in Education Foundation at SCAS in Dallas

1978 The Cecil and Ida Green Fund for Excellence in Engineering and Science Education at MIT

1979 The Cecil and Ida Green Fund for Excellence in Engineering and Science Education at SMU

It is relevant at this point to reemphasize that for more than 50 years Cecil has been a strong proponent and supporter of interaction—cooperation and collaboration—between the sciences and engineering, on the one hand, and the humanities and social sciences, on the other. This attitude harks back to his doodlebugging days when he was learning the great importance of getting people with different skills and attitudes to work together harmoniously. He naturally followed this philosophy of working together when organizing TAGER, and expressed it clearly in a convocation of the SMU Institute of Technology on March 3, 1970, in Dallas:

In a world that is becoming more compact, and also increasingly populated, such serious problems as pollution on a global scale, ecological changes, urban renewal and the evolution of new modes of housing, traffic congestion coupled with increasing difficulties of transportation, inadequate total food resources— all of these cry out for, and indeed demand the technical know-how of the scientist and the engineer for their solutions. It would seem certain that young people, technically trained, have the best opportunity to enjoy the real pleasure and the great satisfaction of knowing that they are really caring for the needs of mankind.

But here again, I feel cooperative thinking has its importance as the social studies students delve also into technical implications, while conversely, the students of science and engineering become increasingly conscious of the far-reaching impact of their future endeavors upon the very patterns of living.

So—in concluding—I would pose the question—what better place to build a modern, sensitively inclined Institute of Technology than within the environment of an excellent liberal arts campus, such as we have here at SMU! Herein might well lie a key for the strongest and therefore the most effective pattern of cooperation for ultimately producing the optimum pattern in future higher education!"

Cecil's foregoing comments in 1970 were surely a clarion call for the development of college and university curricula that would bring science and engineering students into a closer and better understanding with students in the humanities and social studies. His strong support of the cooperative relationship envisioned in the preceding comments is clearly demonstrated by his and Ida's endowments of the three Funds mentioned here.

The Foundation at SCAS provided the means for establishing the kind of curricular relationships that Cecil envisioned.

The Fund at MIT has provided support for research on the learning process and on how to improve teaching throughout the Institute, both subjects of much concern to the current Institute faculty as it again takes up the broader question of curricular modification.

The Fund at SMU has been used to enhance further the impressive improvement of SMU's Institute of Technology which resulted from the Institute's leadership role in the development of TAGER.

APPENDIX G
ENDOWED GREEN STUDENT AWARDS

The Greens first made donations for financial aid to primary and secondary students at St. Mark's School of Texas, in Dallas, beginning in the mid-1950s and continuing into the early 1980s. Inasmuch as their donations were made to a general financial aid fund, they were not identified with any particular students.

CECIL H. GREEN GOLD MEDALISTS AT COLORADO SCHOOL OF MINES

Following is information on the 31 geophysics seniors who have been awarded the Cecil H. Green Gold Medal at the Colorado School of Mines, during the period 1957 to 1987. This information was kindly provided by Mary Spaid, *Mines Magazine* assistant.

Cecil H. Green Gold Medalists, 1957–1987

Date of receipt	Name of recipient	Home town
1957	Donald G. Johnson	Peshtigo, WI
1958	Robert Morrison Hamilton	Tulsa, OK
1959	James Lancaster Payne	Golden, CO
1960	James Robert Heavener	Charleston, WV
1961	Alan Edward McGlauchlin	Golden, CO
1962	Gerald Wayne Hohmann	Harmony, PA
1963	Donald D. Snyder	Center, CO
1964	George Lee Brinkworth	Burbank, CA
1965	Robert James Barday	Monrovia, CA
1966	Angus James MacKay	Laguna Beach, CA
1967	Kenneth Ashley Miller	Denver, CO
1968	Gerald W. Grandey	Downey, CA
1969	Michael Lee Visscher	Arvada, CO
1970	Douglas John Guion	Indianapolis, IN
1971	Robert Alan Crewdson	Rochester, NY
1972	Gregory Evan Evans	Denver, CO
1973	Scott Edward Moravec	Willingboro, NJ

Cecil H. Green Gold Medalists, 1957–1987

Date of receipt	Name of recipient	Home town
1974	James Scott Crompton	Colorado Springs, CO
1975	Anthony John Weber	Arvada, CO
1976	Lawrence Stephen O'Connor	Alhambra, CA
1977	Christopher C. Traeger	Boulder, CO
1978	Stephen Merritt Rasey	Littleton, CO
1979	M. Scott Gillis	Terre Haute, IN
1980	Erik Bartholomew Goodwin	Durango, CO
1981	Mary Elizabeth Edrich	Denver, CO
1982	John Wesley Powell	Grand Junction, CO
1983	Mary Elizabeth Revoir	St. Paul, MN
1984	Terrie Lee Tonkinson	Longmont, CO
1985	John Theodore Etgen	Missouri City, TX
1986	John Andrew Geesen	Hidden Hills, CA
1987	Kurt Ranzinger	Boulder, CO

It may be noted that the above list contains the names of 28 men and three women, who came from the following eleven states: Colorado (15), California (6), Indiana (2), and (1) each from Minnesota, New Jersey, New York, Oklahoma, Pennsylvania, Texas, West Virginia, and Wisconsin.

THE AAUW IDA GREEN FELLOWSHIP FOR TEXAS WOMEN, 1968–1987

The Ida M. Green Fellowship for Texas Women was endowed by Cecil H. Green in 1968 and assigned to the American Association of University Women. The endowment is managed by the AAUW Educational Foundation—the philanthropic arm of the AAUW—which selects one Ida M. Green Fellow every year. The Foundation kindly provided the following information on the 18 Ida M. Green Fellows selected during SYs 1968–69 through 1986–87.

Ida M. Green American Fellowship Endowment Recipients, 1968–1987

Year	Name	Institution	Field
1968–1970	Barbara Page Nettle	U. Texas, Austin	Physics
1970–1971	Kathleen S. Matthews	Stanford Medical Center	Biochemistry
1971–1972	Millie Hughes Wiley	Texas Women's U.	Chemistry
1972–1973	Patricia Anne Tallal	U. Cambridge	Experimental psychology
1973–1974	Judith R. Head	U. Texas Southwestern Medical School	Immunology
1974–1975	Peggy Anne Alley	West Texas State U., Canyon	Organic chemistry
1975–1976	Katherine Anne Green	Texas A&M	Biological oceanography
1976–1977	Martha T. Lubet	U. Texas Health Science Center	Microbiology
1977–1978	Pilla Atkley Leitner	U. Texas-Austin	Physics
1978–1979	Marguerite L. Coomes	U. Texas Health Science Center	Biochemistry
1979–1980	Linda E. McAlister	U. Texas Health Science Center	Biochemistry
1980–1981	Mary Joanne Hintz	U. Texas Health Science Center	Biochemistry
1981–1982	Wendy L. Keeney	Texas A&M	Chemical oceanography
1982–1983	Jo-Ann N. Keene	U. Texas Health Science Center	Immunology
1983–1984	Patricia L. Ashley	U. Texas Health Science Center	Molecular biology
1984–1985	Carol D. Eberhard	U. Texas-Dallas	Physics
1985–1986	Jacqueline Uhr Guise	U. Texas Health Science Center	Neuro-physiology
1986–1987	Amy Leventer	Rice	Geology

AAUW Educational Foundation September 23, 1986—Development
Department

Universities where thesis work done		Fields of specialization	
UTHSCD	7	Biochemistry	4
UT-Austin	2	Physics	3
Texas A&M	2	Biology	2
Rice University	1	Chemistry	2
Texas Woman's U.	1	Immunology	2
UT-Dallas	1	Oceanography	2
UTSMS	1	Experimental psychology	1
West Texas State U.	1	Geology	1
Stanford	1	Neurophysiology	1
U. Cambridge	1		18
	18		

Sixteen of the 18 Ida M. Green Fellows chose to do their thesis
work in a Texas university, and overall showed a strong pref-
erence for research involving some aspect of biology and
medicine.

IDA M. GREEN FELLOWSHIPS AT MIT, 1974–1987

The munificent gift from the Greens that endowed the Ida M.
Green Fellowships at MIT has served its purpose well in that it
has helped an impressive number of outstanding women to
start graduate study in engineering and the sciences. The following list of fellows, prepared by the staff members of the MIT
Graduate School, shows the recipients selected each school
year, the institution from which they received their undergraduate degree, the department or discipline they chose at the time
they entered the Institute, and the graduate degrees subsequently earned. (If degree work was not yet complete at the
time the following list was compiled, no entry appears under
"Degree received.")

Ida M. Green Scholars

Name	Undergraduate university	MIT department	Degree received
1974–1975			
Patricia M. Downey	MIT	Physics	Ph.D. 1980
Randall C. Forsberg	Barnard	Political Science	
Janice I. Gepner	MIT	Biology	Ph.D. 1979
Kathleen Octavio	Clark	Civil Engineering	SM/Eng. 1977
Gail Kendall McCarthy	Berkeley	Mechanical Engineering	Ph.D. 1978
Cheryl Teich	MIT	Chemical Engineering	SB/SM 1975
1975–1976			
Patricia Bjorklund Brennan	Berkeley	Architecture	M. Arch 1977
Alice Cantelow	U. Cal.-San Diego	Earth & Planetary Science	
Caroline Gee	Berkeley	Physics	Ph.D. 1981
Jennifer Gordon	MIT	Nutrition & Food Science	Ph.D. 1981
Marta E. Greenberg	MIT	Electrical Engineering & Computer Science	SB/SM 1977
Joanne M. Higgins	Rochester Institute of Technology	Mechanical Engineering	SM 1976
Marjorie Flavin (Honorary)	Wellesley	Economics	Ph.D. 1981
Kathleen Rupprecht (Honorary)	Notre Dame	Chemistry	Ph.D. 1979

420
Appendix G

Ida M. Green Scholars

Name	Undergraduate university	MIT department	Degree received
1976–1977			
Deborah Cohen	Franklin & Marshall	Management	
Karen Dominguez	UCLA	Architecture	M.Arch 1979
Elizabeth Emerson Hershey	Princeton	Urban Studies & Planning	MCP 1978
Carolyn Schroeder	Brandeis	Mathematics	
Helen Phoebe Sdougos	MIT	Mechanical Engineering	SM 1978
Dianne Simmons	Cornell	Electrical Engineering & Computer Science	SM 1978
Laurel Fisher (Honorary)	MIT	Nutrition & Food Science	Ph.D. 1981
Beth Levin (Honorary)	MIT	Electrical Engineering & Computer Science	SM Ph.D. 1979 1983
1977–1978			
Renee Chow	MIT	Architecture	M.Arch 1980
Frances Hagopian	Brandeis	Political Science	
Sharon L. Haynie	U. Penn.	Chemistry	Ph.D. 1982
Ellen Hildreth	MIT	Electrical Engineering & Computer Science	Ph.D. 1983
Kimberlee Kearfott	St. Mary's U.	Nuclear Engineering	Ph.D. 1980
Barbara Lawrence	Oberlin	Management	Ph.D. 1983

Ida M. Green Scholars

Name	Undergraduate university	MIT department	Degree received
1978–1979			
Nancy Lee Bartlett	Stanford	Chemical Engineering	SM 1979
Dianne Bustamante	Berkeley	Electrical Engineering & Computer Science	
Gerda Endemann	Berkeley	Nutrition & Food Science	Ph.D. 1982
Jennifer Lynn Hall	MIT	Earth & Planetary Science	Ph.D. 1985
Cynthia Horan	Mt. Holyoke	Urban Studies & Planning	Ph.D. 1986
Carol Simon	John Carroll University	Management	SM 1980
Teresa C. Nolet (Honorary)	MIT	Materials Science & Engineering	SM 1979
1979–1980			
Beth Doll	MIT	Mechanical Engineering	SM 1981
Caroline Grant	Queen's U. Canada	Mathematics	
Jean Januszkiewicz Batryn	Penn State U.	Management	
Joo Hooi Ong	MIT	Civil Engineering	SM 1981
Malka Rappaport	Brooklyn College	Linguistics	
Claire Skarda	College of Notre Dame of Maryland	Materials Science & Engineering	SM 1981
Sherry Tvedt	U. Cal.-Irvine	Urban Studies & Planning	MCP 1981

422
Appendix G

Ida M. Green Scholars

Name	Undergraduate university	MIT department	Degree received
1980–1981			
Jan Edwards	Barnard	Linguistics	SM 1981
Kathleen Ginn	U. Virginia	Electrical Engineering & Computer Science	SM 1982
Melinda Hall	Duke U.	Physical Oceanography (WHOI)	Ph.D. 1985
Nancy Jubb	Penn. State U.	Materials Science & Engineering	SM 1982
Jeanmarie Nolley	U. Texas	Nuclear Engineering	Ph.D. 1986
Sheila Sundaram	U. Delhi (India)	Mathematics	Ph.D. 1986
Marilyn Wolfson	U. Michigan	Meterorology	SM 1983
Robin Prager (Honorary)	Harvard	Economics	
1981–1982			
Anne Bickford	Penn. State U.	Civil Engineering	SM 1983
Lynne Butler	U. Chicago	Mathematics	
Susan Dexheimer	MIT	Physics	
Margaret Goud	Stanford	Earth & Planetary Science (WHOI)	
Juliette Levin	Barnard	Linguistics	Ph.D. 1985
Lorraine Olson	MIT	Mechanical Engineering	SM 1983
Susan Muller (Honorary)	Princeton	Chemical Engineering	Ph.D. 1986
Lynn Paquette (Honorary)	Williams	Economics	Ph.D. 1985

Ida M. Green Scholars

Name	Undergraduate university	MIT department	Degree received
Alexana Roshko (Honorary)	MIT	Materials Science & Engineering	
Marguerite Browning	U. Massachusetts	Linguistics	
Anne Connell	U. Glasgow (Scotland)	Mathematics	SM 1983
Kathryn Harrison	U. Western Ontario	Chemical Eng.	SM 1984
Linda Mason	MIT	Materials Science & Engineering	
Jean Schroedel	Washington State	Political Science	
Lisa Urry	Tufts	Joint MIT/ WHOI	
Katherine Cornog (Honorary)	Brown	Electrical Engineering & Computer Science	SM 1985
Elizabeth Hoopes (Honorary)	U. Wisconsin	Economics	
Karin Rabe (Honorary)	Princeton	Physics	
1983–1984			
Phyllis Berg	Cooper Union	Chemical Engineering	SM 1986
Tanya Furman	Princeton	Earth & Planetary Science	
Paula Garfield	U. Delaware	Joint MIT/ WHOI	
Lois Pollack	Brandeis	Physics	
Leslie Ann Sundt	Wesleyan	Economics	

Ida M. Green Scholars

Name	Undergraduate university	MIT department	Degree received
Jennifer Cole (Honorary)	U. Michigan	Linguistics/ Philosophy	
1984–1985			
Martitia Barsotti	MIT	Aeronautics & Astronautics	SM 1986
Wendy Graham	U. Florida	Civil Engineering	
Sarah Kruse	U. Wisconsin	Earth, Atmospheric & Planetary Science	
Gretchen Rohrer	Stanford	Chemical Engineering	
Cheryl Wendelken	SUNY/MIT	Architecture	
Elaine Lee (Honorary)	MIT	Mechanical Engineering	SM 1986
Katherine McCreight (Honorary)	U. Michigan	Linguistics	
Michelle Schulberg (Honorary)	Harvard	Chemistry	
Julie Anderson	Lehigh U.	Chemical Engineering	
Mary Chang	MIT	Applied Biological Sciences	
Annette Huber	Cornell	Civil Engineering	
Rebecca Thatcher	U. Minnesota	Mechanical Engineering	
Karen Walrath	MIT	Electrical Engineering & Computer Science	

Ida M. Green Scholars

Name	Undergraduate university	MIT department	Degree received
Mary Tritschler (Honorary)	Stanford/U. Chicago	Economics	
1986–1987			
Lelani Arris	U. Vermont	Civil Engineering	
Debra Colodner	Yale	Joint MIT/ WHOI	
Christine Govan	NC State	Architecture	
Joanne Liu	U. Cal.	Chemical Engineering	
Judyth Twigg	U. Pittsburgh Carnegie/ Mellon	Political Science	
Mary Manger (Honorary)	MIT	Materials Science & Engineering	

The foregoing list shows that during SYs 1974–75, when the first Fellows were chosen, through 1986–87, a total of 94 women graduate students (76 regular and 18 honorary recipients) entered the Institute—approximately six regular fellows per school year, in addition to the students who though selected as Green Fellows chose instead to accept another fellowship of longer tenure.

CECIL AND IDA GREEN GRADUATE STUDENT FELLOWSHIPS IN GEOPHYSICS AT STANFORD UNIVERSITY, 1977–1987

Following is information about the 21 students who have been selected thus far to receive a Cecil and Ida Green Graduate Student Fellowship in Geophysics at Stanford University:

Year	Student	Home town	Undergraduate degree
1977–78	Einar Kjartansson	Akureyri, Iceland	Denison
	Daniel Moos	Stony Brook, NY	Cornell
1978–79	Sandra Plumlee-Morris	Albuquerque, NM	U. New Mexico
	Davis Engebretson	Port Orchard, WA	Western Washington U.
1979–80	Frederick Schult	Urbana, IL	MIT
	Kathy Koskelin	River Falls, WI	U. Wisconsin
	Kevin Morrill	Aloha, OR	U. Chicago
1980–81	John Garing	Lexington, MA	Princeton
	Eric Peterson	Manhattan Beach, CA	Caltech
1981–82	Douglas Wilson	Wilmington, DE	Stanford
	Andrew Michael	Setauket, NY	MIT
1982–83	Edward LaFehr	29 Palms, CA	U. Utah
	Heidi Scheirer	Allentown, PA	Duke
1983–84	Raphaele Bally	France	Ecole N.S. des Mines
	Isabelle Sanson	France	Inst. Nat. Polytech. de Lorraine
1984–85	Steffen Hodges	Santa Barbara, CA	U. Cal.-Santa Cruz
	Bryan Man	Edmond, OK	Baylor
	Robert O'Connor	Palo Alto, CA	U. Cal.-Davis
1985–86	John Etgen	Missouri City, TX	Colorado Sch. Mines
1986–87	David Goldie	Ottawa, Canada	U. Waterloo
	Richard Saltus	Golden, CO	Stanford & Colorado Sch. Mines

Note: The data were kindly provided by George A. Thompson, chairman, Stanford's Department of Geophysics.

ROSTER OF FULHAM UNDERGRADUATE SCHOLARSHIPS AT SUFFOLK UNIVERSITY, 1981–1987

When Cecil learned from his good friend, Vincent A. Fulmer, that many of the undergraduate students at Suffolk University in Boston deserved special recognition because of their meritorious school record and high motivation toward getting a college education, the Greens endowed a modest Merit Scholarship Program. At the time—it was 1981—Fulmer was board chairman of Suffolk and well knew the quality and needs of many of the students, as did the president, Thomas A. Fulham, for whom the program is named. According to the leaflet describing the program:

The four most qualified freshmen will be awarded a $500 Fulham scholarship upon enrollment, which will be renewed on a yearly basis. One Fulham scholarship will be awarded in each of the following areas:

Humanities

Management

Science

Social Science

The award recipients will be chosen for their academic achievement and academic promise. The criteria include:

High school rank in class

SAT scores

Writing sample

On-campus interview

Recommendations

Fulham scholarships are based on merit, not financial need.

Following are the names of the 21 students who have been selected as Fulham Scholars thus far:

Recipients	First year awarded
Robert E. Bagshaw	1984–85
Nancy D. Bloom	1983–84
Tracey Boisseau	1981–82
Mary Costa	1981–82
Kathleen Crisley	1985–86

Recipients	First year awarded
Rhona Jane Fee	1981–82
Anne Graglia	1985–86
Emily A. Hegarty	1982–83
Robert Johnson	1986–87
Lisa Mafrici	1986–87
Janelle M. Malafronte	1983–84
John Panarese	1985–86
Daniel E. Parks	1982–83
Gabriel Piemonte	1986–87
Maureen Regan	1984–85
Donald J. Robbins	1984–85
Maura D. Robinson	1983–84
Lauri F. Rosenberg	1983–84
Deborah Silva	1985–86
Susan M. Sweeney	1982–83
Barbara Xintaropoulos	1986–87
Total: 21 (15 women, 6 men)	

Cecil and Ida regard this program as one of their most satisfying investments in pleasure. A number of the students have written their appreciation to the Greens, and characteristically, Cecil has responded to each letter with commendation and encouragement.

APPENDIX H
JOURNAL PUBLICATIONS (ARTICLES, CITATIONS, MEMORIALS), ADDRESSES AND SPEECHES, AND MISCELLANEOUS REMARKS BY CECIL H. GREEN (1924–1987)

Cecil H. Green's writings, starting with his SM thesis at MIT, cover a broad range of subject matter and are preserved in a variety of forms—articles in professional journals; institutional publications and public announcements; reports of conventions, special meetings, and the like; and unpublished typescripts of addresses, speeches, comments, and so forth, delivered at many different gatherings.

His writings comprise reports on early research, commentaries on the state of his engineering profession—geophysical exploration; citations of fellow geophysicists for some special honor from a professional society; responses to honors awarded him alone or to him and Ida jointly; memorials to deceased friends; prepared addresses and speeches delivered at dedications, conferences, conventions, symposia, anniversaries, commencements, and the like; and extemporaneous remarks made at many different gatherings. These writings are listed chronologically under a few separate headings for ease of identification as to content and also to enable the interested reader to learn how Cecil first developed ideas and attitudes that were later incorporated in his own special style of exposition and profoundly influenced his and Ida's philanthropy.

SM THESIS (MIT)

1924 A Static Study of the No Load Flux Distribution in a Salient Pole [Synchronous] Alternator. 70 pp., 16 figs. June 1924, Course VI-A.

ARTICLES

1927   Graphical Determination of Magnetic Fields (with E. E. Johnson). Presented at Winter Convention of AIEE, New York, Feb. 7–11, 1927 (5-page preprint).

1938   Velocity Determinations by Means of Reflection Profiles. *Geophysics* 3/4, pp. 295–305, Oct. 1938.

1948   Integration in Exploration.
        Oil and Gas Jour. 46/52, p. 106, 108, April 29, 1948.
        Geophysics 13/3, pp. 365–370, July 1948.
        Amer. Assoc. Petrol. Geol. Bull. 32/7, pp. 1216–1220, July
        1948.

1950   The Relationship of Research and Field Operations in
        Seismic Exploration.
        Colo. Sch. Mines Quart. 45/4a, pp. 1–9, Oct. 1950.

1953   A Cooperative Plan in Student Education. *Geophysics*
        18/3, pp. 525–531, 1953.

1954   Geophysical Management Looks at Safety as a Public
        Relations Factor.
        Geophysics 19/4, pp. 820–824, Oct. 1954. Also see Ab-
        stract with title "Safety & GSI" in *Safety Valve,* May is-
        sue, 1982.

1955   What It Takes to Explore for Oil in Other Lands. *The
        Petroleum Engineer,* p. 368.2, Sept. 1955.

1959   Some Historical Aspects of the Petroleum Industry on
        Its 100th Anniversary (with E. McDermott). (A one-
        page commentary on two articles published in *Oil and
        Gas Journal's* Petroleum Panorama issue of June 1959.
        This comment was also printed in a pamphlet prepared
        by GSI for in-house distribution in Dallas.)

1963   Trends in Geophysics Prospecting for Petroleum [ch. 2]
        (and John W. Wilson) In *Economics of the Petroleum In-
        dustry.* Vol. 1. International Oil and Gas Educ. Center
        and Southwestern Legal Foundation. Houston, Gulf
        Publishing Co., pp. 24–49, 1963.

1972   Protecting and Improving Our Environment—An En-
        gineering Challenge. Colo. Sch. Mines, *The Mines Mag-
        azine,* 62, pp. 13–16, May 1972.

1973–   Importance of the Industry-Education Relationship
1974   (Address before the Chancellor's Council of the Uni-
        versity of Texas System in San Antonio, Oct. 1973, and
        distributed by UTS as a pamphlet.) Also published in
        Colo. Sch. Mines, *The Mines Magazine* 64/5, pp. 6–10,
        1974.

1975 $600,000 endowed professorship in exploration geo-
        physics at CSM named in honor of Charles H. Green

(Cecil's father). Colo. Sch. Mines, *The Mines Magazine* 65/7, p. 4, 1975.

1980 SEG (Society of Exploration Geophysicists) 50th Anniversary Issue—President's Page: "Anniversary Reflections." *Geophysics* 45/11, pp. 1597–1598, 1980.

1981 Comment (On Research and Scholarship at The University of Texas at Austin). In *Discovery* 6/1, pp. 2–5, Autumn 1981.

CITATIONS

1971 A citation for Eugene McDermott for Honorary Membership in the Society of Exploration Geophysicists. Delivered at SEG's 41st Annual Meeting in Houston, Nov. 8, 1971 (4 pp).

1971 A Citation for Francis Anthony Van Melle for Honorary Membership. *Geophysics* 36/1, pp. 248–249, 1971.

1974 Citation for J. Tuzo Wilson for Honorary Membership in SEG. *Geophysics* 39/1, pp. 163, 166, 1974.

1974 Response to his own citation for the 1974 AAPG Human Needs Award. *AAPG Bull.* 58/9, p. 1881, Sept. 1974.

1976 Citation for J. Erik Jonsson for Honorary Membership in the Dallas Geophysical Society, May 7, 1976.

1976 Citation for Fred J. Agnich for Honorary Membership in the Dallas Geophysical Society, May 14, 1976.

1980 Profile on Kenneth E. Burg for SEG Honorary Membership. See also 1981 item.

1981 Honorary Life Membership Citation for Kenneth E. Burg [SEG]. *Geophysics*, v. 46/3, p. 395, 1981.

1981– Maurice Ewing Medal Citation for J. Tuzo Wilson [writ-
1982 ten by Cecil H. Green] (in) Rept. SEG Honors and Awards Committee, 1980–1981, p. 6–7, 1981. *Geophysics* 47/2, p. 303, 1982.

1983 Maurice Ewing Medal Citation for Dr. Frank Press, *Geophysics: The Leading Edge of Exploration* 2/2, p. 36, Feb. 1983.

1984   Citation for Lewis Lomax Nettleton, Sixth Recipient of
        SEG's 1984 Maurice Ewing Medal, at Atlanta, Georgia,
        3 Dec. 1984. Maurice Ewing Medal Citation for L. L.
        Nettleton. *Geophysics: The Leading Edge of Exploration* 4/
        3, p. 25, March 1985.

MEMORIALS

1960   Morris Miller Slotnick (1901–1956)—Memorial presen-
        tation. *Geophysics* 25/2, pp. 547–550, April 1960.

1967   Memorial to Raymond Albert Stehr (1903–1967). *AAPG
        Bull.* 51/11, pp. 2314–2316, portrait, 1967. *The GSI
        Grapevine* 24, p. 23, 1968.

1967   Memorial to Lloyd Viel Berkner (1905–1967). *Geophysics*
        32/6, pp. 1107–1108, December 1967.

1967   In Memoriam—Lloyd Viel Berkner (1905–1967) (and
        R. C. Peavey). *Am. Geophys. Union Trans.* 48/4, pp. 893–
        895, portrait, 1967.

1973   Introductory Remarks at UT-Dallas, on May 14, 1973,
        on the occasion of the Lloyd V. Berkner Memorial and
        Dedication Ceremony (2 pp).

1974   A Memorial Citation for Eugene McDermott (1899–
        1973). *Geophysics* 39/1, p. 137–138, portrait, 1974 (5 pp).

1979   Karcher Memorial. *GSI Grapevine* 35/3, pp. 26–33, 1979.

1979   Memorial—John Clarence Karcher (1894–1978), Father
        of the Reflection Seismograph. *Geophysics* 44/6, pp.
        1018–1021, 1979.

1980   John Clarence "Doc" Karcher (1894–1978), Father of
        the reflection seismograph. Reprint of preceding that
        appeared in *Dallasite* (Dallas) in March 1980.

UNPUBLISHED COMMENTS PREPARED FOR VARIOUS KINDS OF MEETINGS

In the more than forty years since Cecil became part-owner, and
a vice president, of Geophysical Service Inc. in 1941, he has
been called upon repeatedly to speak in some capacity at a wide
variety of meetings. Because these assignments have increased

greatly in number and variety through the years, he has developed his own inimitable style of speaking that has endeared him to his many audiences.

As master of ceremonies, committee chairman, invited speaker, citationist, or host of a luncheon or dinner, he can always be relied upon to conduct his assignment with gracious informality, a sense of humor, and often a touch of sentimentality, while at the same time preserving the dignity appropriate for the occasion. And above all is his evident desire to communicate something of importance to his audience.

In the earlier years, between Pearl Harbor and the 1960s, Cecil's remarks were made extemporaneously to groups that were generally composed of scientists and engineers who shared his professional training and interests. As a consequence he seldom found it necessary to take much time to prepare his remarks because so much of what he discussed he could draw from his own experience as a practical exploration geophysicist. This practice of speaking extemporaneously, however, has left us with essentially no record of what he communicated to his many audiences, though it is evident from his calendar that he spoke frequently.

When the requests for his services as a speaker increased greatly in number, and came from groups with widely different interests, he had to change his way of preparing his speeches. He could no longer depend on spontaneity. After 1960, when his and Ida's program of philanthropy began on a substantial scale, he found it advisable to take more care in preparing what he wished to say, and in preserving a copy of it. As a consequence his papers include seventy or more copies of speeches made between 1960 and 1987. These are of great value to the interested reader because they record the time and situation when he presented innovative ideas, announced stunning surprises, coined a "catchy" phrase, or stated a strongly held conviction or conclusion.

Cecil never indulges in trivia; he makes every phrase count, whether speaking or writing. Though it can be said that on some more informal occasions he may become a bit loquacious, even sentimental, he invariably returns to his principal theme, as the hound to the trail. At formal affairs, however, his carefully prepared and clearly enunciated remarks demonstrate the respect he holds for every such audience he addresses, for on these occasions he is always careful to keep within the time alloted to him. For Cecil, successful communication of ideas and intentions has always been of utmost importance.

Inquiring readers will find the many unpublished copies of Cecil's remarks a mine of information on his personal traits, his views on a wide variety of subjects, and particularly on the importance he attaches to friendliness, cooperation, and collaboration, and to the great satisfaction that he and Ida have enjoyed from their philanthropy.

At my request Cecil consulted his special file containing the typescripts of addresses, speeches, and other presentations that he has delivered orally, and kindly provided the following list—to which I have added supplementary information that further indicates the date, place, and description of each presentation.

CECIL H. GREEN INDEX OF SPEECHES 1948–1987

April 27, 1948	SEG Presidential Address delivered in Denver at the Annual Meeting of the SEG. Also *Geophysics* 13/3, pp. 365–370, 1948.
Dec. 5, 1960	Remarks at the Ground-breaking Ceremony for MIT's Green Center for the Earth Sciences.
March 7, 1961	An Industrialist Looks at Education. Address before the Nuclear Research Foundation of the University of Sydney, Sydney, Australia (16 pp).
Oct. 2, 1964	Remarks at Dedication of MIT's Green Center for the Earth Sciences (in ceremony booklet).

May 7, 1965	Importance of the Industry–Education Relationship. Opening address of 12th Ann. Explor. Meeting of the Permian Geophys. Soc., Midland, TX (11 pp).
Nov. 7, 1966	As he receives the Virgil Kauffman Medal at the SEG Convention in Houston, TX.
March 1967	The Education–Community Relationship. Address at Appreciation Dinner Meeting at Austin College (12 pp).
Nov. 10, 1969	A Bit of Nostalgia Inspired by *The GSI Grapevine*, GSI's 25th Anniversary, in Dallas (10 pp).
Jan. 26, 1970	Remarks at Opening Dinner Session of TCU Development Program, in Fort Worth (4 pp).
March 3, 1970	Remarks at Special Convocation Honoring the Greens at the SMU Institution of Technology (4 pp).
April 10, 1970	Remarks on TAGER COSIP TIES US Workshop, at Austin College, in Sherman (9 pp).
Nov. 18, 1971	Some Current Trends in Science and Engineering Education. Address to MIT Alumni, in Tulsa, OK (16 pp).
Feb. 1, 1972	Remarks at First Council Meeting of the Dallas YWCA, on accepting job of heading 1973 capital fund drive for benefit of YWCA (4 pp).
April 14, 1972	Protecting and Improving our Environment—An Engineering Challenge. Keynote address at CSM's Annual Engineers Day in Golden (15 pp).
April 14, 1972	Remarks at Dedication of the CSM Green Graduate and Professional Center, empha-

	sizing importance of collaboration by private donors with state bureaus in support of education (3 pp).
April 14, 1972	Remarks at Banquet following Dedication of the CSM Green Graduate and Professional Center, emphasizing the importance of enlisting the help of friends in accomplishing goals (3 pp).
April 28, 1972	Response Remarks at Dedication of the Austin College Ida Green Communication Center, emphasizing the important role of private colleges like Austin in developing innovative programs for motivating young people (2 pp).
May 16, 1972	Protecting and Improving Our Environment—A Scientific Challenge. Delivered at the First Annual Spring Honors Convocation at UTD (16 pp).
Feb. 23, 1973	Remarks at Ground-breaking for the Dallas YWCA Main New Facility.
April 13, 1973	Remarks at Ground-breaking for Fine Arts Center at St. Mark's School of Texas in Dallas.
May 14, 1973	Remarks at the Lloyd V. Berkner Memorial and Building Dedication at UTD (2 pp).
July 8, 1973	Remarks at American Alumni Conference in Vancouver, B.C.
Aug. 15, 1973	The Role of Higher Education. Delivered at the Kickoff Dinner for TCU's New Century Program (5 pp).
Sept. 28, 1973	Kickoff talk on Annual Sustentation for St. Mark's School of Texas in Dallas.

Sept., 1973	A Tribute to Eugene McDermott at a St. Mark's School Assembly.
Nov. 11, 1973	Remarks at Ground-breaking for the Maria Morgan Branch of Dallas YWCA.
March 14, 1974	The Importance of the Industry–Education Relationship, delivered at the SEG Mid-Year Meeting at CSM in Golden, CO (13 pp).
April 1, 1974	Remarks as I Receive Human Needs Award. At AAPG Convention in San Antonio, Texas (4 pp). Also *AAPG Bull.* 58/9, p. 1881, Sept. 1974.
April 18, 1974	Remarks at Diamond Jubilee Award Dinner of the National Jewish Hospital and Research Center in Denver, in Dallas.
Oct. 14, 1974	Remarks at Dedication of the H. Ben Decherd Center at St. Mark's School of Texas, in Dallas.
Dec. 3, 1974	Some Lifetime Reflections. Delivered at the Tenth Anniversary Banquet, MIT Earth Sciences Center (10 pp).
Dec. 3, 1974	My Final Report to the Corporation as Chairman of the MIT Earth Sciences Visiting Committee.
Feb. 23, 1975	Remarks at Dedication of the New Facility of the Dallas Central YWCA.
May 24, 1975	Some Lifetime Reflections. Commencement Address at UT-Arlington.
Dec. 3, 1975	Introductory Remarks at a Faculty Seminar, UT Health Science Center in Dallas.
May 16, 1976	Acceptance Remarks for Honorary Doctorate University of Dallas.

Nov. 19, 1976   Investiture of Dr. James Edward White in the Charles Henry Green Chair in Geophysics at the Colorado School of Mines.

May 1, 1977   Dedication Remarks for New Facility YWCA in Garland, TX.

July 19, 1978   My Early G.E. Recollections. A letter to a Dallas manager whom he liked.

Sept. 12, 1978   Importance of the Industry–Education Relation—Case History, Stanford University.

Oct. 26, 1978   Staff Address on Case History—Importance of the Industry–Education Relation, Scripps Clinic & Research Foundation.

Feb. 2, 1980   Dedication Remarks for the M/V *Edward O. Vetter* at Port Hueneme, CA.

Feb. 20, 1980   Dedication Remarks for the M/V *Fred J. Agnich* at Christening Ceremony.

May 5, 1980   Remarks at MIT Leadership Campaign Dinner at Dallas Petroleum Club.

May 13, 1980   Remarks at St. Mark's School of Texas Trustee Dinner at Dallas Country Club.

May 21, 1980   Remarks at Faculty Celebration Honoring MIT President Jerome B. Wiesner (5 pp).

June 19, 1980   Commencement Address at Graduation Exercises of Woods Hole–MIT Joint Program at Woods Hole, MA.

Aug. 6, 1980   Remarks at My 80th Birthday Celebration UTHSC in Dallas.

Nov. 3, 1980   Nostalgic Remarks on Topic—Some Longtime Reflections. Dallas Society of Exploration Geophysicists (16 pp).

Nov. 16, 1980   Opening Remarks at National Conference on "Biological Imaging: Contributions from

	Contemporary Physics and Engineering," at SCRF.
March 19, 1981	A Case History—Importance of the Industry–Education Relationships (An Alternate Title: The Practical Importance of People and Human Relationships). Delivered at a Dinner Celebration of the Addison and Randolph Association, TCU, Fort Worth, TX (14 pp).
April 26, 1981	Remarks at the Dedication of the Allen Electronic Organ in CSM's Green Graduate and Professional Center, Golden, CO.
July 16, 1981	Remarks at Gound-breaking for N.W. Garden Annex, SCRF, La Jolla, CA.
Nov. 3, 1981	Remarks at 50th Year Campaign Victory Dinner, St. Mark's School of Texas, Dallas.
Feb. 15, 1982	Remarks at Christening of the M/V *Cecil Green II*, at Panama City, FL.
April 23, 1982	A Constant Challenge. Delivered at the Dedication of the Student Union at UT-Dallas (5 pp).
Sept. 22, 1982	Acceptance Remarks upon receiving the 1982 Philanthropy Award to Cecil and Ida Green from the Dallas Historical Society (2 pp).
March 9, 1983	Remarks when unveiling Portrait of Richard W. Lyman in Green Library, Stanford University.
April 18, 1983	Remarks upon receiving Honorary Membership from the Institute of Electrical & Electronic Engineers.
June 10, 1983	Remarks for Ida at Dedication of MIT's Ida F. Green Hall.

June 13, 1983   Remarks at 70th Birthday Celebration Dinner for Sir Richard Doll, Radcliffe Observatory, Green College, Oxford University, England.

Oct. 5, 1983   Remarks at 10th Anniversary Luncheon of MIT's Ida M. Green Fellows.

Oct. 16, 1983   Opening Remarks at Second National Conference on Biological Imaging, at National Academy of Sciences, Washington, DC.

Aug. 1984   Preface (with A. Suttle, Jr.) for NAS Report on Second National Conference on Biological Imaging.

Sept. 24, 1984   The Importance of People. Delivered at the SMU Computer Conference, Dallas (8 pp).

Oct. 2, 1984   Remarks at the 20th Anniversary Celebration of the Green Center for the Earth Sciences at MIT (3 pp).

Nov. 7, 1984   Introductory Remarks at a Luncheon for Frank Press and Fred Seitz in the Dallas Petroleum Club.

Mar. 29, 1985   Introduction of Frank Press at Dallas Petroleum Club Dinner following Dedication of the Patrick E. Haggerty Building at the University of Dallas.

May 25, 1985   Remarks at the Dedication of the New SEG Headquarters Building in Tulsa, OK (3 pp).

June 29, 1985   Remarks as a Gold Plate Honoree at the Annual Meeting of the American Academy of Achievement in Denver, CO.

July 18, 1985   Interview Questions for Helen Monroe, Executive Director, Community Foundation, San Diego, CA.

Aug. 22, 1985	Remarks at MIT-TI-GSI-VI-A Luncheon at Dallas, Petroleum Club (delivered for Cecil, who had to be absent).
Oct. 15, 1986	Remarks at the Dedication of the new Headquarters Building (Green Tower) of the Society of Exploration Geophysicists in Tulsa, OK (3 pp).
Aug. 9–22, 1987	An Excursion to China. A brief report on an invited visit to China to deliver one of the keynote addresses at the International Conference on Geophysical Exploration held in Zhou-Xian (3 pp).
Aug. 9–22, 1987	Remarks by Cecil H. Green at the Joint International Conference of the Chinese Petroleum Society and the U.S. Society of Exploration Geophysicists in Zhou-Xian (4 pp).
Oct. 22, 1987	Remarks by Honorary Chairman Cecil H. Green, at the formal announcement of MIT's Campaign for the Future (3 pp).
Nov. 13, 1987	Remarks by Cecil H. Green on the Early History of TAGER on Its 20th Anniversary, at the Infomart in Dallas (7 pp).
Nov. 17, 1987	Cecil's Response after receiving the Philanthropist of the Year Award by the Dallas Chapter of the National Society of Fund Raising Executives, at Loews Anatole Hotel, Dallas (4 pp).
Mar. 11, 1988	Remarks by Cecil H. Green at 50th Reunion Celebration by ARAMCO and GSI in Dhahran, Saudi Arabia, March 11–16, 1988 (5 pp).

April 14, 1988   Dedication of the Ida M. and Cecil Green
Faculty Club at the University of California,
San Diego.

April 29, 1988   Response by Cecil H. Green on Reception
of The Freedom of the City Award, from
Mayor Gordon Campbell in Vancouver,
B.C.

SUMMARY
The source materials used in this biography, including copies of
Cecil's and Ida's published and unpublished writings, will be
preserved in due course in appropriate archives. These, it is expected, will be made available eventually to biographers, historians, and other scholars interested in the Greens.

APPENDIX I
SOURCES OF INFORMATION ABOUT CECIL AND IDA GREEN AND ABOUT GEOPHYSICAL SERVICE INC. AND TEXAS INSTRUMENTS INC.

1950   Kelly, Sherwin F., "The Rise of Geophysics." *Canadian Mining Manual* 1950, pp. 1–7.

1955   "They started in 1930—25th Anniversary." *GSI Grapevine* 11/8 Sept. 1955, pp. 1–40. (The story of Geophysical Service Inc.)

1956   "The Story of Texas Instruments Incorporated." Presented by J. E. Jonsson, president, before the New York Society of Security Analysts, Inc., Thursday, Dec. 13, 1956. Pamphlet published by Texas Instruments Inc., Dallas.

1959   Eubanks, Bicknell. "Young Firm Successfully Challenges Giants in Transistor Field. Texas Instruments Gets in on Ground Floor." *Christian Science Monitor,* Boston, June 12, 1959, p. 18.

1961   McDonald, John. "The Men Who Made TI (Part 1)." *Fortune,* Nov. 1961, pp. 117–119.

1963   *The Cecil and Ida Green Building Dedication, Massachusetts Institute of Technology.* Comments by participants in the dedication and numerous photographs. MIT, Oct. 2, 1964, 48 pp.

1966   Shrock, Robert R. *A Cooperative Plan in Geophysical Education. The GSI Student Cooperative Plan—The First Fifteen Summer Programs, 1951–1965.* Published by Geophysical Service Inc., Dallas, 1966, 144 pp.

1966   Sweet, George E. *The History of Geophysical Prospecting.* Vols. 1 and 2. Copy 504 of the Everette Lee De Golyer Memorial Edition (1886–1956). Los Angeles, Science Press, Oct. 1966, 326 pp. (See especially Part 27 on Geophysical Service Inc.; Ch. 81 on McDermott and Green.)

1966   Shrock, Robert R. "The GSI Student Cooperative Plan. Representatives of Industry, Education and Government Work Together in Summer Geophysical Training." Amer. Geol. Inst. *Geotimes* 11/5, Dec. 1966, pp. 15–20.

1967   "Seismograph." *Bulletin*, Standard Oil Company of California, 14/4 Oct. 1967, pp. 12–17. (An excellent description of the activities of a geophysical exploration seismic crew.)

1969   "The Management Style of Patrick Haggerty. Can a System Perform Like a Hero? (by DA)." *Innovation*, New York, no. 8, 1969, pp. 16–31.

1970   "Globe-Girdling GSI Had Dramatic Start." *Dallas Morning News*, May 17, 1970, p. 11B.

1970   Tinkle, Lon. *MR. DE: A Biography of Everette Lee De-Golyer*. With Foreword by Norman Cousins. Boston, Little Brown, 1970, 393 pp.

1973   Davenport, Arthur W. *A Great History of the Great Class of 1923, M.I.T.* Norfolk, VA, Liskey Lithograph Corp., 1973, 352 pp. (See pp. 190–191 for comments about Cecil Howard Green.)

1974   Scott, Otto J. *The Creative Ordeal: The Story of Raytheon.* New York, Atheneum, 1974, 429 pp.

1974   Shrock, Robert R. *The Ida Green Room in the Cecil and Ida Green Building, Center for the Earth and Planetary Sciences, Massachusetts Institute of Technology.* Cambridge, MA, MIT Graphic Arts Service, Dec. 1974, 20 pp.

1975   Owen, Edgar W. *Trek of the Oil Finders: A History of Exploration for Petroleum*. Semicentennial Commemorative Volume, American Association of Petroleum Geologists Publication, Tulsa, OK, March 1975, 1,647 pp.

1975   Caruth, Donald. "Interview [taped] with Cecil H. Green, March 10, 1975." Transcribed Feb. 1, 1977. North Texas State University Oral History Collection (Business Archives Project), No. 7. Typescript, 106 pp.

1976   Lord, Frank. "Cecil H. Green Honored at Petroleum Club Luncheon." *GSI Grapevine* 32/3, 1976, pp. 3–13.

1976   Petty, O. Scott. "Seismic Reflections: Recollections of the Formative Years of the Geophysical Exploration Industry." Houston, TX, *Geosource Inc.*, 1976, 81 pp.

1977,   Shrock, Robert R. *Geology at MIT 1865–1965*. Vol. 1. The
1982   Faculty and Support Staff, 1977, 1,032 pp.; Vol. 2. Departmental Operations and Products, 1982, 762 pp.

1978   Fulmer, Vincent, A., et al. "An International Tribute to Cecil and Ida Green, Washington, D.C." Printed by MIT for distribution at the gala affair honoring the Greens held at the National Academy of Sciences, Washington, D.C., on Nov. 9, 1978, 48 pp.

1980   Atchinson, Dave, and MacKie, Sandy. "Sleuths: Oil Scouts Keep Tabs on Competitors' Drilling." *Standard Oiler* (Standard Oil Company of California), Sept.–Oct. 1980, pp. 13–15.

1982   Bates, Charles C., Gaskell, Thomas F., and Rice, Robert B. *Geophysics in the Affairs of Man: A Personalized History of Exploration Geophysics and Its Allied Sciences of Seismology and Oceanography*. Oxford, England, Pergamon Press, 1982, 492 pp.

1982   Harrington, Theodore. "Cecil Green and Gerald Westby: Lifetimes of Excellence." (Profiles of two of North America's most highly regarded exploration geophysicists.) *Geophysics: The Leading Edge in Exploration 1/1*, June 1982. (Green's profile, with photos on pp. 16–23.)

1983   Fulmer, Vincent A., et al. *Dedication of Ida Flansburgh Green Hall*. Prepared by MIT to commemorate the dedication held on June 10, 1983, 40 pp. (An excellent biography of Ida.)

1983   Fulmer, Vincent A., et al. *An International Tribute to Cecil and Ida Green, Washington, D.C.* Prepared by MIT for distribution at the gala affair honoring the Greens and held at the National Academy of Sciences on Nov. 9, 1978, 48 pp. (An informative biography of the Greens.)

1983   Smith, Helen C. "Profile: Cecil and Ida Green." *Universal Medical* (quarterly newsletter of the Division of Institutional Services, University of Texas Medical Branch, Galveston) 15/1, Fall 1983, pp. 34–35.

1983   Seal, Mark. "Cecil and Ida Green: They've Used the Wealth Amassed from Founding Texas Instruments to Become 'Super Santas' for Dozens of Schools and Research Institutes." High Profile in *Dallas Morning News*, Dec. 25, 1983, pp. E1–E5.

1984   Triolo, James S. "Partners in the Enterprise (The Story of Cecil and Ida Greens' Association with Scripps Clinic)." Recorded in an inhouse memorandum by one of the Green's closest friends at SCRF. Typescript, June 22, 1984, 21 pp.

1984   King, Carol L. "Cecil Green; pioneer in exploration geophysics." Stanford Earth Scientist Section, *Stanford Observer*, Dec. 1984, pp. 3 and 9.

1985   Sharp, Anne. "The Natural." A profile of Cecil H. Green. Prepared in connection with presentation of The Great Trek Award to the Greens by UBC's Alma Mater Society in 1984. The Alumni UBC *Chronicle*, Spring 1985, pp. 14–16.

1985   Passino, Karen. "The Pioneering Spirit of TI's Founding Fathers." *TI Engineering Journal* 2/5, Sept.–Oct. 1985, pp. 2–3.

1985   Christopher, Sandy. "Cecil Green: Looking Back." An interview. *TI Engineering Journal* 2/5, Sept.–Oct. 1985, pp. 10–19.

1986   Proubasta, Dolores. "Erik Jonsson." A profile. In *Geophysics: The Leading Edge of Exploration* 5/6, June 1986, pp. 14–23.

1986   "America's Tailor-Trousered Philanthropists." A review of Nielsen's. *The Golden Donors* (see preceding reference). In *The Economist*, Aug. 23, 1986, pp. 73–74.

1987   Millar, D. D., ed., et al. *The Messel Era—The Story of the School of Physics and Its Science Foundation within the University of Sydney, 1952–1987.* Australia, Pergamon Press, 1987, 157 pp.

1987   "The 20th [1967–1987] Anniversary [of] TAGER." *Bulletin* (Association for Higher Education of North Texas), Fall 1987. Dedicated in memory of Ida Green. The program prepared and distributed during the Anniversary contains a brief, informative essay, "The TAGER Story," which describes the crucial role that Cecil and Ida played in the creation and funding of TAGER.

1987   Schmuckler, Eric, King, Ralph, Jr., and Lataniotis, Dolores, A. "The 400 Richest People in America," *Forbes* 140/9, Oct. 26, 1987. (Index of names, p. 320 ff, does not include the Cecil H. Greens.)

1987   Karcher, J. Clarence. "The Reflection Seismograph: Its Invention and Use in the Discovery of oil and gas fields." *Geophysics: The Leading Edge of Exploration* 6/11, Nov. 1987, pp. 10–19.

1987   Sweet, George E. "Comment on J. C. Karcher's The Reflection Seismograph." *Geophysics: The Leading Edge of Exploration* 6/11, Nov. 1987, p. 20.

1987   Golden, Gayle. "Cecil without Ida." *Dallas Morning News*, Dec. 23, 1987, pp. 1C–3C.

INDEX

Page references in *italics* refer to illustrations.